The Physical Environment

Series Editor

Bob Digby — Lecturer in Education/Geography, Brunel University, London

Authors

Bob Digby — Lecturer in Education/Geography, Brunel University, London
Jane Ferretti — King Edward VII School, Sheffield
Ian Flintoff — The Heathland School, Middlesex
David Kinninment — Kenton School, Newcastle upon Tyne
Rachel Lofthouse — Prudhoe County High School, Northumberland
Graham Ranger — Geography and Environmental Education Adviser, Derbyshire Local Education Authority
Graham Yates — Ryton Comprehensive School, Gateshead

Heinemann

Heinemann Educational Publishers
Halley Court, Jordan Hill, Oxford OX2 8EJ
a division of Reed Educational & Professional Publishing Ltd

MELBOURNE AUCKLAND FLORENCE PRAGUE
MADRID ATHENS SINGAPORE TOKYO SÃO PAULO
CHICAGO PORTSMOUTH (NH) MEXICO IBADAN
GABORONE JOHANNESBURG KAMPALA NAIROBI

First published 1995

99 98 97
10 9 8 7 6 5 4 3

ISBN 0 435 35227 X

Designed and typeset by Pentacor plc,
High Wycombe, Bucks (*Russell Horton*)

Illustrations by Arcana, Stefan Chabluk, Karl Cooper, Tracy Hawkett, Russell Horton, Kathy Lacey, Mark McLaughlin, Keith Smith, Sandra Storey

Index compiled by Dr F. E. Merrett, member of the Society of Indexers

Cover design by MCC
Printed in Spain by Mateu Cromo Artes Graficas SA

Acknowledgements
The authors and publishers would like to thank the following for permission to use photographs/copyright material:

American Geographical Association, T. J. Bassett, 'Fulani herd movements' in *Geographical Review*: maps 208, 211, 212, 213, J. J. Pigram, 'Salinity and basin management in south-eastern Australia', in *Geographical Review*: graph 219; Association of London Authorities/London Borough's Association/South East Institute of Public Health, *Air Quality in London*: table 142; Australian Bureau of Statistics, *Yearbook of Australia*: maps 82; R. G. Barry and R. J. Chorley, *Atmosphere, Weather & Climate*, Methuen: diagram 97; Blyth Valley Borough Council: rose diagrams 40; Bureau of Meteorology, Department of the Environment, Sport and Territories, Australia: diagrams and graph 115; Chapman and Hall, World Conservation Monitoring Centre, *Global Biodiversity Status of the Earth Living Resources 1992*: map 201, 206, 207, tables 202, 204; *Collins-Longman Atlas for Secondary Schools*: basis of map 224b; Elsevier Science Publishers, from Rossignol-Strick 1985: diagrams 106; English Nature: maps 60B; Reprinted by courtesy of *The European*: extract, 4; *Financial Times*, extracts 147, 214, map 148; Friends of the Earth: table 141, extracts from 'Transport-related atmospheric pollution in London', *Summary of Evidence 2*, 142, 143; Frosted Earth, *The Surrey Weather Book*: extract 93; R. M. Fuller, 'Lowland grassland in England and Wales: a review of grassland survey 1930–1984', *Biological Conservation*: graph 245; Geographic Association, Lamb 1972 in *Geography*: table 87, G. O'Hare and J. Sweeney, 'Lamb's circulation types and British weather: an evaluation' in *Geography*: maps 87; Geography Teachers' Association Victoria, A. Fajan, 'Salinity in irrigated and wetland areas', *Interaction*: maps 221; ©Alan Godfrey 1989: map 32T; Guardian Newspaper Group: extracts 29T, 80, 101MB, 144, 244; HarperCollins, V. Bishop and R. Prosser: T*he Environment*, table 61, C. Krebs, *Ecosystems and Human Activity*: table 152; HMSO, after Pedgley 1962, ©Crown Copyright, reproduced with the permission of the Controller of the HMSO: *A Course in Elementary Meteorology*, and Bennetts *et al* 1988: diagrams 99; HMSO ©Crown Copyright, reproduced with the permission of the Controller of the HMSO: maps 90–91; Dr Mike Hulme, Climatic Research Unit, UEA, 'Causes of climate and variability in the African Sahel', SAGT Journal: graphs 111, 116; Hydraulics Research, 1989: table 60T; *The Independent*: extracts 46, 129; Israel Government Tourist Information: extract 51; Kilum Mountain Forest Project: map 191, extracts 194, 196; *Lake District Herald*: extracts 7, 8; G. Lean and D. Hiarschsen, *Atlas of the Environment*, Banson Marketing/WWF: maps 176BL/BR; WWF graph 177, diagram 178; Longman, G. O'Hare, 'Biomass on four selected ecosystems', in *Soils, Vegetation, Ecosystems*: table 151, P. J. Gersmehl, 'An alternative biogeography' in *Annals of the Association of American Geographers*: diagrams 154, *Core Geography*: basis of map 224a; Chris Madden, *When Humans Roamed the Earth*, Earthscan/WWF: cartoon 174; MAFF Publications, *Environment Matters – Our Living Heritage 1994*: map 235; Methuen 1988, Chorley, Schuman and Sugden, *Geomorphology*: graph 26T; The Estate of L. Musk for kind permission to reproduce diagrams from *Weather Systems*, Cambridge University Press: 98R, 120BL; *National Geographic*, Karl, 'The global warming debate': graphs 133T/M; *National Geographic*: map 74; National Rivers Authority: graph 10, map 60T, table 60B; National Trust, graph 157, *Kinder Scout 10 Years On*, graph 171; Methuen 1988, Chorley, Schuman and Sugden, *Geomorphology*: graph 26T; Nature Conservancy Council: map 155; Nelson, M. Jones and B. Walsh, *Assignments in Physical Geography*: diagrams 162; *The New Internationalist*: 'Saints and Sinners', extract 135; *New Scientist*: extract 101; *New York Times*, extract 79; *Off-Licence News*: extract 101; Ordnance Survey ©Crown Copyright reproduced with the permission of the Controller of the HMSO: basis of map 8, 32M/B, 33, 56, 236; Paladin, Rosenblaum and Williamson, *Squandering Eden*: extract 116; Philip Allan Publishers Ltd, M. Penny, 'Investigating the productivity of upland forests', *Geography Review*: diagram 198, graphs 198, table 199; Prentice Hall, T. McKnight, *Physical Geography: a landscape appreciation*: map 71; *Richmond and Twickenham Times*: extract 94; Roundhouse Publishing, A. Neuman, T*ropical Rainforest*: diagram 182; Routledge 1992, M. Newson, *Land, Water and Development*: basis of map 22, table 30; Royal Meteorological Society, Landsea *et al*, in *Weather*: maps 114, 117, graphs 117, 118, extract 119, map 120; Solo Syndications, *Evening Standard*: extracts 80, 101; Reproduced by kind permission of Tees and Hartlepool Port Authority Ltd, map 66; Teesmouth Field Centre, Hartlepool Power Station, Tees Road, Hartlepool, Cleveland: map 62; *The Shepparton News:* extract 220; UK Meteorological Office ©Crown Copyright reproduced with the permission of the Controller of the HMSO: tables 85, 94, 95; extract from The *Meteorological Magazine* 100, Hadley Centre: map 129, graph 133; US Department of the Interior, *The Story of the Hoover Dam*, 1976: extracts 25; Walter, H., *Vegetation of the Earth*, EUP: diagram 155T; John Wiley, J. P. Grime, *Plant Strategies and Vegetation Processes*: diagram 246; WMO/UNEP, *Scientific Assessment of Climatic Change*: graphs 125, 126; World Wide Fund for Nature, Korup Project: extract 183, maps 184, 190, table 185, diagram 188; YDNPC, working paper 21, Ken Willis and Guy Garrod, *Landscape Values: a Contingent Approach and Case Study of the Yorkshire Dales National Park*, basis for scenarios, 232–33.

Photographs
Airphotos: 44, 58, 64, 236; Associated Press: 79; Australian Picture Library/John Carnemolla: 215; Ian Berry/Magnum Photos: 29; Alan Blackburn: 35T/M/B; Bruce Coleman: 24, 179T, 180B, 183, 241; Bob Digby: 45L, 76, 214, 216, 217, 218L/R, 228T/B, 231; Frosted Earth Publications, *Surrey Weather Book*, 93; Halton Deutsch: 5; Mike Hulme: 111L/R; 113; David Kinninment: 16, 18, 20, 21, 31, 34, 41; *Lake District Herald*: 7; Paul Morrison: 45R; The National Trust: 164, 165, 169, 172; NERC: 89, 90, 92L/R; NHPA: 54, 74, 150, 155, 201, 202, 206, 207T/M, 247; NPA: 110; OXFAM: 226; Panos Pictures: 209/Jeremy Hartley 208; Photoair: 197; Popperfoto: 4, 132; Graham Ranger: 180T, 186T/B, 189, 192, 193; Robert Harding: 50, 140, 142; Roger Scruton: 227; Solo Syndication: 80; SPL: 69, 120, 181; Teesside Development Corporation: 65; Telegraph Colour Library: 83; Warren Spring Laboratory, 144.

Contents

SECTION 1 Managing landform systems

SECTION 2 People, weather, and climate

SECTION 3 Ecosystems and human activity

Throughout this book you will find:

Theory boxes which explain geographical processes (e.g. page 9)
Techinque boxes which explain geographical techniques (e.g. page 86)
Activities to help you explore and understand geographical ideas (on green tints e.g. page113
Section summaries which list key ideas and are useful for revision (e.g. page 146)

1

Managing landform systems

Holding back the tide

In 1953, the Dutch suffered a nationwide disaster. Flooding swept through the country, caused by a 'deadly combination of spring tide and gales'. A low-pressure system brought heavy rain, while a surge in the high tides from the North Sea prevented most of the water from escaping out to sea. The waters breached and destroyed many of the protection barriers of the Netherlands, flooding thousands of square kilometres of countryside. The human death toll was 1853. The Dutch swore it would never be allowed to happen again.

Netherlands floods are the worst since 1953

THE Netherlands was hit by the worst flooding since 1953, when more than 1800 were killed.

At least one person died and 250 000 people had to leave their homes as emergency services warned that waterlogged flood defences were close to bursting point.

More than 1550km^2 of agricultural land were submerged and hundreds of homes ruined just 13 months after floods last swept the country.

Farmers, many of whom refused to leave their holdings, used tractors and trailers to move millions of animals out of danger or sent livestock to market early.

In the largest evacuation this century emergency services cleared the prison at Maastricht and dozens of hospitals, and carried thousands of books to safety.

Emergency workers toiled all night, piling sandbags on to sodden dykes around Tiel, the critical area, lying at the confluence of the Maas and Waal rivers.

Police and soldiers patrolled by boat and helicopter to discourage looting while people struggling to move possessions to the safety of upper storeys criticized those offering 'flood cruises' to sightseers as voyeuristic money-grabbers.

Some blamed the damage – put at up to 40 billion guilders ($23.5b) – on environmentalists who campaigned against raising, reinforcing, and extending the dykes which contain the Rhine, Waal, and Maas.

Prime Minister Wim Kok declared a state of emergency on 31 January and pledged a nationwide disaster.

Flood victim Truus de Ruiter said: 'My husband was stuck all night in a convoy of trucks and cattle trailers. Then the authorities banned the transport of animals to make way for people so my husband has stayed behind.

'Having to leave behind everything we have worked so long for is bad enough; what we will face if the biggest dykes break and the polder becomes a bathtub . . . I cannot bear to think about.'

▶ **Figure 1** From *The European*, 3–9 February 1995.

▶ **Figure 2** The Dutch floods of 1953 – many people died, and many hundreds of thousands more were evacuated from their homes.

In 1995 this promise proved to have been in vain. Despite the fact that the Dutch had become world leaders in coastal protection, much of the Netherlands was flooded in the early part of 1995.

About one-third of the Netherlands has been reclaimed from the sea. Large barriers, or 'dykes', have been constructed both to keep out the sea and to drain water away from the land into the sea. As land has been drained of water, it has shrunk, lowering the level of the land. This reclaimed land is below sea-level. Transport links and some settlements are built along the dykes, but pressure on space means that farmland and some towns and cities are built on land below sea-level. The dykes remain at the former level of the rivers draining the land, and are higher than the surrounding land.

Figure 1 is a report from *The European*, and reveals the impact that the 1995 flood had on people. The flooding was not merely the result of strong tides; additional water came from the river systems behind the dykes. The wettest January on record resulted in threatening water-levels in river systems such as those of the Rhine and the Maas. Heavy rain was made worse by premature melting of Alpine snow in Switzerland two months earlier than normal. Much of France, Belgium, Germany, and the Netherlands was subjected to serious flooding.

Effects of the flooding in the Netherlands

1 What were the consequences of the flooding in the Netherlands? Think about the effects:
  a) locally
  b) nationally
  c) internationally.

2 What are the short and long-term implications for the Dutch?

▲ **Figure 3** Flooding – a European or national problem?

Why do we study landform systems in Geography?

The study of landform systems such as coastlines, river basins, deltas, and estuaries is of practical as well as educational relevance to people. If we do not understand how these systems operate, then how are we to prevent or avoid disasters happening in the future? These events are entirely natural; it is only the fact that we live in vulnerable areas that makes them hazardous. A study of landform systems also helps us to understand the processes that affect the Earth.

Many of the causes of flooding in the Netherlands are natural. However, many people believe that it has been significantly influenced by human activities. For example, in eastern France the removal of hedges and woodland to intensify farming practices has meant that rainwater reaches river channels more quickly.

Environmental groups have suggested that modifications to the main channels of the river Rhine, such as straightening meanders and dredging, have turned the river into a massive 'funnel' capable of channelling water quickly to the north of Europe. As a result of these changes, this channel is now 80km shorter. Flowing at an average speed of 2–3km per hour, water from melting snow now reaches northern Germany and the Netherlands 30 hours earlier than it used to.

Yet others blame the environmental groups themselves. An argument raged in early 1995, with engineers complaining that environmentalists had delayed dyke strengthening and reinforcement. Reinforcement plans, they argued, had been sacrificed for the sake of a couple of trees.

Perhaps in the past the Netherlands has invested money in coastal defence at the expense of the flooding potential of rivers. The Dutch Prime Minister Wim Kok said that in the past the Netherlands had a magnificent reputation for defending itself against the threat posed by the sea, but 'now that the threat from the rivers seems greater than many had thought, we will have to prove our worth there too'.

Despite the views of Wim Kok and of many of the Dutch people, it was stated in *The Observer* that the immediate effects of flooding were much worse in France, Germany, and Britain than in the Netherlands, where flooding was the greatest. Dutch preparation for emergencies had allowed them to evacuate 20 000 people with little trouble or dissent. Two months later, the after-effects of the flooding became apparent. In March 1995 Dutch farmers were counting the costs of the damage to soil resulting from lead and other chemicals which had been deposited on the land as a result of the flooding.

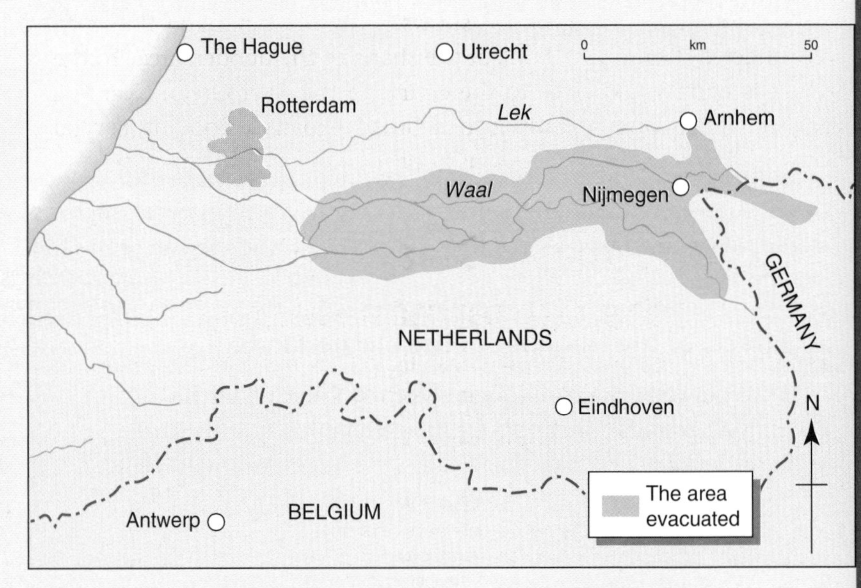

▲ **Figure 4** Area evacuated during the flooding of the Netherlands 1995.

1 How much should people attempt to modify landforms for their own benefit?

Do some research, looking at old newspapers, textbooks, or CD-ROM for information on floods in the past. Suitable examples include the Lynmouth flood of 1953, the floods along the east coast of England in 1953, and the floods of Bangladesh in 1988.

2 How much should the Dutch treat flooding as:

a) their problem

b) a problem involving the Germans and the French

c) a problem involving the Swiss?

3 Who should manage investment for such flood protection? Should it be:

a) the Dutch themselves, alone

b) a wider body such as the European Union?

4 Landform systems are often studied separately as 'coastlines' or 'river basins'. Should we consider rivers and coasts to be part of a global system, of which we are all a part? Use the case study above to argue your view.

1 Managing river basins

Flooding in Keswick

Figure 1.1 is a news report from the weekly *Lake District Herald*, 28 December 1985. It describes the floods which affected Keswick on Saturday 21 December 1985, and tells the story of families whose homes were flooded. Their homes were flooded to a considerable depth, as the photograph in Figure 1.2 shows. Only some parts of the town were affected (Figure 1.3). Look at the shape of the flooded area. To the west it is narrow and close to the course of the river Greta. Near Fitz Park, it widens and spreads into a broad expanse, flooding several houses. These were the first serious floods in Keswick for eight years.

'Operation Clean-up' after the Christmas flood

DOZENS of Keswick families had their Christmas holiday ruined when flood water poured into their homes on Saturday.

People living in High Hill, Crosthwaite Road, and Limepotts Road were worst affected by the water which rose to five feet in places after the River Greta burst its banks.

Mr Sydney Thorley's home in Crosthwaite Gardens was one of the worst affected by the flood.

BUNGALOW HIT

'We had about five inches of water throughout our bungalow and about 18 inches in the garage,' Mr Thorley told the *Herald*. 'The water was coming in so fast that we had to call the police and fire brigade. They managed to commandeer a boat and came right up to the front door to get us.

'It has undoubtedly spoiled our Christmas. Fortunately, we have no relatives or friends coming this year but we were hoping for a quiet Christmas,' said Mr Thorley.

Mr Barry Abbott's bungalow in Limepotts Road was also under several inches of water, and furniture, carpets, and beds were damaged.

DIESEL OIL

'All sorts of things came in with the water, even diesel oil. It all happened very quickly and the water went into every room.'

'We have been busy mopping up and we are all waiting for the insurance men to come.'

Many people living in Crosthwaite Road found their cars damaged by the flood water. In some cases it rose as high as the windscreen and spoiled upholstery.

Ravensfield was not flooded, though residents were evacuated as a precautionary measure because only a wall was holding back the raging River Greta. Some residents were taken to Keswick Cottage Hospital and some to other homes in the county.

Police stacked sandbags against the riverside wall which held and prevented much worse flooding.

The force of the extra water in the River Greta washed away part of a bridge at The Forge under the Keswick by-pass. A bridge, which links The Forge with Old Windebrowe, became impassable and was closed by the police.

Keswick Rugby Club was under four to five feet of water and players and officials spent the whole of Sunday mopping the premises out.

ROADS CLOSED

The squash club and the town's bowling green were also hit and some roads out of the town, including the Keswick to Borrowdale and Keswick to Thirlmere routes, were closed.

However, police reported no casualties as a result of the floods.

Police paid tribute to the way people at Keswick aided others in trouble.

'When we were filling sandbags a tremendous number of local people turned out and did their share,' said Superintendent Davidson.

◀ **Figure 1.1** From the *Lake District Herald*, 28 December 1985.

▲ **Figure 1.2** Limepotts Road, Keswick during the floods of 21–22 December 1985.

List all the effects of flooding reported in Figure 1.1. Categorize these effects into *tangible* (i.e. those that can be measured) and *intangible* (i.e. those that cannot be measured but which are still important). Are there other effects you can think of which the newspaper should have reported?

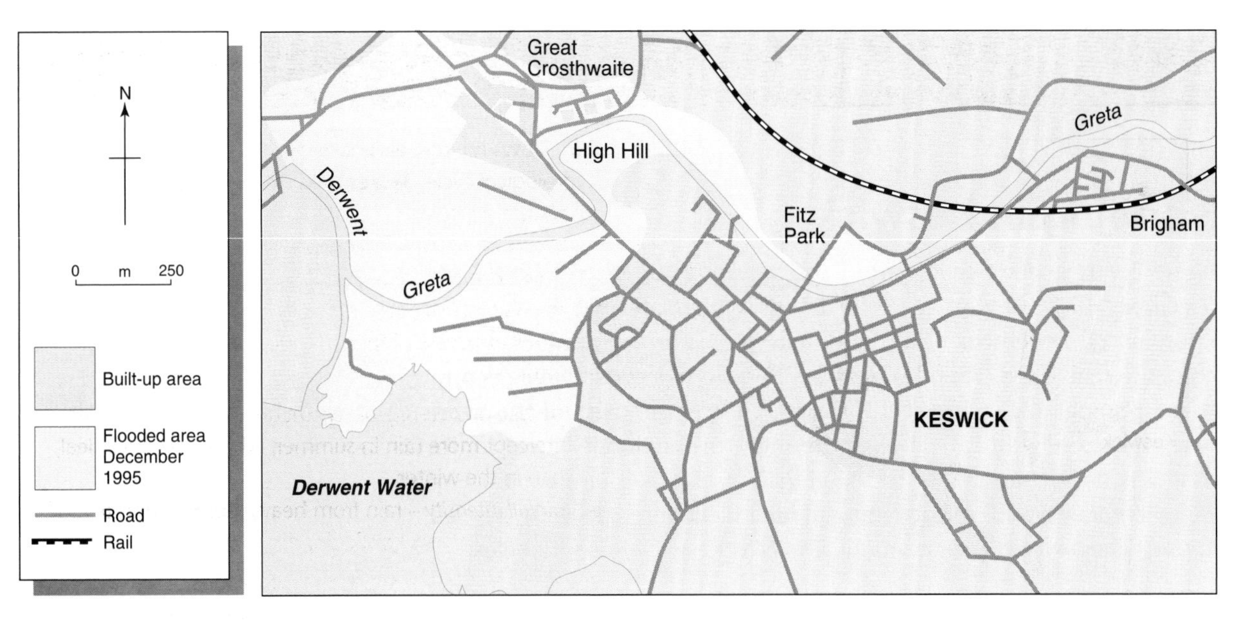

▲ **Figure 1.3** The extent of flooding in Keswick. Notice the shape of the flooded area – the floods affected housing only in the western part of the town.

1 a) How large was the area affected by flooding? Describe the location of the affected area.

b) Using Figure 1.3 and an atlas map of the Keswick area, or an OS map, explain why the flooded area is this shape.

2 In Figure 1.4, how did the residents of Keswick react to the flood? Were they right to feel this way?

3 This section of the book is about managing landform systems. Who is responsible for managing flood systems in the Keswick area? What are their responsibilities, implied in Figure 1.4?

4 Does the evidence about the flood so far suggest that 'somebody was not doing their job'?

Keswick flood victims voice their anger at meeting with NWWA

OVER fifty people turned up at a meeting between Keswick flood victims and officials of the North West Water Authority.

The meeting, set up by Workington MP Mr Dale Campbell-Savours, was intended for those affected by the pre-Christmas deluge, but others attended to voice their anger.

The Water Authority are accused of allowing the Thirlmere Reservoir to remain too full, making the river Greta unable to cope with the overflow.

One member of the public who attended, Mr Eric Skilton, said he was 'far from satisfied' with the Water Authority's response to the allegations.

'We tried to get the Water Authority to immediately reduce the depth of Thirlmere to avoid another catastrophe but they didn't give much hope of this. We would be more satisfied if the lake was reduced straight away. This is the first thing to do as a safety precaution.

'Until the Authority reduces the level of the lake, the dangers are still here.'

Mr Skilton said Thirlmere's level needed to be reduced for only three months of the year when the lake was at its highest. At present, it was 4 m above danger level, he said.

There was general agreement from the Water Authority that they would pursue the matter with instruments and continue to investigate.

However, the Authority gave no offer of compensation to people whose homes and property had been damaged.

▲ **Figure 1.4** From the *Lake District Herald*, 1 February 1986.

Flooding occurs when a large volume of water enters a river system and cannot be contained within the river channel. By studying floods, we can learn a lot about how a river basin works. By studying a river basin closely, we can learn more about it and predict how likely it is to flood.

The hydrological cycle of a drainage basin

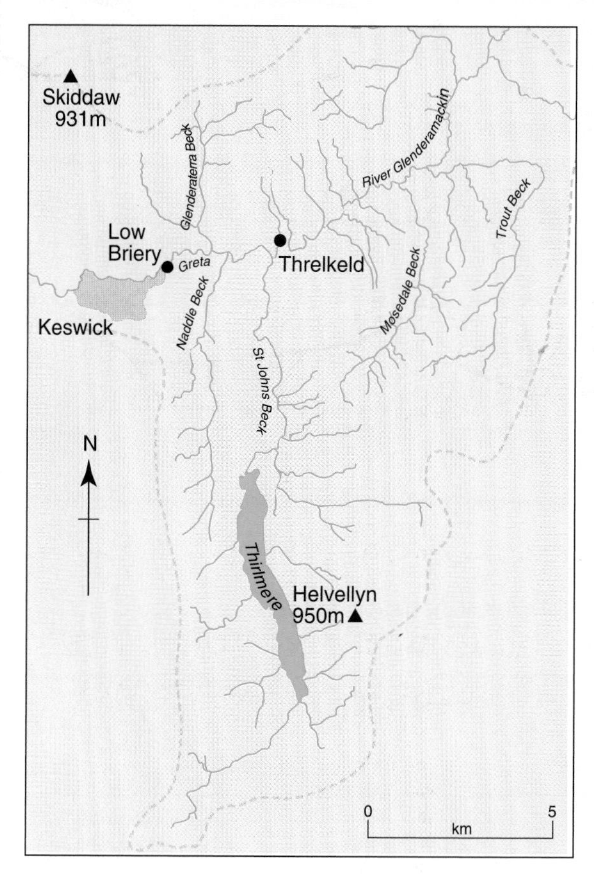

▲ **Figure 1.5** The drainage basin of the river Greta. This is a tributary river basin of the river Derwent in north-west England.

The hydrological cycle is the circulation and transfer of water in its various forms within and outside a river basin. River basins are also known as **drainage basins**, because they form the area from which rainfall runoff is drained and transferred into river channels. Figure 1.5 shows the drainage basin of the river Greta. The basin has a border called the **watershed**. A number of small streams or tributaries flow from the watershed, and these join to form larger streams and the main river course.

The hydrological cycle may be observed within river basins and can be shown as a systems diagram, as in Figure 1.6. **Systems** consist of four main parts: inputs, stores, flows, and outputs. An input is anything that is added to the system; an open system is one in which inputs are added from outside. The hydrological cycle within a drainage basin is an open system, as it gains precipitation from the atmosphere and energy from the Sun. **Flows** are movements of water within the hydrological cycle. **Stores** hold the water temporarily within the system.

When it rains, very little rainfall reaches the river channel directly. A large proportion is held and stored in the leaves and branches of plants – this is called the **interception zone**. The amount intercepted by plants depends on:

- *the type and density of vegetation* – deciduous plants intercept more rain in summer, when they are in leaf, than in the winter
- *rainfall intensity* – rain from heavy storms saturates soil more easily than gentle drizzle.

Water drips from leaves or flows down stems and reaches the ground surface. Some is evaporated straight into the atmosphere. Water reaching the surface, either straight away or via plants, moves into the soil. This process is called **infiltration**. The ability of soil to absorb water before saturation is reached is called its **infiltration capacity**. If infiltration capacity is exceeded then surface runoff occurs. The amount of infiltration depends on certain factors.

- *Antecedent rainfall* This is the amount of rainfall which has fallen recently. A spell of wet weather may saturate soil and prevent further infiltration.
- *Permeability of the soil* Sandy soils absorb water easily, and runoff rarely occurs. Clay soils are more closely packed, contain smaller air spaces, and saturate quickly.
- *Intensity of the rainfall* Rapid downpours are unlikely to be absorbed, and run off the surface more quickly.
- *Surface cover* Some surfaces, such as forest, have a large surface area to intercept rain. The passage of rain is therefore delayed.

Once water enters the soil, some is taken up by plant roots and transpired into the atmosphere. **Transpiration** is the process by which water vapour passes through leaves. Evaporation and transpiration are usually referred to together as **evapotranspiration**, as they are very difficult to measure separately. Once water enters the soil by infiltration, gravity causes it to move downslope through soil air spaces. This is known as **throughflow**. Water continues to infiltrate into solid rock below the soil. Rocks with joints or cracks allow water to move more easily to the groundwater area. The upper limit of this saturated area is known as the **water table**. Water from this area seeps slowly towards the river channel as groundwater flow, or **baseflow**.

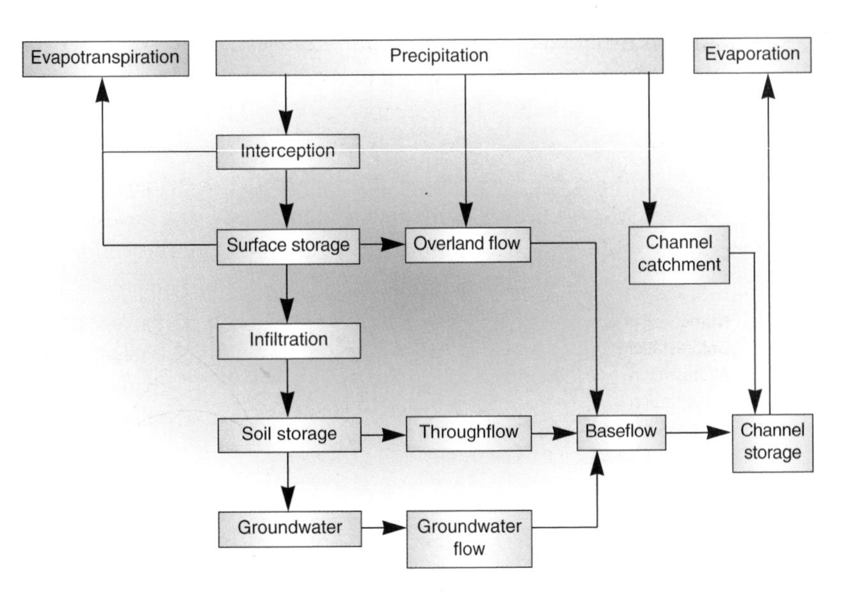

▲ **Figure 1.6** A systems diagram of the hydrological cycle.

▶ **Figure 1.7** The regime of the river Greta in 1985. Compare this graph with the rainfall pattern shown beneath the graph.

Mean daily flow m³/s

Year from 1 January 1985 at 0900 GMT

	Jan	Feb	Mar	Apr	May	Jun	Jul	Aug	Sep	Oct	Nov	Dec
Rainfall monthly total (mm)	85.4	15.8	137.9	153.1	102.0	80.7	147.2	407.8	190.4	146.6	193.5	348.2

1. On Figure 1.7, identify the day of the flood in Keswick described in Figures 1.1–1.4. What was the likely cause of the flood?
2. What can be deduced from Figure 1.7 about seasonal weather conditions in 1985?
3. In 1985 there were occasional heavy summer rainstorms, but this did not seem to affect the river flow much. Why?
4. Even in lengthy rainless periods the river Greta does not dry up. Why might this be?

The regime of the river Greta

A river regime shows how the pattern of discharge in a river varies. Discharge is the volume of water in a river channel. It is measured in cubic metres per second (m^3/sec, or cumecs). Look at Figure 1.7, the regime for the river Greta. Discharge varies over the year. Many factors affect the level of discharge in a river channel, including intensity and duration of rainfall, gradient or steepness of the slopes in the drainage basin, soil type and thickness, drainage basin density (the relationship between the number of tributaries and the area of the drainage basin), and shape and size of the drainage basin.

Factors affecting flooding

Flooding is usually the result of a combination of factors. Figure 1.8 is a Venn diagram which shows factors in a drainage basin, such as slope form, human activity, soil properties, and the storms that may affect it. Some of these produce conditions which could lead to flooding in the river basin and should be shown in the shaded part of the Venn diagram, where the factors intersect. Others are less likely to do this, and belong in the clear areas of the Venn diagram.

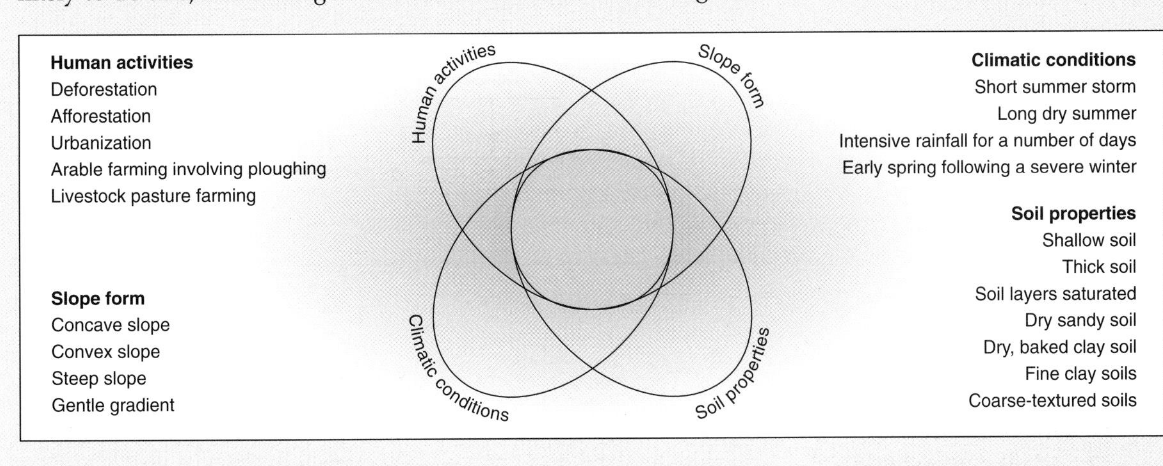

▲ **Figure 1.8** Venn diagram to show factors in a drainage basin that could affect flooding.

Copy the Venn diagram in Figure 1.8 onto a large sheet of paper. Think about what each factor involves. Will it lead to an increased likelihood of flooding? If it will, place it in the central blue area. Or will it reduce the likelihood of flooding? If so, it belongs in one of the outer areas. Complete the diagram by adding factors to it. You will notice that many conditions produce flooding. Sometimes several occur together; at other times it only takes one or two.

The flood hydrograph

Study the flood hydrograph in Figure 1.9a. A **hydrograph** shows river discharge plotted against time. It is different from the river regime because it shows discharge over a much shorter period, usually only a few hours. A hydrograph consists of two different parts: rainfall, and river discharge. Bars indicate rainfall, and the line graph represents river discharge. Rising and falling (receding) lines on the discharge graph are referred to as 'limbs'.

Many factors affect how quickly runoff reaches a river channel, and these in turn dictate the steepness of the rising limb. The faster that runoff reaches the river channel, the steeper the rising limb is, and the more likely the river is to flood. The hydrograph for a forested river basin, for example, where

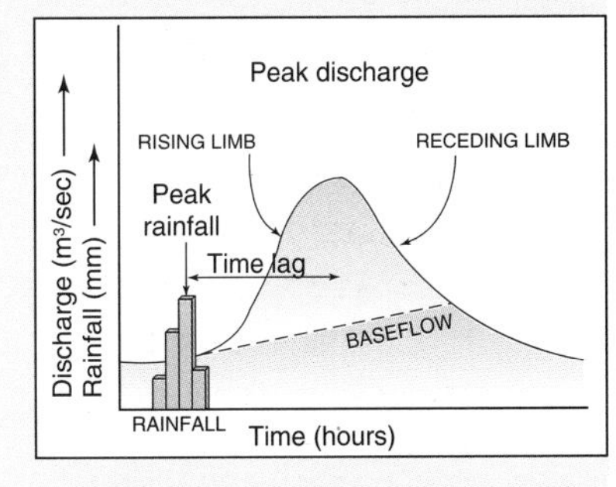

▲ **Figure 1.9a** A typical flood hydrograph.

Continued on page 12

Continued from page 11

there is good interception, might have a gently rising limb (as in Figure 1.9b), while the one for a deforested basin is likely to have a steep rising limb (as in Figure 1.9c). The time between the point of maximum rainfall and maximum river discharge is the **time lag**. Time lag depends on factors such as interception. For example, a deforested basin with steep slopes is more likely to have a short time lag than a thickly vegetated basin with gentle slopes.

The receding limb shows how quickly the river recovers from the runoff that results from a storm. A dense vegetation cover in a river basin increases the chance that the rainfall will be intercepted. In this case, there will be a long period before water reaches the river, and the receding limb of the hydrograph will be gentle (see Figure 1.9b). Usually, most river water is from baseflow. Baseflow is water that the river channel receives from seepage through the bedrock. This water takes a long time to reach a river. Any rise and fall in the river following a storm is likely to be the result of throughflow and surface runoff. Underground water (baseflow) does not appear in the river until much later.

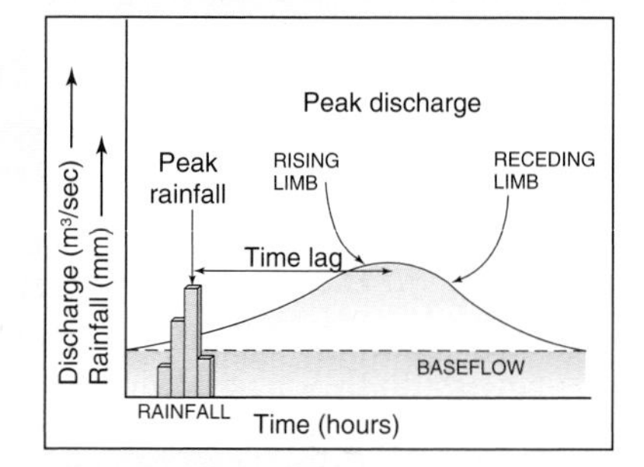

▲ **Figure 1.9b** Hydrograph for a gently rising and receding river. Note how the river rises more slowly and time lag is longer between peak rainfall and peak discharge.

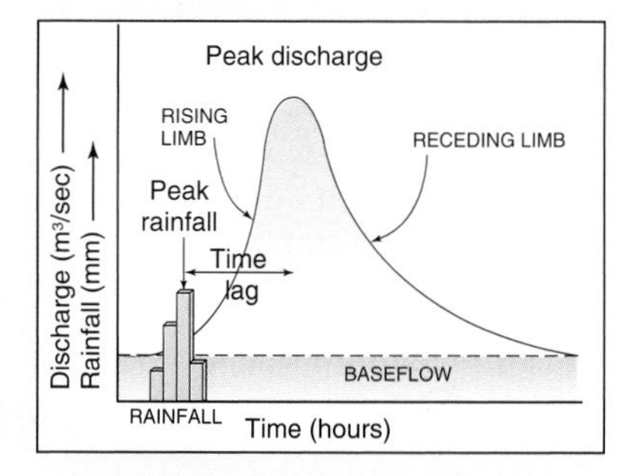

▲ **Figure 1.9c** Hydrograph for a fast-rising and receding river. Note how the river rises more quickly and time lag is shorter between peak rainfall and peak discharge.

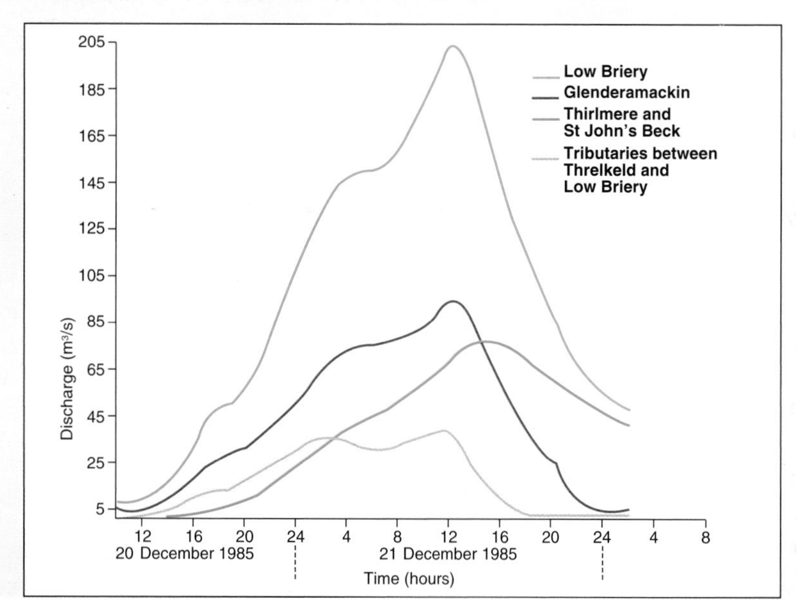

◀ **Figure 1.10** Simulated flow hydrographs (based on rainfall and flow data) at Low Briery showing flow contributions from the main tributaries of the river Greta for the flood of 21 December 1985.

Read the theory box on the flood hydrograph. Figure 1.10 shows computer models of the hydrographs for the river Greta and selected tributaries of the Greta for the storm of 21 December 1985, which caused the flood of the river Greta in Keswick.

1 a) Which of the main tributaries contributed most to the flooding of the river Greta? Why? Study Figure 1.5 on page 11, which shows the locations of these tributaries.

b) Which of the main tributaries contributed least to the flood? Why?

c) Which of the main tributaries peaked latest? Why?

2 a) Use an Ordnance Survey map of the area to trace the river Greta and its tributaries.

b) On your map, mark the watershed of the river Greta's drainage basin.

c) Annotate your map to show factors that influence the flow of the water in this drainage basin, referring to Figure 1.10.

3 In what ways might the flood in Keswick be linked to the shape of the drainage basin?

How should the flood problem be solved?

This exercise is about possible solutions to the flood problem in Keswick, through a 'balloon debate'. A number of people have to argue their case to stay in a hypothetical balloon that is losing height. The balloon stays in the air by losing people. People stay in the balloon by the strength of their arguments.

1 Five speakers, or pairs of speakers, are allocated roles, with viewpoints which they must defend.
2 Speakers have two minutes to justify their argument.
3 When everyone has presented their view, the audience votes to eject one viewpoint from the balloon.
4 The discussion continues in five-minute rounds. At the end of each round one viewpoint must be ejected.
5 The most persuasive argument wins.

The five roles and viewpoints are:

1 *Thirlmere is a flood control device* In fact, Thirlmere is a reservoir built to supply Manchester with water, but people in Keswick believe that one way to control excess runoff should be to store it in lakes, so that no flooding will occur. Thirlmere could be used to regulate the flow of the river Greta, particularly during peak runoff periods.
2 *The engineers' view* Engineering solutions provide the best means of controlling floodwaters through Keswick. This assumes that technology can control flooding, and that money spent is money well spent. Such schemes meet with local approval because they are visible and show that something is being done.
3 *The opposite view to (2) above* Floodwaters can be controlled through engineering, but only so far. Any means of controlling flooding can only be as good as its design, and the standards to which it has been built. If its design standards are exceeded by subsequent floodwater levels, damage may be even greater than if there had been no attempt to control floods in the first place.
4 *Flood abatement schemes are necessary* By changing land use upstream, such as afforestation, the regime of the river is altered. Increased afforestation would increase interception, and delay the time taken for rainwater to reach the river. This view attempts to analyse the causes of flooding through river basin processes.
5 *National Rivers Authority (NRA)* The NRA argues that Thirlmere is not there to control flooding. It believes that Thirlmere has been successful in supplying water to Manchester for over 50 years. The NRA also believes that Thirlmere is part of one tributary, St John's Beck, in a river basin that contains many others. Flooding could never be controlled by Thirlmere alone.

Flood alleviation

Flood alleviation is concerned with reducing the effects of flooding on the human environment. The National Rivers Authority (NRA) is responsible for flood alleviation in England and Wales. There are several possible approaches – these are shown in Figure 1.11. Study this carefully. Possible strategies range from engineering modifications to river channels and protection of buildings from flood damage, to doing nothing and bearing the cost of any flood that occurs. The four main strategies are:

- structural protection measures through engineering solutions
- river basin management
- modifying the burden of loss
- bearing the loss caused by flooding.

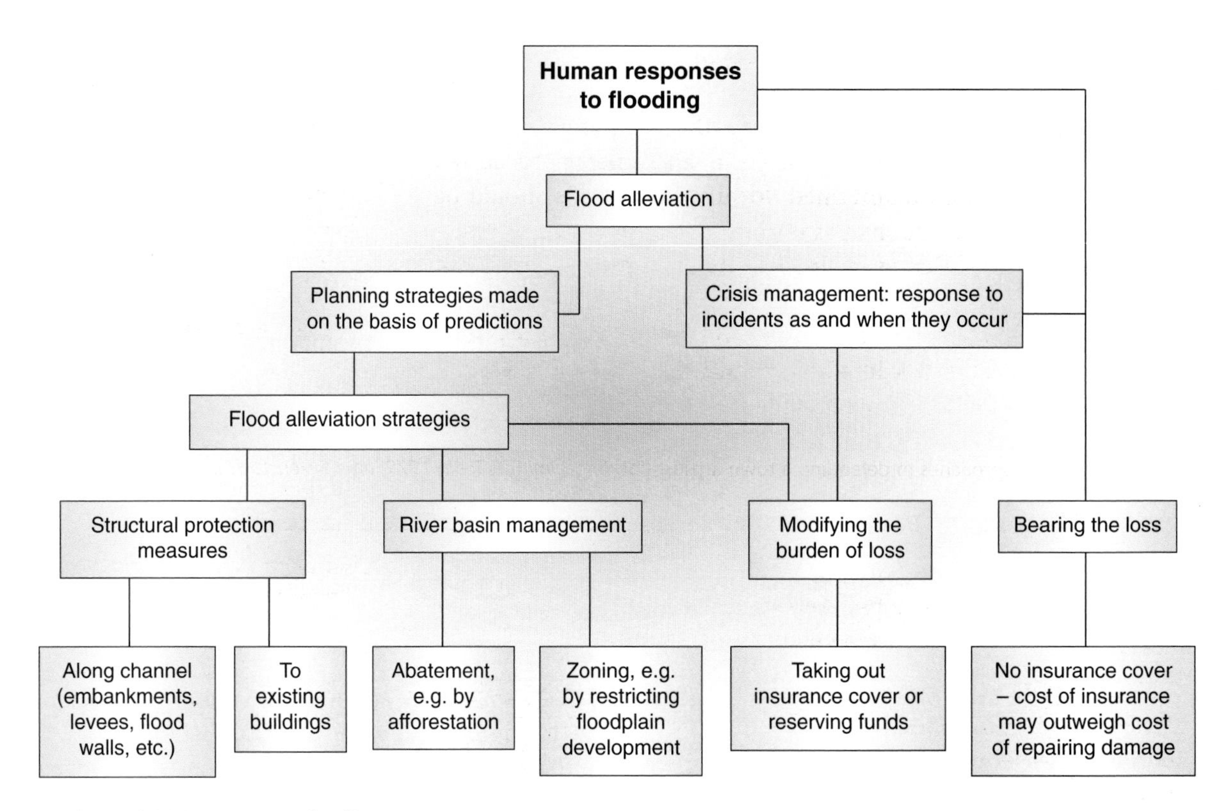

▲ **Figure 1.11** Responses to flooding.

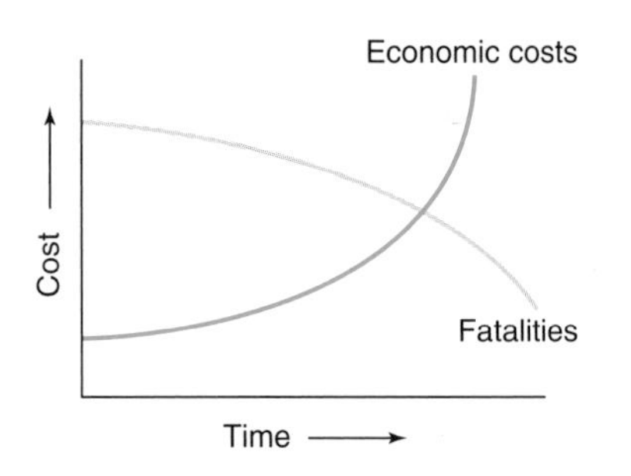

▲ **Figure 1.12** Economic costs and fatalities caused by flooding in developed countries. Notice how although fatalities are falling, economic costs are rising sharply. Think about why this might be.

For economic reasons, bearing the loss caused by flooding is most likely to be found in developing countries. It is not likely to meet with support in a developed country, as at Keswick. The graph in Figure 1.12 shows that in developed countries the loss of life due to flooding is falling with time, but economic losses are increasing. Why is this?

Structural protection measures through engineering solutions

Engineering structures represent the most publicly acceptable response to the flood hazard. 'Flood protection' is a misleading term, as protection is only as good as the design of the structure in question. But since it is the most publicly accepted response, residents may believe it to be infallible. When a protection scheme is breached, as it eventually will be, damage is likely to be greater than if there had been no scheme in place at all. This is because of the attitude of complacency that seeming safety can bring.

Education programmes need to work alongside structural schemes, to encourage flood-proofing. River management is increasingly more concerned with trying to reduce flood losses than preventing flooding altogether.

Embankments, or levees, and flood walls (Figure 1.13b)

These are designed to restrict flooding to defined limits on the floodplain and to allow controlled flooding of certain areas. They are a relatively cheap form of flood protection and operate to design standards if they are well maintained. Ideally, flood walls should be located as far away from the river as possible to increase the potential storage of flood water. However, flood walls and embankments can increase flooding both upstream and down-stream – upstream because they may constrict flow, downstream because they may discharge water more quickly into an area that is less able to absorb it.

Figure 1.13 shows the basic flood problem and five approaches to structural defence measures.
Study the diagrams carefully as you read the text.
Consider which approach you feel is the best.

▼ **Figure 1.13** Five approaches to defending a town against flooding *(Smith & Tobin 1979, after Nixon 1966).*

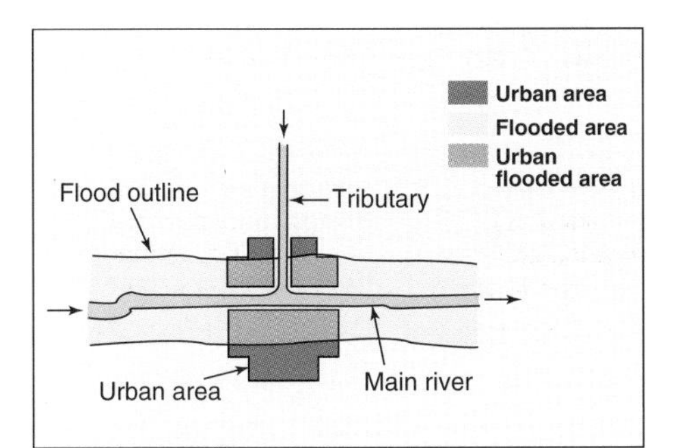

(a) The basic flood problem.

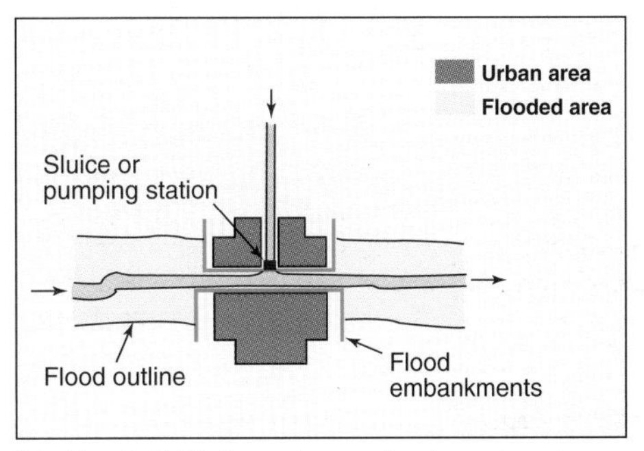

(b) Flood alleviation using embankments or levees.

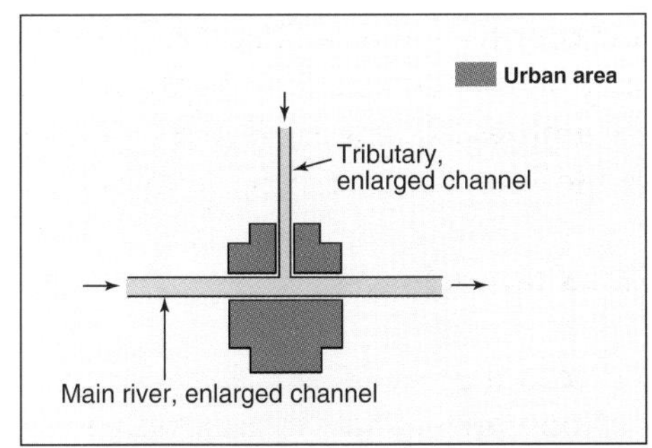

(c) Flood alleviation using channel improvements.

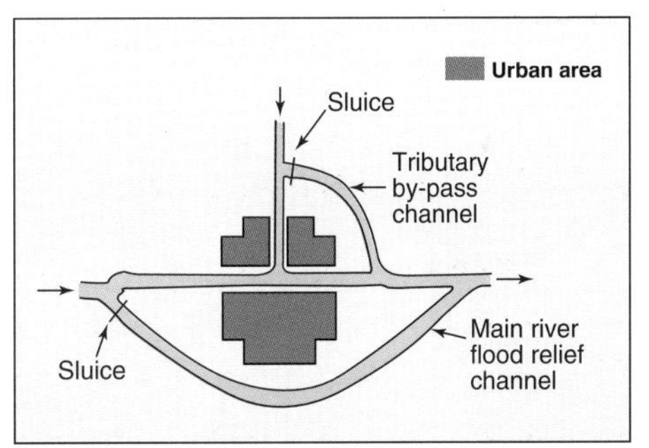

(d) Flood alleviation by means of relief channels.

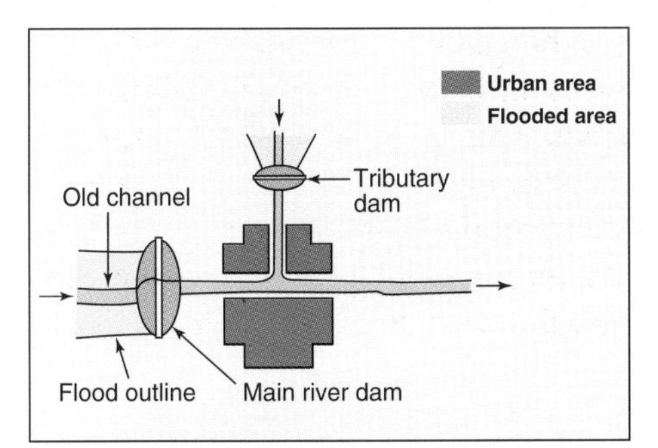

(e) Flood alleviation achieved by means of reservoir storage.

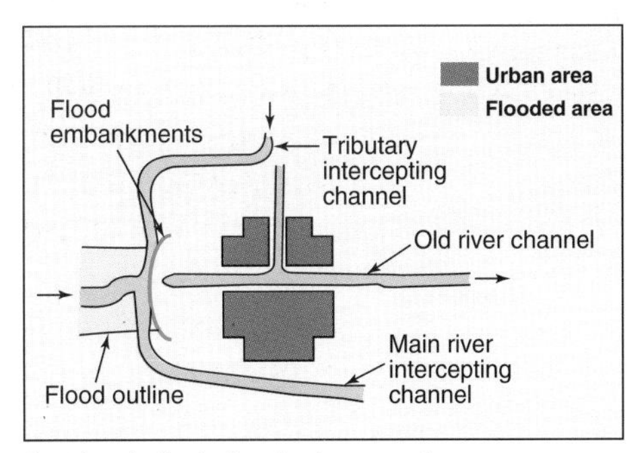

(f) Flood alleviation by interception.

Channel improvements (Figure 1.13c)

Channel improvements are designed to confine floodwaters to the river channel. There are two main ways in which this can be done.

1 Channel roughness can be reduced by clearing vegetation and other obstacles or by lining the channel with a smooth surface such as concrete. This reduces friction and allows faster discharge. Look at Figure 1.14. The river bed is strewn with stones and boulders. These are rough surfaces which increase friction, slow the river down, and reduce its channel capacity. The wall dividing the river from the road is much smoother, and enables the water to flow more quickly. Measures like this help water to flow away more quickly and reduce the risk of a flood, but can increase flooding downstream.
2 The channel can be widened or deepened by dredging. This increases the capacity of the river to hold water.

Relief channels (Figure 1.13d)

The channel can be shortened by cutting meander loops and steepening the gradient in order to allow water to discharge away from the area at risk. These schemes are initially very attractive and extensively used but have a number of disadvantages. Faster discharge may increase the likelihood of flooding downstream. In addition, straightening the channel may be self-defeating, as the river reverts back to its natural state of dynamic equilibrium (this is explained in the theory box opposite). This means that costly maintenance work is frequently needed.

▼ **Figure 1.14** River bed and wall in Keswick. Note the roughness of the river bed and the smoothness of the wall which divides the river from the road. The wall enables water to flow more quickly than it would if the whole channel were rough like the river bed, as it is smooth and offers less friction.

Equilibrium in river systems

Rivers reach a state of balance with processes that form them, called **equilibrium**. Rivers can be seen as systems. Balance within systems is complex, as the flow of energy and materials passing through may vary. If rainfall decreases then a river system has less energy and is less able to carry sediment. Sediment is therefore deposited; over time, it builds up and increases the gradient of the river channel, causing the river to speed up as the gradient increases. Sediment is therefore removed again and equilibrium is restored, as the river returns to a state of balance. This is known as **self-regulation**. Any changes to the river channel create conditions in which the river will attempt to re-create its original channel.

Sometimes conditions change so that the river is unable to recover its original equilibrium, and a new equilibrium has to be found. For example, when Mount St Helens in Washington State, USA, erupted in 1980, a massive amount of sediment was released into the neighbouring river systems. The original balance could not be recovered because the inputs into the system were so great. A new balance had to be found by the rivers.

Flood storage reservoirs (Figure 1.13e)

These store excess water in the upper reaches of the catchment, which is gradually released by careful regulation. Because of the high costs involved, reservoirs have never been built in Britain for the sole purpose of flood alleviation, though some have contributed to flood control. In this country, only upper river courses offer suitable sites away from lowland areas where most flooding occurs, so such reservoirs would have minimal effects downstream.

Flood interception schemes (Figure 1.13f)

These involve changing the course of the river, and work in three different ways:

- the river is re-routed to by-pass settlements under threat
- new channels can be used in addition to the natural channel to store water
- flood embankments help to contain floodwater well away from places under threat.

River basin management

River basin management means reducing the harmful effects of a flood, while accepting that a flood may happen. Three methods are well tried:

- flood abatement
- flood-proofing
- floodplain zoning.

Flood abatement

Flood abatement aims to reduce flooding downstream by changing land use upstream. The method most frequently used is afforestation, or planting additional trees. Afforestation delays the passage of water into river channels through increased interception. Also, greater evapotranspiration by the new vegetation reduces the amount of water in a river channel. This method can be effective but requires

Flood management – a decision-making exercise

Your role: You are on work experience with a firm of consultant engineers. While you are there, you are to produce a flood management plan for Keswick for North West Water. You need to refer to all the data in Figures 1.1–1.16.

Your report should include:

- a statement of the flooding problem at Keswick
- a table showing alternative types of flood alleviation scheme, with the benefits and problems of each type
- an annotated copy of the maps in Figures 1.3 and 1.5, to show the details and location of your suggested schemes
- a justification of the schemes that you recommend.

Flood frequency analysis

It is useful to estimate the chances, or **probability**, of a flood of a certain size occurring in a given period. Flood management depends on such estimates.

Probabilities are estimated as follows.

1 The size of the largest flood for every year is arranged in rank order, 1 being the largest, for the time during which records are available.

2 The following equation is then used to calculate the interval in time between floods of the same size:

$$T = (n + 1) / m$$

where T is the interval, n is the number of years of observation, and m is the rank order. The figure you get will tell you in how many years you can expect a flood of this size.

large areas of land, and time for forests to become established. For example, channels dug for plantations before tree planting on the slopes of Ingleborough and Whernside in the Yorkshire Dales National Park, actually caused an increase in runoff and flooding in the initial stages. Therefore this method is not suitable for all catchments.

Flood-proofing

Flood-proofing involves designing new buildings or altering existing ones to reduce damage that would be caused by flooding. These structures may be temporary or permanent. Flood-proofing tends to be less effective against high-level, fast-flowing, and longer-lasting floods.

▲ **Figure 1.15** A mix of older and newer buildings in Keswick. Notice how all the buildings are situated well above the level of the river. This is a good example of flood-proofing (avoidance).

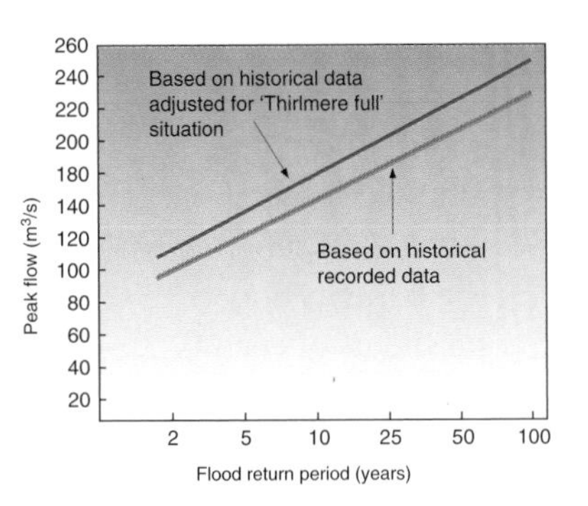

▲ **Figure 1.16** Flood frequency distribution at Low Briery gauging station.

Floodplain zoning

This is a management strategy whose aim is to reduce floodplain development. It suggests that floodplains are divided into zones.

Zone 1 The *prohibitive zone* (nearest to the river) where no further development is allowed except for essential waterfront facilities.

Zone 2 The *restrictive zone*, where only essential development and recreational facilities are permitted. All buildings should be flood-proofed.

Zone 3 The *warning zone* (farthest from the river) where inhabitants receive warnings of impending floods and are reminded regularly of the flood hazard.

What measures were proposed for Keswick?

Consulting engineers Babtie, Shaw, and Morton produced a report which attempted to consider the different options for Keswick.

1 Thirlmere was considered and found to give some protection against floods. It would not be possible to rely on Thirlmere, however, because it is not designed for this purpose. Also, it is not in a suitable location to control the flow of water in the Greta basin. Why do you think this is? Refer to Figure 1.5.
2 Increasing floodplain storage capacity upstream from Keswick would reduce downstream flood flows. This could be achieved by constructing a flood control weir which would raise water-levels upstream during flood conditions, or by constructing an embankment around the edge of the floodplain. This would create a storage reservoir to hold water for longer in the natural floodplain. Farmers whose land would be affected opposed this.
3 The cost-effectiveness of local flood alleviation schemes in Keswick were considered. The channel between Great Crosthwaite and Greta Bridge, and the bridge itself, have insufficient capacity to maintain floodwaters. Therefore the capacity of both the channel and the bridge have to be improved.

The following five basic schemes have been investigated.
1 The construction of flood walls and embankments at Crosthwaite, near the pencil factory, and in the vicinity of Greta Bridge.
2 Removal of the weir at Greta Bridge, with local regrading.
3 Removal of the weir at Greta Bridge, and regrading of the channel for approximately 400m upstream.
4 Removal of the weir at Greta Bridge, and widening of the channel for approximately 350m upstream.
5 Removal of the weir at Greta Bridge, with both regrading and widening of the channel. This design has been based on a 50-year flood flow at 210m^3/sec.

Scheme 1

Scheme 1 was used as a base with which to compare other schemes. It is probably unrealistic as it is likely to result in peak river levels upstream of Greta Bridge up to 450mm higher than those during the flood of 21 December 1985.

Embankments and flood walls would be built out of stone with reinforced sheet-piling at certain locations. They would be designed to contain floods to a level predicted to occur only once in 100 years. The flood would be contained within the channel through Keswick with the exception of the Upper and Lower Fitz Park areas.

Scheme 2

Scheme 2 concerns Greta Bridge (Figure 1.17). At present, the bridge does not allow sufficient water to pass under it at peak flow, which encourages flooding upstream of the weir. The capacity of the bridge would be improved by removal of the weir and lowering of the river bed under the arches. This scheme would be insufficient in itself but would enhance benefits from other schemes. Replacing the original bridge with a new single-span bridge was also considered, but the cost was likely to be in the region of £0.5 million.

1 Figure 1.16 shows the frequency with which flood levels are likely to be repeated at Low Briery in Keswick. This is called the **flood return period**. Estimate from the graph what the peak flows are likely to be for Keswick at:
a) 25 years
b) 50 years
c) 100 years.

2 Compare your answers with the level of the river Greta in the 1985 flood, shown on Figure 1.10. What is the return period for a flood of this level? By which year, therefore, is it likely that a flood of this level will have been repeated?

3 How should engineers use such data?

Scheme 3

This involves regrading the channel and removing the weir. Dredging the channel would give the river a greater capacity and allow a faster discharge. Between Lower Fitz Park and Greta Bridge – a distance of about 400m – the level of the river bed would be lowered by approximately 0.5m, and the pencil factory bank of the channel trimmed to a gentler slope.

Scheme 4

Under scheme 4, the river channel is widened and the weir is removed. Though costly, it is thought to be more reliable than regrading the channel as it is less likely to change with time. The initial degree of flood protection is likely to be maintained. Channel widening would be carried out along the pencil factory bank for a distance of about 350m. Embankments and flood walls would be used in conjunction with this scheme.

Scheme 5

This combines schemes 3 and 4. The additional channel capacity created by regrading and widening the existing wall along Main Street would mean that no increase in height is required. In the long term, it is likely that maintenance of the regraded river would be required.

▲ **Figure 1.17** The Greta bridge in Keswick. Notice the arches and central support. This support takes up space and prevents the fast flow of water in times of flood. However, the bridge is of historical interest, and instead of removing it and constructing a new bridge, the plans are to deepen the river bed and increase water capacity in that way.

Assessing the effectiveness of flood alleviation schemes – a cost–benefit analysis

Five schemes for Keswick were considered by the NRA. A means of analysis and evaluation would help to identify the best scheme. One technique frequently used is cost–benefit analysis, in which the costs of a scheme are weighed against the potential benefits.

Costs and benefits can be divided into primary and secondary categories. Primary costs and benefits are directly concerned with the scheme. Primary costs in a flood alleviation scheme would be the purchase of materials or purchase of land. A primary benefit would be saving the cost of cleaning up following a flood. Secondary costs and benefits are related to the scheme, but not so directly. A secondary cost might be educating people about the possibility of a flood exceeding the design capacity of the scheme. Secondary benefits might include increased value of property that was previously prone to flooding.

Public projects of this kind have some costs and benefits which are difficult to measure. Placing a value on the costs and benefits of a scheme can be very difficult. Some things are easy to cost, such as building materials. Others, such as loss of trade caused by flooding, are more difficult. A reduction in human stress caused by flooding is even more difficult to value. Can stress be valued in terms of the costs of a therapist to help deal with that stress, for example? Such analysis is bound to be highly subjective, and means that figures vary according to personal opinions of individuals or groups.

What was eventually decided?

In Keswick it was finally decided to:

- increase the capacity of Greta Bridge by removing the weir and deepening the channel (shown in Figure 1.17)
- increase the capacity of 450m of channel upstream of the bridge
- construct 1300m of earth embankment, and modify and construct flood walls along 300m of the river.
- improve the wall along different stretches of the river in Keswick.

There were two phases to the scheme. The first included the construction of flood defence embankments and walls, and sewer outfall improvements. The second phase included river works and regrading upstream of High Hill Bridge, retaining walls, deepening under the bridge, and constructing a flood bank alongside the pencil works.

You will have your own views about whether or not these were the best solutions. The question for the borough engineers who planned the solutions will always be: 'Were they the best possible solutions?' Until the next large flood comes, we won't know. If the improvements do their job, will people congratulate the engineers? If they don't work, the newspapers will no doubt 'point the finger'.

1 Draw a table with the following headings: Scheme, Tangible costs, Intangible costs, Tangible benefits, Intangible benefits. Using two colours, list the primary and secondary costs and benefits involved in the proposals for Keswick. Distinguish between *tangible* and *intangible* costs and benefits.

2 Do the benefits of doing something about flooding outweigh the costs? Or is there a case for the approach which refers to 'Bearing the loss' (see Figure 1.11), where nothing is done, there is no cost, and flooding is dealt with as and when it happens?

▼ **Figure 1.18** Gabions used to strengthen the river bank. Gabions are metal cages which hold stones in place, and these are used to protect and build up river banks. Small tributary streams are lined with concrete to allow faster movement of flood waters.

Large river basins – the Colorado

Figure 1.19 shows the world's largest river basins. We look here at the Colorado river in the USA. The Colorado ranks as one of the world's major rivers. Its basin contains some of the most spectacular scenery in the USA and, in such an arid part of the country, the river is vital for the water it provides. The Colorado has been described as 'the most legislated, most debated, and most litigated river in the entire world' (Reisner, 1990). Why should a river be a subject for the law, and court cases?

Compared with other river basins of a similar size, the Colorado is both more exploited and more managed. This is because it is used to support a large population and industrial base in part of the USA, in many cases outside its own basin. The Colorado is vital to the states of south-west USA, supplying water to 17 million people in seven states and Mexico. Using its water, farmers produce \$500 billion worth of farm sales, and electricity companies generate \$21 million of electricity supply. It is also used as a public water supply, for flood control, and for recreation. Its drainage basin is 632 000km^2, or one-twelfth of the USA – an area larger than France – and most of it is semi-arid land. It falls more than 3000m from its source, through the Rocky Mountains, to its mouth in the Gulf of California. Included in its basin is the Grand Canyon (see Figure 1.26).

The Colorado is one of the first rivers in the world to have its entire flow used up before it reaches its estuary in the Gulf of California. During most years, its delta is dry. However, there is only so much water, and demand for it is rising every year. A study of the Colorado raises many issues about the management and use of river basins. Before large-scale management, the Colorado was subject to considerable seasonal fluctuations in discharge. There was often widespread flooding from April to June, caused by snowmelt in its Wyoming headwaters. These floods were part of the Colorado system; they brought fresh sediment to the floodplains, and recharged the area's groundwater.

▼ **Figure 1.19** The world's largest river basins

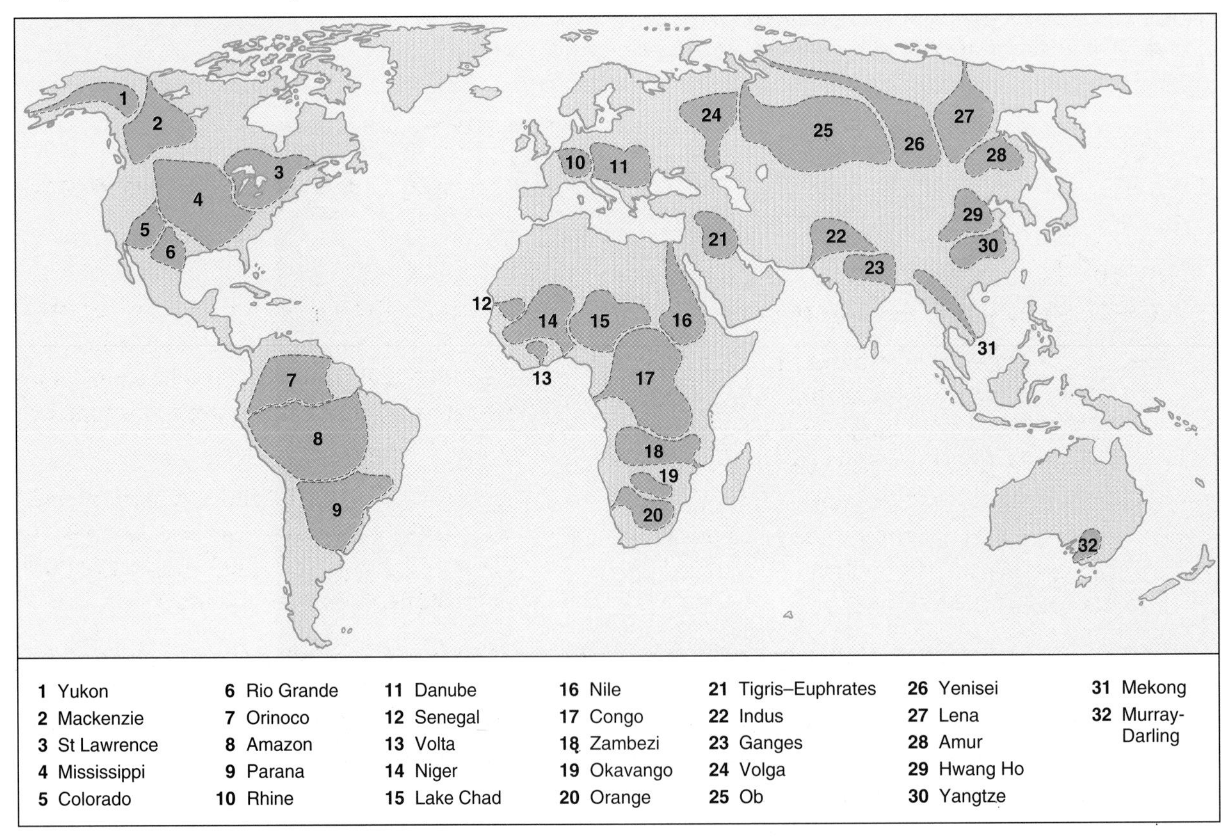

Figure 1.21 lists the major dams on the Colorado system. The first of these was the Hoover Dam (Figure 1.22), which was approved for construction in 1928, and was in operation by 1935. Figure 1.21 shows that since the Hoover Dam was built, more and more dams have followed. This has prompted the view that the Colorado is 'the world's largest plumbing system'. In fact the Colorado system is often referred to as a cascading system which is controlled by this series of dams and reservoirs. Many of the lakes behind the dams attract recreational visitors, and thus income for the area. The Colorado river schemes have been carried out with wider purposes in mind.

River	State	Dam	Reservoir	Date	Height (m)	Capacity (million m³)
Colorado	AZ, CA	Parker	Lake Havasu	1938	98	995
Colorado	AZ, NV	Davis	Lake Mojave	1950	61	2795
Colorado	AZ, NV	Hoover	Lake Mead	1936	221	45 748
Colorado	AZ	Glen Canyon	Lake Powell	1964	216	41 513
Colorado	CO	Granby	Lake Granby	1950	91	829
Los Pinos	CO	Vallecito	Pine Lake	1941	49	199
Blue	CO	Green Mountain	Green Mountain Lake	1943	94	191
Aqua Fria	AZ	Waddell	Lake Pleasant	1927	78	251
Gila	AZ	Coolidge	San Carlos Lake	1928	76	1 843
Verde	AZ	Horseshoe	Horseshoe Lake	1949	43	214
Verde	AZ	Bartlett	Bartlett Lake	1939	87	277
Salt	AZ	Stewart Mountain	Saguarto Lake	1930	63	107
Salt	AZ	Mormon Flat	Canyon Lake	1938	68	90
Salt	AZ	Horse Mesa	Apache Lake	1927	93	377
Salt	AZ	Roosevelt	Roosevelt Lake	1911	85	2 125
Green	WY	Flaming Gorge	Flaming Gorge	1964	153	5 826
Strawberry	UT	Strawberry	Strawberry Lake	1913	22	353
Price	UT	Scofield	Scofield Lake	1946	38	851
San Juan	NM	Navajo	Navajo Lake	1963	123	2 606
					Total storage capacity	106 928

AZ Arizona, CA California, NV Nevada, CO Colorado, WY Wyoming, UT Utah, NM New Mexico

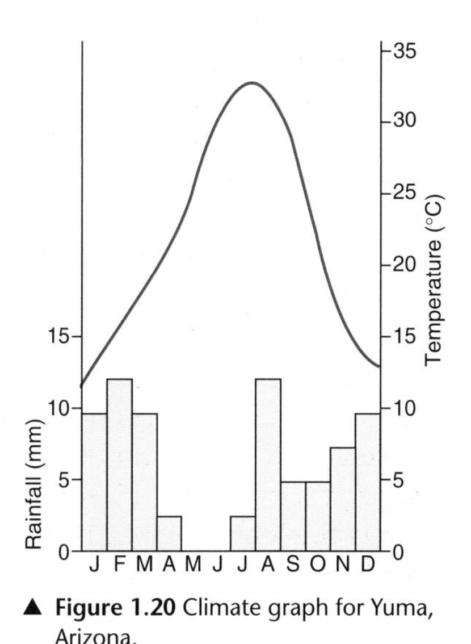

▲ **Figure 1.20** Climate graph for Yuma, Arizona.

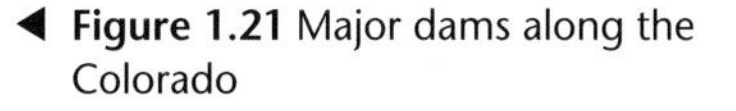

◄ **Figure 1.21** Major dams along the Colorado

Issues facing the Colorado river

1 a) Locate the Colorado basin in your atlas. Draw its basin on an outline map of the USA, including its major tributaries. Locate their sources and label these on your map.

b) Identify and label the states of the USA through which the Colorado and its tributaries flow. Show the Mexican border.

2 a) Using an annual rainfall map of the USA, show those areas of the basin that receive less than 500mm of rain per year.

b) Identify and locate Yuma, Arizona on your map. Annotate your map with details about the Arizona climate, such as annual rainfall and temperature patterns, using Figure 1.20. What are the implications of a climate such as this for water management in south-western USA?

3 Find a population distribution map in your atlas. Identify and label the largest cities of Arizona, Nevada, and California, and other areas of high population density.

4 What is suggested by your map about the imbalance between people and water distributions? What questions need to be asked and answered?

1. Find up-to-date data on the populations of the seven states in the Colorado basin.
2. In US elections, senators are elected on the basis of populations of different states. The greater the population of a state, the more senators represent that state in Washington. What is the geography of political strength within the seven states you have located? Where is the political 'centre of gravity'?
3. In whose interests do you think the Colorado schemes were first developed? Who stood to lose from these proposals?
4. What appear to be the main costs and benefits of the Hoover Dam as shown in Figure 1.23? Who stood to have the greatest gains? Who stood to have the greatest losses?

▲ **Figure 1.22** The Hoover Dam.

Look at Figure 1.22. Although flood control became the prime factor by which to gain public support for the dam, the river has a variety of uses that exploit its potential as a resource. Storage and water supply for agriculture and industry are the main reasons for the schemes. Public water use is also important, as is the generation of hydro-electric power (HEP). Recreation is now also very important, creating almost five times as much income as public water supply.

Flood control

The US Bureau of Reclamation is responsible for controlling the flow of the Colorado and for ensuring that there is storage space in the reservoirs for flood control. This is done through a complex set of regulations. Each winter the Bureau of Reclamation must make important decisions on how much space is to be left in each dam. This is calculated by taking into account the amount of snow that has fallen and settled and guessing how early the spring is going to be, based on temperature patterns. It can take weeks to release sufficient excess water to create storage capacity in the reservoirs, because eac¹ of the sections of river between each dam has only a certain discharge capacity. If the amount of discharge from each dam is underestimated, the capacity of the reservoirs is exceeded. This might cause flooding, which could create huge costs in damage. If an overestimate is made, too much water is released from the reservoirs, and there will be shortages later in the year. Remember that all of the water in the Colorado system is used and that demand is increasing.

a

Farmers, tempted by fertile desert soil, tapped the Colorado for water to irrigate and create rich gardens – but annually the river rose in destructive flood-tide to destroy their crops, and annually the river's flow dwindled so that all living things were faced with water shortages.

b

The proposed [Hoover] dam would be so high that its reservoir could store the entire flow of the Colorado river – including all average floods – or two whole years. Furthermore, it would be located below the large tributaries and thus would provide for their control. The dam would create a power head within transmission distance of the power markets of southern California. And it would be in the midst of a heavily mineralized region in Nevada and Arizona where low-cost power could be a boon to strategic metal production.

c

All the states could see, theoretically, the advantage of a great dam in the lower basin; but in all cases they were concerned about its effects on their individual fortunes. They were haunted by a bad dream of seeing 'their' water leave their borders, committed to a state whose better fortunes had enabled it to make use of the water.

d

The basic doctrine of water law recognized in all the basin states except California was that of prior appropriation and use. In other words, the person complying with . . . legal formalities . . . thereby secured a first right to its use. California had a dual system of water rights. In addition to appropriation rights, the State [of California] also recognized riparian rights – the right of a landowner on the bank of a stream to the water flowing past his property. Other States . . . [had given this up].

▲ **Figure 1.23** Extracts from *The Story of the Hoover Dam,* a guide for those visiting the Hoover Dam Visitors' Center.

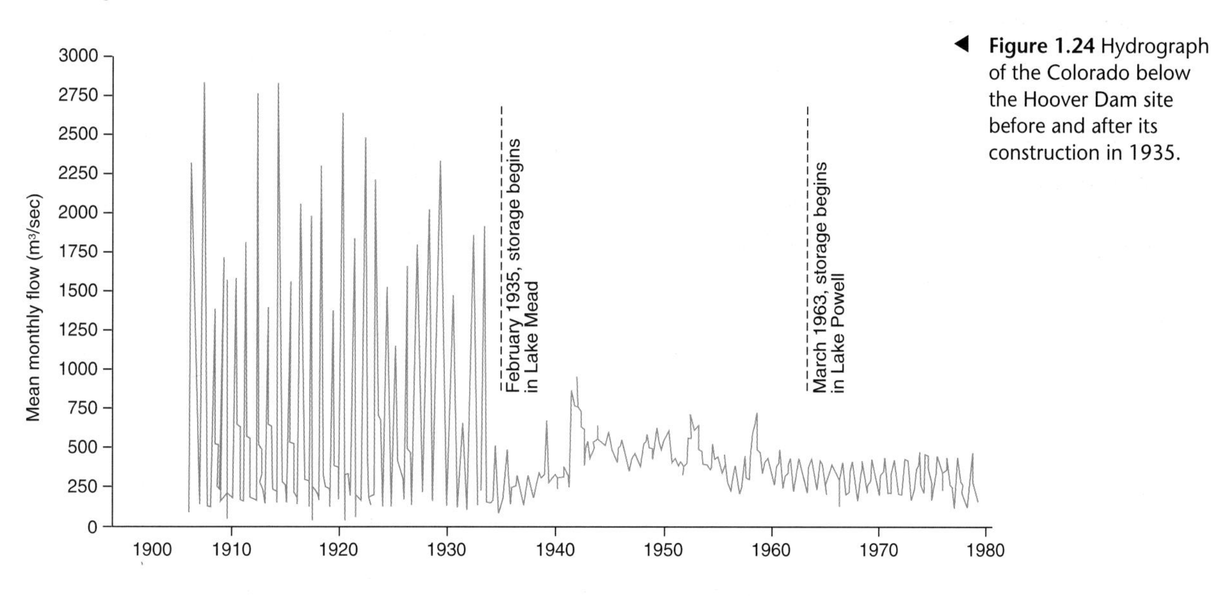

◀ **Figure 1.24** Hydrograph of the Colorado below the Hoover Dam site before and after its construction in 1935.

Look at Figure 1.24.

1 What does this hydrograph suggest about the effects of the Hoover Dam?

2 In the spring of 1983, there was serious flooding along parts of the Colorado. Construct a model like the one in Figure 1.6 on page 10 to show the factors that might have led to this flooding.

3 Construct a similar model for the Colorado during times of drought, or an Arizona summer. Use Figure 1.20 to help you.

1 What effects will dams built as part of the management of the Colorado have on sediment transport and flow along the river? Refer to the theory box on dynamic equilibrium on page 17.

2 To what extent is the present-day allocation of the Colorado's water between states fair and even? Does this uneven distribution matter?

The effects of management on the Colorado river system

The release of large quantities of water from a dam often results in heavy scouring immediately downstream of the dam. This water contains little or no sediment, as it has been deposited in the reservoir behind the dam, so it has more energy and a great capacity for erosion and transportation. Figure 1.25 shows the changing profile of Lake Mead behind the Hoover Dam because of sedimentation. It has been estimated that on average the Colorado carries 140 million tonnes of material through the Grand Canyon in a year. Most material is moved when the river is in spate or in flood in late spring, when snow melts in the upper parts of the basin. In the past, the Colorado and its load helped to create the Grand Canyon, and provided the area around its delta, in Mexico, with valuable recharges of fertile sediment. Management of the Colorado has upset the equilibrium of the system.

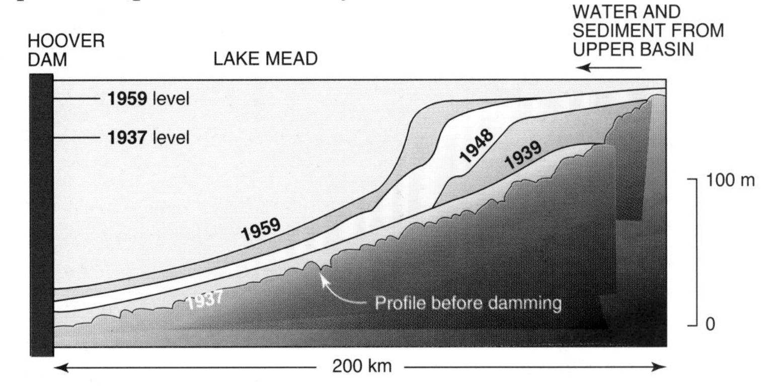

▶ **Figure 1.25** Sedimentation within Lake Mead behind the Hoover Dam.

▼ **Figure 1.26** Lower and upper basin entitlements to the waters of the Colorado.

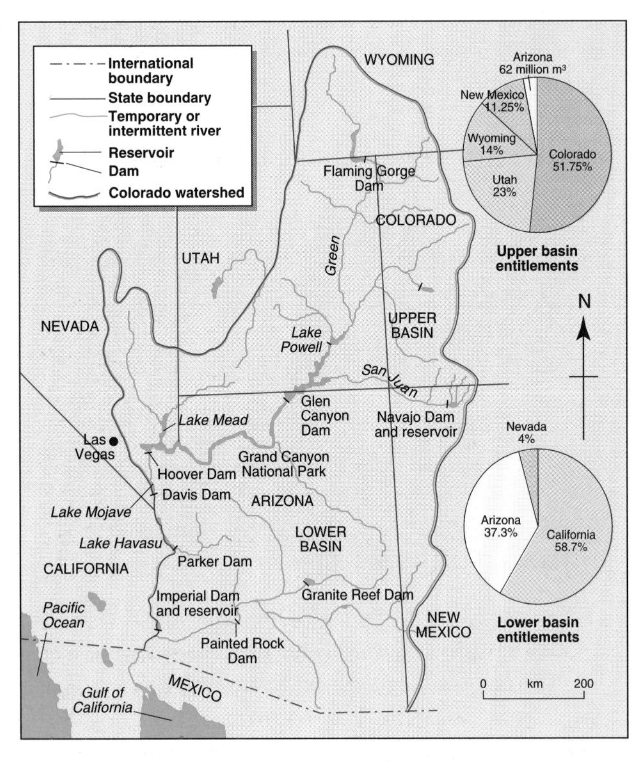

Using dams and reservoirs for water supply

Perhaps the most important reason for control of the Colorado, and certainly the most lucrative, is that of water supply for agriculture. Water is a vital natural resource, often seen as so important that war and other hostilities have broken out over it. In the film *Chinatown*, water is hoarded by the rich of California at the expense of the poorer farmer – a scenario which was based on real situations. The water of the Colorado is an international resource, because not only does the basin occupy seven states of the USA, but the river also enters the sea through Mexico.

Demand for water is so great and is such a point of conflict, that each state through which the Colorado flows is entitled to a certain amount of its waters – this is called a state's **entitlement**. Figure 1.26 shows lower and upper basin entitlements to the Colorado's waters, by state.

'The state of California has so far bullied itself into a position of strength against all other states in the Colorado basin. It is time to allocate water fairly between the states. If California cannot provide its people with sufficient water without threatening one of the world's major basins, then that is California's problem and no-one else's.'

'The provision of water is fundamental. Decisions about water provision and basin management can only really be made at a national level. California's need for water is essential for the national economy. No other interests matter, least of all those of states whose water is used elsewhere.'

'Whatever the pros and cons of arguments between California and other states of the USA, Mexico is the true loser. The USA should pay compensation to Mexico immediately and restore the flow of the Colorado to the Mexican people within five years.'

▲ **Figure 1.27** Options for the future of the Colorado.

Discuss each of the statements for the future of the Colorado outlined in Figure 1.27.

Should large dams be built? The case of the Narmada river and the Sardar Sarovar Dam

This chapter has so far explored river basin management in the economically developed world. Water resources and drainage basins need to be managed with some care. One way in which this can be done is to build large dams to store and regulate the use of water. In the late 20th century some very large dams – super dams – have been built, of which the Hoover Dam was the first. Are the issues in managing river basins the same in an economically less developed country?

This case study looks at one of the world's largest projects – managing the basin of the Narmada, one of India's major rivers, in western India. Figure 1.28 is a map of the Narmada river basin. In India, 14 per cent of public expenditure is spent on super dams – 1500 large dams have been built in India, 14 of which are super dams. However, this has not been achieved without criticism. These schemes have met with much controversy from within and outside India, especially for the proposal to dam the Narmada river.

Look at the photograph of the Sardar Sarovar Dam in Figure 1.29. What are your reactions to this photo? River basins, and human features within them, can be looked at from an aesthetic – or artistic – point of view. People's opinions and viewpoints depend upon what they already know, however. Though the Hoover Dam may be a remarkable sight, we may see other considerations behind this. Knowing that the waters of the Colorado are allocated unequally may affect our attitude towards the Hoover Dam. Sometimes, when we know more, we look at something with a different viewpoint. This is true for the ways that some people view the schemes along the Narmada river.

▼ **Figure 1.28** The Narmada river in west India.

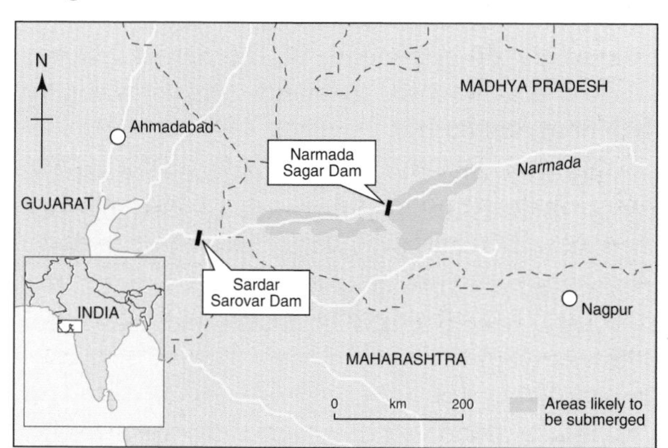

▼ **Figure 1.29** The Sardar Sarovar Dam on the Narmada river.

The Narmada is the largest west-flowing river in the Indian subcontinent. The states of Gujarat and Madhya Pradesh in north-east India suffer from a general shortage of rainfall, and the Indian government claims that £20 billion is lost in agricultural production because of the recurring drought problem. The government has started work on a huge scheme to harness both the power and water resources of the Narmada drainage basin. This involves a series of more than 3000 major and minor dams. The whole scheme could take up to 100 years to complete.

Project	Sardar Sarovar	
Villages	234	
Population	100 000	
Project	Narmada Sagar	
Villages	254	
Population	170 000	
Project	Omkareshwar	
Villages	27	
Population	13 000	
Project	Maharashtra	
Villages	58	
Population	14 000	
Total number of villages affected		573
Total number of population affected		297 000

▲ **Figure 1.30** Villages and populations that will be submerged by the Narmada dams.

Work has started on the two biggest dams. These are the Sardar Sarovar in Gujarat, and the Narmada Sagar Dam in Madhya Pradesh (see Figure 1.28). It is estimated that the Sardar Sarovar Dam will irrigate 1.8 million hectares, supply 3.5 billion litres of drinking water a day, and produce 1450 megawatts of hydro-electricity which would be enough to support an industrial boom. The Sardar Sarovar will take 22 years to complete at a cost of £3.5 billion.

It would appear that the benefits of the scheme are enormous. Why then is there such massive opposition to the scheme in India, and indeed world wide? Figure 1.30 lists the villages and populations that will be submerged by the Narmada dams – 100 000 people will have to be moved because of the Sardar Sarovar Dam alone.

Many villages and religious and historical sites will be submerged, 600km^2 of valuable agricultural land and 3,500km^2 of forest will be lost. It is thought by protesters that more land will be submerged by the dam than will be irrigated by it, and that 80 per cent of villages in the Sardar Sarovar area will receive no irrigation benefits from the scheme. Much of the land to be submerged is good-quality agricultural land, while a lot of the land to be irrigated is black clay soil which becomes waterlogged easily and is unsuitable for farming.

There is a problem in finding, and funding, sufficient land for resettlement of those people who are forced to move because their villages are to be submerged by the reservoirs. Many have taken drastic action to protest about the schemes – see Figure 1.31. It is not just the fact that many people are going to lose their homes and their land that is causing such consternation. It is believed that the building of the Sardar Sarovar and other dams will not only be a massive social disaster but will also be an environmental, ecological, and political disaster. Environmentalists believe that an irrigation project on this scale will cause waterlogging of soils and an increase in soil salinity. This happens when poor drainage prevents the removal of excess salts, which then accumulate. Moisture in the soil evaporates, and salts crystallize on the surface.

Siltation is another worry of opponents of the Sardar Sarovar scheme. Rivers transport and deposit sediment, which naturally creates problems when dams are built.

- Silt carried by the river will collect behind the dam. The reservoir eventually silts up, as there is no economic way of dredging out the sediment.
- Water emerging from the dam will carry little or no sediment. The river increases in energy, which results in accelerated erosion of the river channel downstream.
- Normally, sediment carried by a river in suspension is deposited on the floodplains of the Narmada and is generally very fertile. Farmers will therefore lose the means by which their soils are replenished with nutrients each year.
- Farmers whose land is submerged will lose their best-quality soils. With large numbers of people relocating and much farmland being submerged, new land must be found. This is likely to lead to deforestation, which will result in increased soil erosion rates and therefore more rapid siltation of the reservoir.

▶ **Figure 1.32** Two contrasting views of the Narmada scheme. From the *New Internationalist*, December 1990.

'The government officials want to drive us away. If we were not struggling we would have been pushed out by now. They say the dam will benefit other people. We are not against benefits reaching other people but that should not be at the cost of destroying our lives. The government's policies benefit other people but it is the tribal people who bear the costs. We know that we will get arrested and beaten, but we also know that worse things can happen.'

SARLI, age 35, Kakrana Village

'Our track record in this is not necessarily good but we are learning from this. I want to make this resettlement package that has to go to the people as transparent as possible by involvement of NGOs (Pressure Groups) to see that the people get what they are supposed to get – not only land but also the land that was earmarked for them, the right amount. And that they get the full amount of money.'

KAMAL NATH, Minister of Environments and Forests

Anti-dam villagers prepared to make the ultimate sacrifice

Unless the Indian government keeps its promise of setting up a team to review the Sardar Sarovar dam project, a group of anti-dam activists called the 'Save or Drown Squad' will sacrifice themselves in the Narmada River on, or shortly after, Friday August 6. We appeal to the Indian government to save the lives of these people by immediately setting in motion a genuinely independent review of all aspects of the project, as agreed with anti-dam campaigners at the end of June.

We also appeal to the British government to ensure that the technical studies funded by the British Overseas Development Administration are made public as soon as they are completed, and urge the UK not to give any more support to the discredited Sardar Sarovar project or to any related projects. The British Government should also ensure that the World Bank, which was forced to terminate its loans for the dam in March this year, does not renew funding.

If the dam in Western India is completed, 200,000 people will be displaced by its long reservoir. Hundreds of thousands more will lose their livelihood. A growing body of evidence shows that the project will not deliver its claimed benefits in terms of electricity generation and the supply of water for agricultural, industrial, and domestic use, and that its costs are prohibitive.

Two weeks ago, the first 50 houses behind the dam were submerged. Many tribal villagers, who have for years said that they would drown rather than allow the dam to be built, were dragged by the police from water which had risen above their waists.

Six hundred villages from the Narmada valley are staging a week-long protest in Delhi. At the beginning of August, supporters from all over India will converge on Manibeli, the second village behind the dam and the centre of the people's resistance to it. If the government has not started the review by August 6, the 'Save or Drown Squad' will face the Narmada waters, believing that the ultimate sacrifice is the only non-violent means left to show their opposition to the Sardar Sarovar dam, and the anti-people and anti-nature development which it symbolizes.

Dr Gautam Appa and Others

▲ **Figure 1.31** From *The Guardian*, 2 August 1993.

Another major problem is the increased risk of earthquakes. It is thought by many seismologists that the building of large-scale dams could trigger major earthquakes, a factor referred to as 'reservoir-induced seismicity'.

However, the main problems relate to the effect on people. Up to 1.5 million people will lose their homes and have to be resettled elsewhere. The majority of these people are from tribal communities such as Bhib, Pardhans, and Kols who have created a subsistence way of life based on farming, fishing, and forestry. Families who have to move have been promised five hectares of land each but most will receive poor-quality grazing land in exchange for their rich land close to the river.

Much of the funding for the project has come from the World Bank which approved of the scheme in the 1980s. After much lobbying in which the World Bank came under great pressure, it finally withdrew its support in 1994.

Read the two comments in Figure 1.32. Form groups of three or four. Each group will reply to one of the comments. Discuss each argument.

Essay titles

1 What similarities and differences are there between the way in which the Colorado river has been managed, and the proposals for the Narmada?

2 'No matter what we like to say, at the end of the day management of a river basin is a matter of politics; in all matters of politics, the big players always win at the expense of the small ones.' Discuss.

Ideas for further study

Select one of the world's largest river basins (see Figure 1.19). Carry out a small group enquiry into this basin. Below are some suggestions for a route of enquiry.

1 Where is the river basin? What is its size, and what is the length of the river?
2 What are the key features of the river basin, for example its valley form, its landscapes?
3 What pressures are affecting the river basin? What kinds of pressure are these? (You could include industrialization, water supply, energy, clearance of vegetation, etc.)
4 Who is responsible for the pressures placed on the basin? Is anyone responsible for managing these?
5 How are these pressures affecting the basin?
6 What strategies exist for managing the basin? How can these be compared? Whose interests does each strategy protect?
7 What is the likely outcome? Can all the competing pressures be accommodated or resolved? Who gains? Who loses?

The Narmada scheme is only one large dam scheme and is not even completed yet. The effect of the Narmada can be assessed by comparing it with existing super dams, such as the Aswan High Dam, the Tennessee Valley scheme, and the river Volta.

Summary

- There are many pressures on drainage basins. Some of these arise from natural incidents such as flooding, and human attempts to manage them.
- Flooding and the movement of water within drainage basins can be better understood through models such as the hydrological cycle and storm hydrographs.
- Changes within drainage basins may take place as a result of human activities. These include small-scale activities, such as floodplain usage, or large-scale schemes such as water provision for a large catchment.
- Conflicts may arise as people attempt to manage drainage basins. These may range from local disagreements about the best way to manage a flood problem, to international disagreement over water provision and entitlement.
- Sometimes decisions about the way to manage river basins are made a long distance from those people and places that are affected by the decision.

References and further reading

J. R. Bevan 'Water resources and river basin management', in M. Barke (ed.) *Case Studies of the Third World*, Oliver and Boyd, 1991.

A. Clowes and P. Comfort, *Process and Landform*, Oliver and Boyd, 1987.

A. Goudie, *The Human Impact on the Natural Environment*, Blackwell, 1986.

K. Hilton, *Process and Pattern in Physical Geography*, Unwin Hyman, 1985.

N. Howes, 'Equilibrium in physical systems', *Geography Review*, September 1989.

M. Newson, *Land, Water and Development*, Routledge, 1992.

R. Prosser, *Natural Systems and Human Responses*, Nelson, 1992.

M. Reisner, *Cadillac Desert. The American West and its Disappearing Water*, Secker and Warburg, 1990.

F. Slater, *People and Environments: Issues and Enquiries*, Collins Education, 1986.

2 Managing coastlines

Why do we manage coastlines?

A coastline is the boundary where the land meets the sea. This definition is not as simple as it appears, because the boundary may be different at low tide and high tide, or during storms and in periods of calmer weather. It is therefore more usual to refer to a **coastal zone**, a wider part of the coastline which may be affected by a variety of processes. The coastal zone is marked by a variety of landforms, such as beaches, cliffs, and sand dunes, and it adjusts to a range of natural processes, from intense wave energy and storm-force winds to periods of calm. It is also a focus for human development; housing, commerce, recreation, and tourism all create additional pressures on the coastline. Coastlines throughout the world are increasingly under pressure, for example, as leisure interests and increasing affluence lead to global tourism. The need for careful management of coastlines is becoming clear, as the studies in this chapter show.

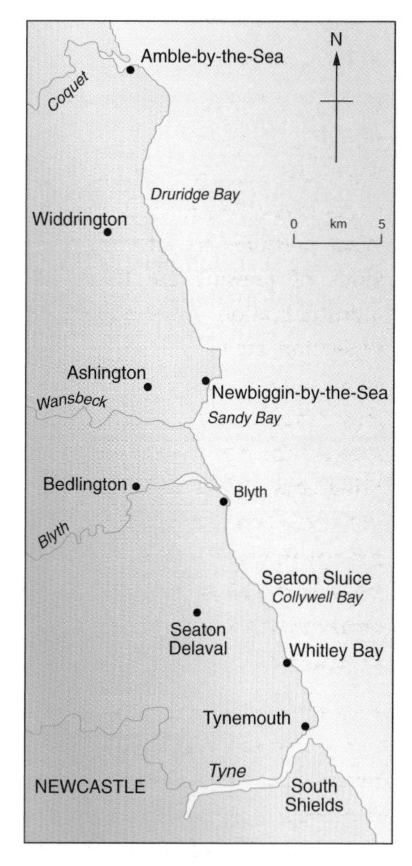

▲ **Figure 2.1** The south Northumberland coast.

The Northumberland coastline from Collywell Bay to Druridge Bay

This study is designed to show you the many different issues and processes affecting a small stretch of coastline in Northumberland. This section runs from the southern edge of Collywell Bay, 15km from Newcastle-upon-Tyne, to the northern end of Druridge Bay (Figure 2.1). In total, this stretch is about 25km from south to north. We look here at the issues that concern people and the environment they live in or visit, and at who is responsible for managing the coastline.

Collywell Bay

Figure 2.2 was taken on the clifftop near Collywell Bay, along the southern part of this stretch of coastline. This section is to the south of the village of Seaton Sluice (see Figures 2.3 and 2.4). The area has been made a Site of Special Scientific Interest (SSSI) because of its special geological features. There is a typical erosion feature – a **stack** – just north of Collywell Sands (see Figure 2.4). The geology within this small bay helps to account for its shape and form.

▲ **Figure 2.2** Warning sign on the clifftop at Collywell Bay.

What part has geology played in the formation of this coastline?
Study Figure 2.5, which shows the geology of Collywell Bay. Consider the role of geology in the development of today's coastline. The geology is complex, but broadly consists of sandstones, shales, mudstones, and coal seams at right-angles to the coast. Sandstone is the most resistant of these, but you will see on Figure 2.5 that there is more than one kind. The Crag Point sandstone has created the headlands of Crag Point and Rocky Island. Is it possible to rank the

order of resistance of the other rock types shown, from most to least? Clearly, the Charley's Garden sandstone has eroded more than the Crag Point sandstones, and is weaker. However, it has produced a small stack, Charley's Garden (see Figures 2.3 and 2.4). None of the other rock types still remaining has produced a similar feature, so we can deduce that sandstones are more resistant than other rocks.

1. What does the photograph in Figure 2.2 suggest about the geology in this area? Why do you think that removal of material from the beach is forbidden?
2. Draw a sketch map of the Collywell Bay area of Seaton Sluice. Annotate your sketch map to show different coastal features such as the bay and headlands, but also pay attention to smaller-scale features shown in Figure 2.4.
3. Read about cliff erosion in the theory box on page 34. Annotate your map further to show changes that have occured in the bay between 1887 and 1994. Changes to the physical and human geography of the area should be noted.
4. Why do you think that erosion is considered to be a serious problem at Collywell Bay?

▼ **Figure 2.3** Map of Collywell Bay, 1887.

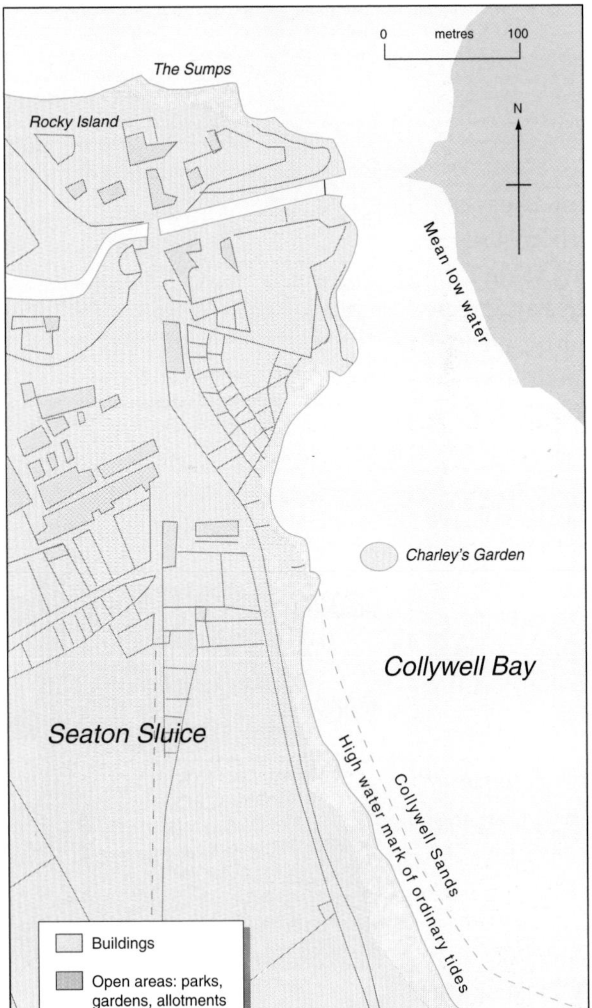

▼ **Figure 2.4** Map of Collywell Bay, 1994.

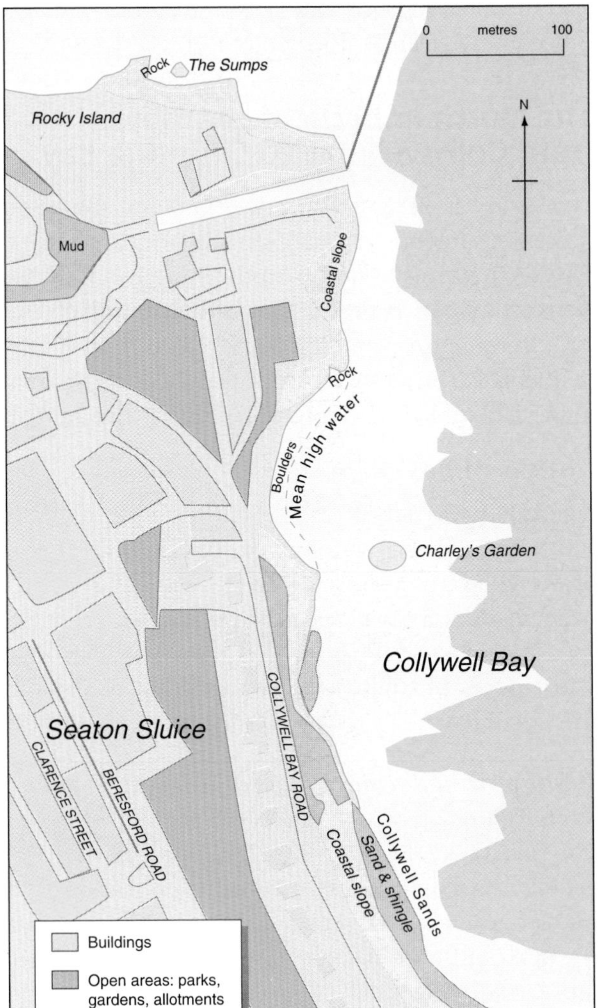

The sandstones are thickly bedded, well jointed, and have a 'blocky' appearance. The debris created by erosion at the cliff foot and on the shoreline is also of large blocks of sandstone, as these fall away from the cliff. The rocks were all formed between 220 million and 190 million years ago, during the Permian period.

Why is there a problem of erosion at Collywell Bay?

Look at Figure 2.6 on the next page. Find this area on Figures 2.3 and 2.4. How rapid do you think the rate of cliff erosion is here? Why is erosion a problem in Collywell Bay? Geographers are concerned with these questions because they are questions that link people and their environment. Figures 2.3 and 2.4, which show maps of the same area in 1887 and 1994 respectively, can help to provide some answers.

Collywell Bay does not suffer so badly from erosion as, for example, the Holderness coast on North Humberside, where cliff recession may exceed 2m per year. However, although it is slower, erosion in Collywell Bay is still a serious issue, and expensive measures have been taken to protect the cliffs. The exposure of rock strata, or layers, along this coast means that it is of particular interest to geologists, who are keen to protect it from further erosion – which is why it has been designated an SSSI. Pressure from residents has also played a significant part in attracting investment to the area, and providing protection measures.

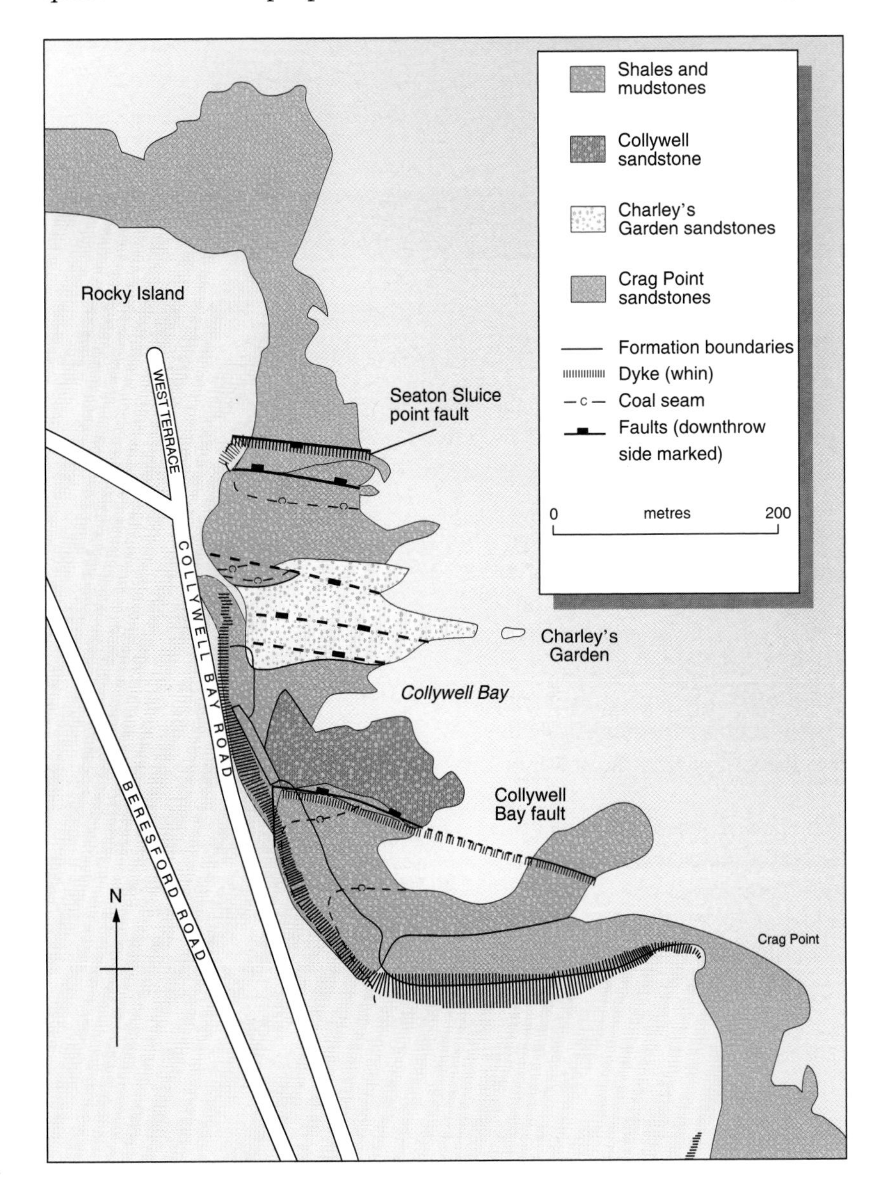

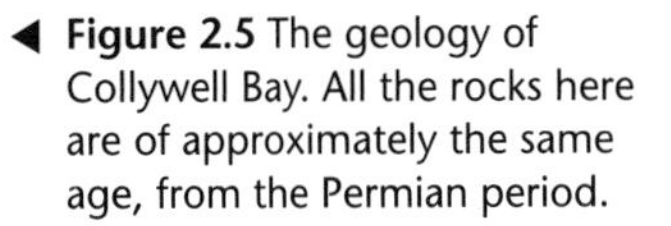
◀ **Figure 2.5** The geology of Collywell Bay. All the rocks here are of approximately the same age, from the Permian period.

▲ **Figure 2.6** Collywell Bay from the south, showing houses along Collywell Bay Road.

How do cliffs erode?

Cliff processes can be subdivided into two categories:

- those that operate at the foot of the cliff
- those that operate on the face of the cliff.

Cliff-foot processes

The main agent of erosion at the foot of a cliff is wave energy. Waves are a direct result of the action of wind which creates friction on the water surface. The greater the wind strength, the greater the friction and the greater the size of the wave. Waves also grow with time. They may be generated a long distance from where they eventually reach a shoreline; the distance travelled is known as the **fetch**. The fetch is the distance over which the wind blows largely uninterrupted. The greater the fetch, the greater the wave size. The greatest wave sizes are therefore likely to be on coasts facing major oceans, such as the east coast of Australia, or the Atlantic coast of Ireland.

Figure 2.7 shows how a cliff foot may be undercut at Collywell Bay. As waves approach the shore, friction between the water and the sea bed slows them down. The lower layers of water are slowed down, and upper layers override them, creating a build-up of water which eventually spills over as a wave. At the foot of a cliff, this may have several effects.

▲ **Figure 2.7** Wave attack at the foot of a cliff. Notice how the water attacks the cliff foot head-on at this point. This particular attack is against weak muds and clays which erode easily.

- *Hydraulic pressure* The force of a wave against a rock face creates a great deal of pressure within the air spaces in fractures and joints or between rock strata. The approach of water compresses air within fractures, which then explodes as water recedes and is released. The repetition of this explosive effect causes further fracture of the rock, which then falls away from the cliff.
- *Abrasion or corrasion* As waves approach, wave energy enables material to be picked up and thrown against the foot of the cliff. The mechanical action of material against the cliff foot is known as **abrasion**. It is easy to imagine how, in Figure 2.7, any debris carried to the cliff foot can quickly abrade such weak material.
- *Corrosion or solution* This is the chemical effect of salts or acids held within sea water. Weak acids dissolve or corrode alkaline rocks containing calcium carbonate, such as limestone or chalk. Salt crystallization occurs on all rocks facing the sea, as salt rehydrates on the surface. The stresses created by crystal growth and chemical effects of salts weaken many rocks, particularly those that are already weak and fragmented.

▲ **Figure 2.8** Slip lines along a cliff, showing mass movement along the cliff face. Stress lines are set up, caused by structural failure of the rock. These are easily lubricated with water during wet periods. As a result, the cliff edges slip, still intact.

Cliff-face processes

Cliff-face processes involve the action of weathering and mass movement of rocks on the face of the cliff. These are collectively known as **subaerial processes**, and consist mainly of weathering and mass movement.

Weathering

Weathering is the disintegration or decay of rocks at the point where they are located. Mechanical weathering processes include freeze–thaw, where water within rock joints or spaces freezes, expands by up to 10 per cent of its volume, and forces fragments of rocks apart. Chemical weathering processes include solution and corrosion, which are described above. Together, these act on the cliff face and weaken rocks that are already exposed to the sea.

Mass movement

Mass movement is the downslope movement of rock material under gravity. Gravitational stresses overcome the resistance of the rock. It is significant in the way that it causes cliff collapse or recession. The simplest example is cliff collapse. Other processes include landslips or landslides, where a weakened mass of cliff material slips intact along a line of weakness, often following heavy rain or a storm, during which the weight of the mass increases as a result of the rainwater (see Figure 2.8). Slumping is similar but involves disintegration of the whole structure. Figure 2.9 shows a slump, which appears to have been caused by a spring, which added water and weight to a part of the cliff face, causing the cliff to fail.

▲ **Figure 2.9** Slumping along a cliff. This material has become saturated with water, probably from the underground spring shown in the photograph. The weight of saturated clay has caused the cliff to fail, and the movement has brought the material down rapidly, mixing as it moves so that it is no longer intact.

How does erosion affect people's attitudes?

How might living close to an area where erosion is taking place affect people's attitudes and feelings? You have already seen how housing developments have continued at Collywell Bay in the past 100 years, in spite of erosion. The evidence seems to suggest that, so far, erosion is slow, but that remedial action is needed and people need to be warned (see Figure 2.2). Do people see erosion as a threat? If they do, what is it like to live in a place at risk and under threat? Do local councillors believe that the opinions of people who live in Collywell Bay are significant? The issues of erosion, managing the coastline, and natural processes have a major impact on people's lives. However, there are costs and benefits for each situation; protecting the coastline in one area can merely move the problem further along the coast.

There are many situations where people's opinions help to explain links between people and the environment. The following techniques box shows you how to construct questionnaires to help you in your own research.

Designing a questionnaire

This section is designed to give you some pointers about using questionnaires or interviews to collect information from people. Designing a questionnaire is time-consuming and needs to be undertaken with care. Questions need to be tested before being used. Consider the following issues when you design questions.

Closed or open questions?

A *closed question* is one that seeks a single closed response. For example:

- In which age category are you:
  0–9, 10–19, 20–29, 30–39 . . .
- How did you travel here today?
  car, bus, train, motorbike, bicycle . . .

Closed questions are useful, particularly when you want to design a database, as the results are easy to key in and analyse.

An *open question* is one that seeks a free response and does not try to influence the answers. These are useful in gaining insights into people's opinions, or where there is likely to be a wide range of possible answers. You may start with a closed question, and ask for a YES/NO answer, and then follow with an open request for explanation. An example of a questionnaire is given in Figure 2.10.

Wording

Be careful with the wording of questions. Avoid technical language which might not be understood. For example:

- Which would be the best strategy for coping with the erosion problem here?

This might be more simply put:

- What do you think could be done to stop the cliff falling into the sea here?

Beware, too, of asking people to tell you what they *think* they do, rather than what they *actually* do. For example:

- Which beach do you normally visit?

This may give you a different answer from:

- Which beach did you last visit?

Sample size

Sampling is useful and necessary when the population is too large to be covered in total. You may decide to see who comes along the street and ask whoever looks friendly enough. This is called **random sampling**. During the day, you may find more women than men on a shopping street. You might want to balance this. If you know that 46 per cent of the population is male in your study area, you might want to make sure that 46 per cent of your sample are male. If one-quarter of all men are over the age of 65, you may want to make sure that of the men you interview one-quarter are men over 65. This is called **stratified sampling**. To obtain reliable results, you need at least 100 responses. If the population is small, for example 20 households along a stretch of coast, then sampling is not necessary.

Interviewing

At Collywell Bay, there are about 20 houses on the clifftop. This may be too small a sample for questionnaires, but could provide you with a rich source of people to give you detailed information in an interview. Here, open questions are vital if you want people to talk. Listen carefully to what they say. You might ask if you can tape the interview, which means you don't have to write down what they say as they talk. You can then listen more carefully again later.

1 Are you aware of the problem of coastal erosion here? YES ☐ NO ☐

2 Are you concerned about the problem of coastal erosion here? YES ☐ NO ☐

3 a) Does coastal erosion affect you in any way? YES ☐ NO ☐
b) If 'Yes', can you say how it affects you?

4 a) Do you feel anything should be done about coastal erosion? YES ☐ NO ☐
b) If 'Yes', can you say what should be done?

5 a) Are you aware of anything that has been done already about coastal erosion? YES ☐ NO ☐
b) If 'Yes', can you say what has been done?

6 Who do you think should be responsible for financing any work on the coast?
a) District Council
b) County Council
c) Central Government
d) Charity or private donations
e) Private individuals

7 Who do you think should be responsible for managing any work on the coast?
a) District Council
b) County Council
c) Central Government
d) Charity or private donations
e) Private individuals

8 Would you be prepared to pay higher council rate to finance such work?

▲ **Figure 2.10** Example of a questionnaire on coastal erosion.

How can Collywell Bay be protected?

Look at Figure 2.11, which shows one aspect of coastal management, that of coastal defence. The map shows annual spending on coastal defence in Britain by section of coast. The amount varies considerably in different parts of the country. This money is spent by local councils, who are responsible for coastal defence work. Once local borough engineers have identified areas where they believe work is necessary, funding is sought from council budgets and from the Ministry of Agriculture, Fisheries, and Food (MAFF). Decisions are made by elected councillors acting on the advice of engineers.

Like many studies in Geography, you will find that people who decide whether or not to take action and spend money, are affected by their own perceptions and preferences. Imagine, then, how borough engineers will feel when, having identified work which they believe is necessary, funding is reduced or even turned down by councillors who prefer to spend money in other ways.

There are two broad options for Collywell Bay:

- 'hard', or engineering, solutions
- 'soft' approaches.

1 Why should the Ministry of Agriculture, Fisheries, and Food be responsible for coastal defence and protection?
2 Look at Figure 2.11. The investment in coastal defence is not evenly distributed around Britain. Why?
3 How might people whose property is affected by coastal erosion try to get support from their council?

Hard approaches to coastal defence

'Hard' solutions are those that involve engineering work. Figure 2.6 shows the broad sweep of Collywell Bay and the sea wall constructed at the foot of the cliff. This is to protect the cliff foot from further erosion. However, although effective in many ways, there are some problems with this method of defence. There is evidence on Figure 2.6 that, while the cliff foot has been protected, the cliff face has continued to be active. Figure 2.12 also shows that there can be undercutting of the sea wall itself. Although wave energy no longer attacks the cliff foot, it is only deflected, and the energy is often consumed by scouring of the bed at the foot of the wall. This eventually leads to collapse. A further effect is that the energy of the sea is focused on the beach, leading to beach erosion and removal.

Stone revetments protect the beach through the use of stone boulders, which act by absorbing much of the wave energy and increasing the surface area exposed to wave attack (see Figure 2.13). Although effective, these are regarded as ugly, and are therefore unpopular. Another alternative is to drain those parts of the cliff that are likely to become saturated, thus preventing slumping. This is done by boring into the cliff and pumping out water at the base.

Beaches are fundamental to cliff defence. The energy of approaching waves is absorbed, or dissipated, by friction. The best source of friction is a wide beach, since most wave energy is then lost before a wave reaches the cliff foot. You may have noticed in comparing the two maps of Collywell Bay in 1887 and 1994 that the beach has diminished. The effect of the sea wall, therefore, has been to remove the very feature which would have helped to protect the cliff foot in the first place!

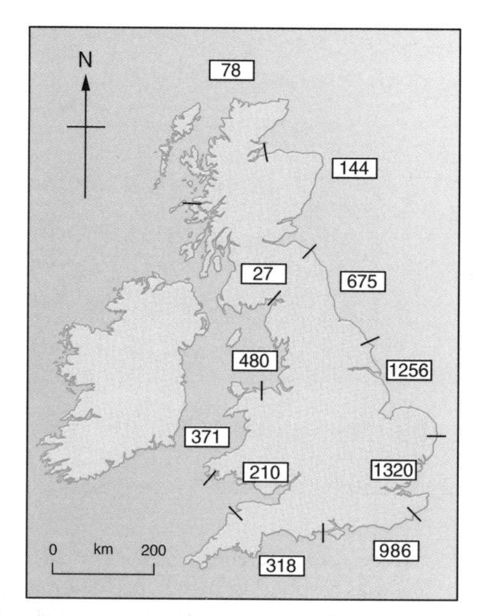

▲ **Figure 2.11** Annual cost of coastal defences in Britain (thousands of pounds).

▲ **Figure 2.12** The effect of wave attack on a sea wall.

Other ways of designing cliff-foot defences have been attempted. Groynes help to retain beach sediment. These are usually wooden structures, built at right-angles to the beach. They are designed to trap sediment which would otherwise be removed by longshore drift (Figure 2.14). However, by preventing removal of beach material from one point, other locations may suffer. Figure 2.15 shows how the construction of groynes may protect the cliff at point P, while threatening further erosion at point Q. Point Q becomes starved of sediment, thus reducing beach size and increasing wave energy at the cliff foot.

▲ **Figure 2.13** Section through a revetment. A revetment is a structure consisting of large angular boulders, put in place to protect a cliff. The surface area of the boulders and the air spaces between them absorb wave energy and reduce the potential rate of erosion.

▼ **Figure 2.14** Longshore drift. As waves break at an angle, the swash carries beach material towards the cliff from A^1 to A^2. Wave retreat, or backwash, carries material back down the slope from A^2 to A^3.

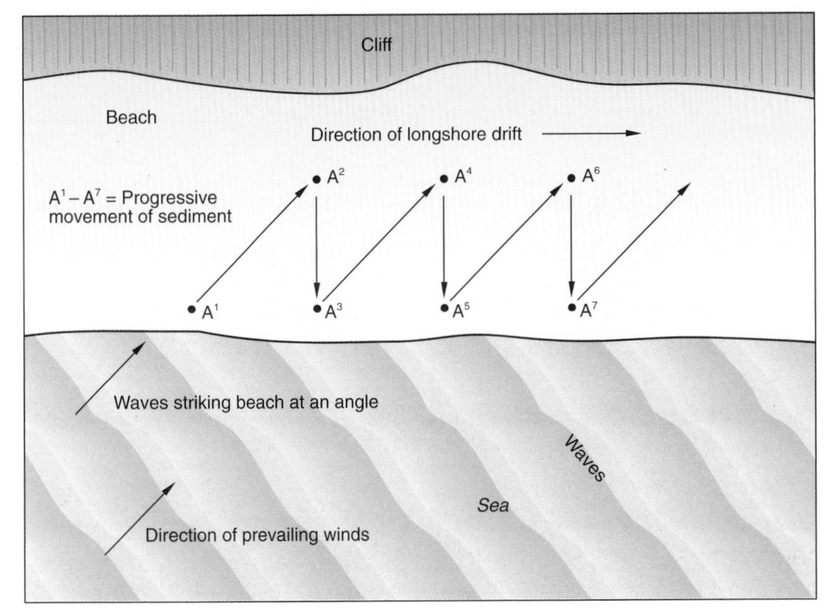

▼ **Figure 2.15** The effects of constructing groynes on a beach. Beach accumulation at P has resulted in beach starvation at Q. As a result, wave energy is now increased closer to the cliff, and the threat of cliff erosion here is greater.

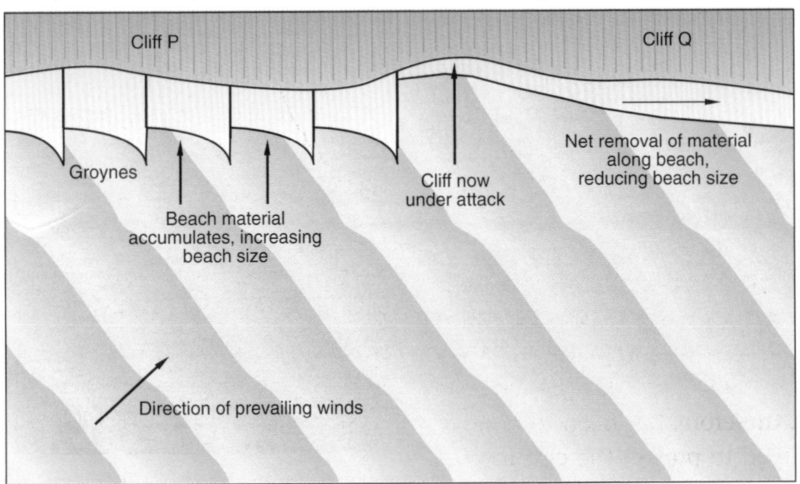

Designing improved defences

The old part of the wall at Collywell Bay is deteriorating and needs repair. You are working for a firm of engineers contracted to advise on the best way to protect the cliff at Collywell Bay. Write a report about the best strategy for renewing the defence, using the data in Figures 2.16–2.20 and referring to any of the information in this chapter so far. Aim for 750–1000 words, with diagrams, showing which designs you consider most effective, and why. State:

- the problems faced at Collywell Bay
- the effects of previous attempts to protect the cliff foot. Comment on the different designs of the sea wall.
- your evaluation of their success
- possible alternatives to protect the cliff which take into account the coastal processes at work in the bay
- reasons why you accept some of these alternatives, but not others
- your choice of strategy for cliff management at Collywell Bay.

▼ **Figure 2.16** Wind speed and direction along the south Northumberland coast.

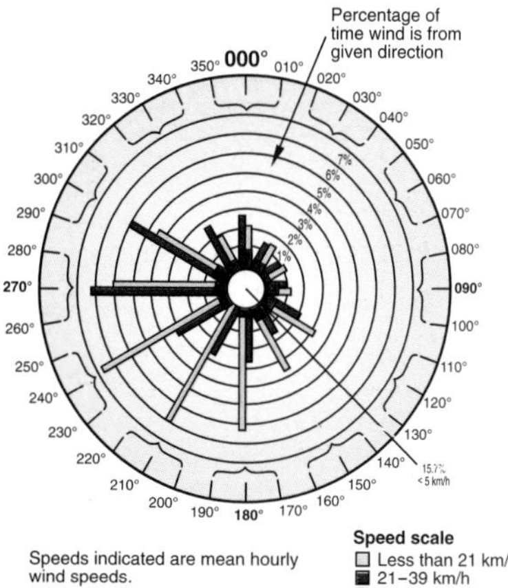

▼ **Figure 2.17** Gale-force wind speed and direction along the south Northumberland coast.

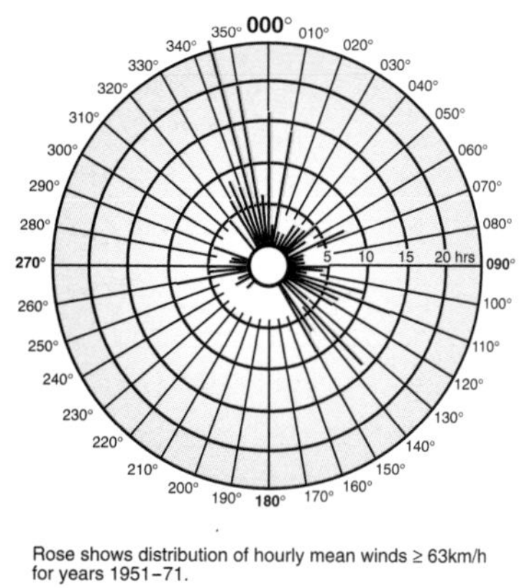

▼ **Figure 2.18** Section through the original sea wall at Collywell Bay.

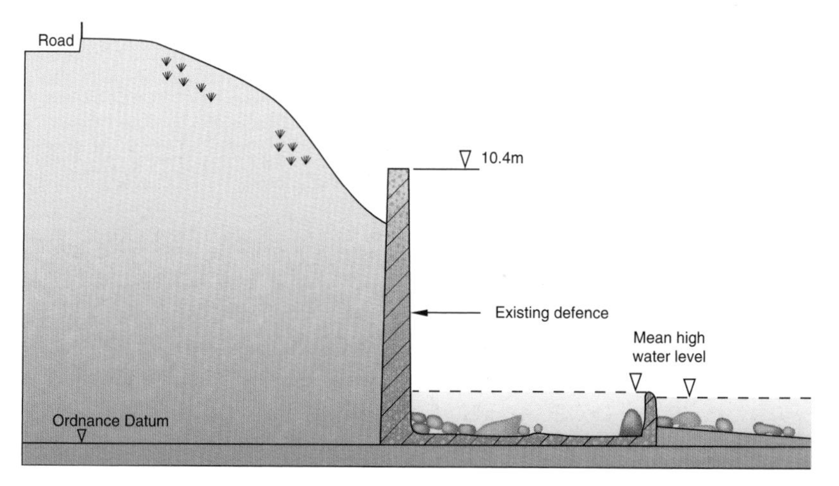

▼ **Figure 2.19** Section through part of the northern wall.

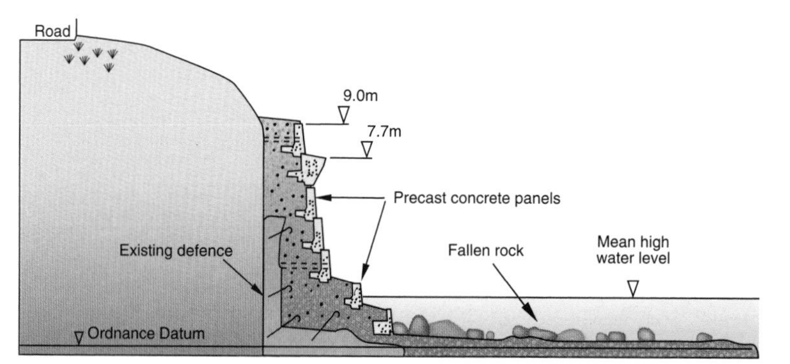

▲ **Figure 2.20** Part of the original sea wall at Collywell Bay. More of this wall is visible in Figure 2.6. It is the highest wall in the centre of the bay, and is showing signs of ageing.

'Soft' solutions to sea defences

The cost of 'hard' engineering solutions is high. Attempts to reduce central government expenditure, and also new thinking on the problems of coastal defences, have led the MAFF to re-assess its preferences for coastal defence. 'Soft' options for coastal defence are those that adapt and supplement natural processes such as beach nourishment (see below). In effect, this means that some cliff-lines might be left to retreat, or at least be protected by cheaper, more sustainable methods.

Offshore breaks involve the creation of an artificial reef at sea, which reduces wave energy before it reaches the shore zone. This can be made of a range of materials, from dredged sediment to old car tyres which are dumped and anchored strategically.

Figure 2.21 shows a beach nourishment scheme in Portobello on the Firth of Forth, east of Edinburgh. During the 19th century, extraction of sand from the beach provided raw material for a local sand and brick works. This depleted the beach so much that by 1926 it had almost disappeared. Storm waves were severe and caused much damage, as waves reached as far as the houses on the promenade. In 1970, a beach nourishment project was introduced to import sand from Fisherrow Sands and release it from offshore (see Figure 2.21). Reducing the gradient of Portobello beach, together with groyne construction, and the use of coarser/heavier sand (in order to improve beach stability), has proved successful in keeping the beach intact since 1970.

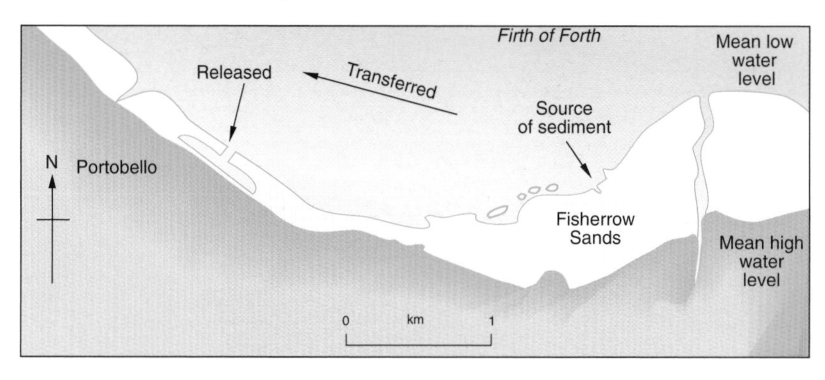

▲ **Figure 2.21** Beach nourishment scheme at Portobello, near Edinburgh.

Reaching policy decisions

Divide into four groups. Each group should take a different statement to discuss from the list below. Justify the statement to the rest of the class and be prepared to answer questions.

- '"Managed retreat" is just another way of cutting essential government costs. It leads to the abandonment of land.'
- 'There is no point in doing anything to defend coastlines against a sea that will always have the last word.'
- 'People who purchase property close to receding cliffs know exactly what they are doing. It's no one's responsibility but their own to see that their home is defended.'
- 'Hard defences will continue to be the only option where important natural or human assets are at risk. But "managed retreat" is suitable where sea walls protect land of low value or quality. Hard defences would only prevent valuable sediment from reaching sand dunes, shingle beaches and salt marshes elsewhere.' (After Claire Hutchings, *Geographical Magazine*, March 1994)

1. What evidence is there of sediment movement in the Sandy Bay area? In which direction does it seem to be taking place?
2. How far does the direction of movement correspond with what you would expect from the rose diagrams in Figures 2.16 and 2.17? How would you explain this?
3. Where and what is the source of the sediment?
4. Using Figure 2.22, compare the land use in the different sections of the Bay.

Sandy Bay

Sandy Bay is located to the north of Blyth, between the estuary of the river Wansbeck and Newbiggin-on-Sea (Figure 2.22). The caravan park at the southern end of the bay brings a seasonal influx of tourists who stay in the area, and Newcastle-upon-Tyne is only 20 minutes' drive away for daily visitors.

Why is erosion such a problem at Sandy Bay?

Study Figure 2.23. It shows a succession of shore lines at Sandy Bay since 1853. In the last 150 years, the coast has receded rapidly, much faster than at Collywell Bay. The geology of the area is significant. Here, as at Collywell Bay, the main rock types are sandstone, shales, and coal seams. The sandstone has been undercut, and repeated collapse has caused the upper parts of the cliff to retreat. Further south, the cliffs consist of boulder clay, the resistance of which is low. (Both Figures 2.7 and 2.9 show boulder clay.) Once removed, the boulder clay particles remain suspended in sea water rather than being deposited on the beach; some are carried south towards the spit.

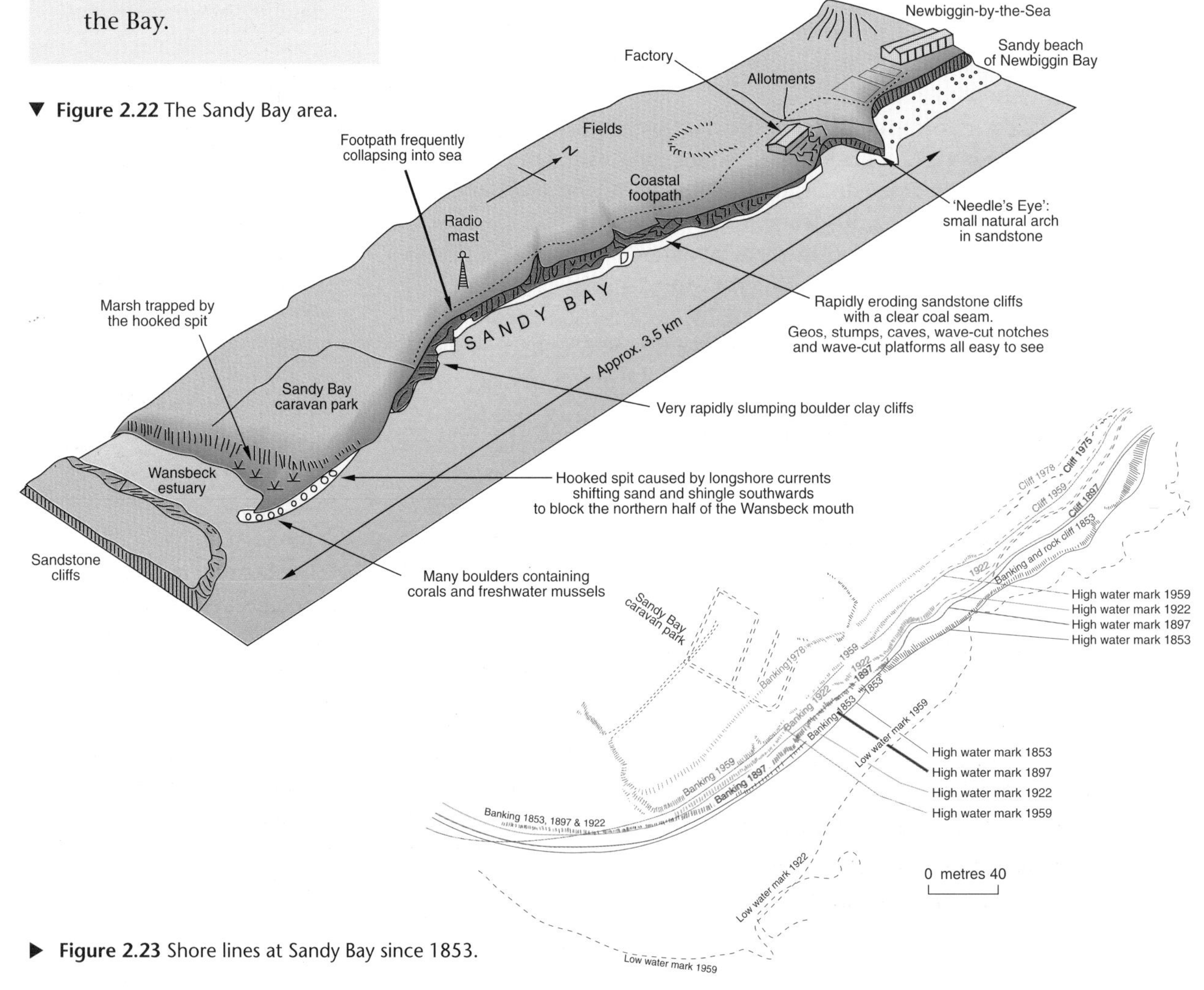

▼ **Figure 2.22** The Sandy Bay area.

▶ **Figure 2.23** Shore lines at Sandy Bay since 1853.

1 Study Figure 2.23. Measure the distance eroded on the cliffs between:
a) 1853 and 1897
b) 1897 and 1922
c) 1922 and 1959
d) 1959 and 1978.

2 Find the average rate of erosion per year in each of the four periods. Plot this on a graph.

3 Is erosion constant, speeding up, or slowing down?

4 What rate of erosion would it be reasonable to assume for the next ten years?

5 Predict the location of this coastline by the time you are aged 40.

Erosion and human activity

Large areas of Northumberland have been subject to coal mining during the last 200 years. Sandy Bay has been especially affected. There is a high degree of subsidence here, caused by the removal of coal from deep underground seams, which have then been left to collapse into each other. Some of the surface subsidence has accelerated the rate of erosion. Sea water is thus able to invade the beach and reach the cliff foot. Field drainage on farmland has also contributed to the problem. Where drains reach the cliff, the outflow of water from the fields has run down the cliff face and caused the cliff to retreat. The situation has been made worse by saturation of clays which have slipped when lubricated by the runoff.

Little has been done to combat erosion. Some boulders have been placed, or have fallen, at the cliff foot, and these help to protect the cliff. However, longer-term solutions are necessary to protect the cliffs at Sandy Bay. A proposed scheme is illustrated in Figure 2.24. It consists of a technique known as 'rip-rap'.

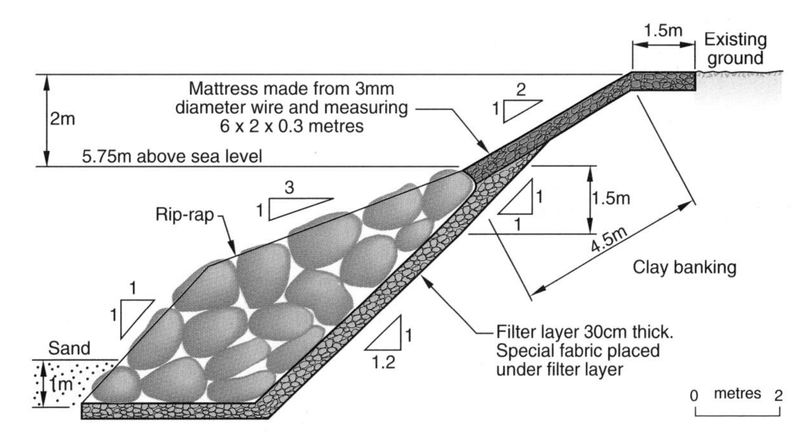

▲ **Figure 2.24** Possible protection scheme for Sandy Bay.

6 Study Figure 2.24. Describe the 'rip-rap' method.

7 Identify its component parts. How does it help to solve the problems of:
a) cliff-foot erosion
b) storm-wave energy
c) excessive land drainage and saturation of clay
d) cliff slumping?

8 The cost of this scheme is about £600 per metre at 1995 prices. There are many possibilities but the most likely are:
- protect the whole 3.5km stretch
- protect certain parts of it
- protect none of it.

Form groups of two or three people. Carry out an analysis, such as a cost–benefit analysis (see page 20), of the various possibilities. Which option seems most feasible? How do your recommendations compare with those made by other groups? How might the owners of the caravan site react to your recommendation?

9 The local council were advised that defence of this stretch of coast was economically unviable. How might the caravan site owners defend their case for protection work to be carried out, at a time of spending cuts in local government?

Druridge Bay – part of our heritage?

Druridge Bay is the largest and most northerly stretch of this part of the Northumberland coast. It lies at a point where the relatively undeveloped coastline of Northumberland meets the more developed and industrialized coastline of the Tyne and Wear conurbation. The bay consists of a long, mainly sandy stretch of beach and dune coastline between Low Hauxley to the north and Cresswell to the south. The coastline includes sand dune reserves owned by the National Trust, Druridge Bay Country Park, six SSSI, and two nature reserves. The number of features of ecological interest seems to indicate that Druridge Bay is an area to be valued and possibly conserved. At present, there are plans to designate it as part of Britain's Heritage Coast and thus to protect it. Should it become a part, or are there pressures which might prevent this?

Human activities at Druridge Bay

The case for making this bay a part of our Heritage Coast is far from clear, however. There are significant pressures on the Bay. These include pressures from visitors, agriculture, coal extraction, sand extraction, and nuclear energy.

Visitor pressure

The Bay is only 25 minutes' drive from Newcastle-upon-Tyne. Figure 2.26 shows Druridge Bay on a typical day in mid-June. While the beach does not attract the crowds of people who might visit a major resort, the numbers of people are significant. Car parks vary in distance from the main beach, but involve a walk of some 200–300m through the sand dunes. This has resulted in the erosion of dunes by trampling.

Draw a sketch of Druridge Bay from Figure 2.25. Label your sketch with the features listed in this section. You will add more features to this sketch later, so make it at least A4 size.

▼ **Figure 2.25** Druridge Bay.

Dune erosion is cumulative: that is, once a section of dune is exposed, the removal of sand by wind threatens the survival of other plants nearby as the surface of the sand is lowered. Thus dunes quickly become scarred by large patches of bare sand (Figure 2.27). Though these can be managed, and paths can be fenced off during re-growth, dunes become highly mobile once vegetation cover is lost, as Chapter 7 explains.

▲ **Figure 2.26** Tourist pressure during the summer at Druridge Bay.

▲ **Figure 2.27** Dune erosion at Druridge Bay. The bare exposed patches of sand typify serious erosion at a fairly early stage. What begins as small isolated patches of bare sand soon expand as sand is removed by wind, and other species of plant are threatened.

Agriculture

Most land adjacent to Druridge Bay is fairly marginal in terms of agricultural quality. However, overstocking of cattle and sheep on grazing lands adjacent to the dunes has affected the plants in the area. The use of hay as a winter feed has introduced grass seeds of varieties not previously found in Druridge Bay, and these are becoming dominant in some grazing areas.

Coal extraction

The decline of the coal industry in north-east England during the mid-1980s and early 1990s has resulted in the closure of all the deep mines within 25km of this area. Opencast extraction of coal continues, however. East Chevington site is the largest adjoining the Bay. Opencast mining always has an impact on the landscape. Ladyburn Lake is a former site, and is now a habitat for 38 species of water fowl. On the other hand, site restoration also affects soil characteristics; restored farmland is never the same as it was before mining began.

Sand extraction

Study Figure 2.28. It describes an issue involving Druridge Bay which seems to threaten the beach and dunes.

One of the most beautiful and unspoilt beaches in northern England is disappearing – carted away by the lorryload to be turned into roads, shops and buildings.

The pure white sand of Druridge Bay, an eight-mile [12.8km] stretch of Northumberland coast lined with dunes, lagoons, and nature reserves, is the victim of a 30-year-old planning decision described as legal nonsense.

In the early 1960s, the government overruled local planning authorities and gave the go-ahead for sand extraction at Druridge on condition that only a 'small mechanical digger' was used. The permission was granted to a small firm for local building needs. But in the 1970s, the rights were bought by Northern Aggregates, a subsidiary of Ready Mixed Concrete (RMC), and the rate of extraction accelerated.

Planners blame the loose wording of the original permission, typical of the days when environmental awareness was low. Today's 'small mechanical digger' can scoop up two tonnes in its bucket. And the demand for building materials has grown substantially. According to planners at Castle Morpeth Borough Council, the condition gives RMC 'carte blanche to take as much sand as they like'. At a rate of 40,000 tonnes a year, about 1.5 million tonnes of sand have been extracted from Druridge, causing 'extreme' erosion. Beach levels have dropped, the dunes are narrowing by a metre every year, and underlying clay, rock, and a fossilized forest are regularly exposed as the sand thins.

The erosion also jeopardizes plans by Northumberland Wildlife Trust to create new habitats for otters, marsh harriers, and now the rare bittern. One of the threatened dunes, an SSSI, has a colony of scarce marsh helleborine orchids.

Ironically, RMC prides itself on its environmental record and is a corporate member of the Yorkshire, Cleveland, and Durham wildlife trusts, though not of the Northumberland trust, which rejected its application. The firm argues that its contribution to erosion is 'minimal'. Local councils say they cannot afford the estimated £500 000 to compensate RMC for revoking the permission.

Last year RMC agreed to cut extraction by a quarter and has said it will stop using Druridge if similar sand can be found elsewhere. An alternative site has been earmarked but will not be available for at least a year.

Many protesters claim this will be too late for Druridge as the bay is a 'closed' system and cannot generate its own sand. Ian Douglas, reserves manager for the wildlife trust, said the risks were pointed out in the 1960s. 'Thirty years later, we are still talking.'

▲ **Figure 2.28** From *The Independent*, 3 January 1995.

The issue is complex because sediment movement is complex, as Figure 2.29 shows. Sediments have sources – cliffs from which they originate – and sinks, where they are deposited. Beaches, coastal spits, dunes, and offshore bars all act as points where sand is deposited. The depletion of one, such as the beach, has an effect upon the others. Beach erosion, or sand extraction from it, lowers the beach level, which brings the sea further inland, threatens the dunes, starves the spit of sand, and thus depletes the offshore bar. These actions both develop and affect water movement and sediment transfer around Druridge Bay.

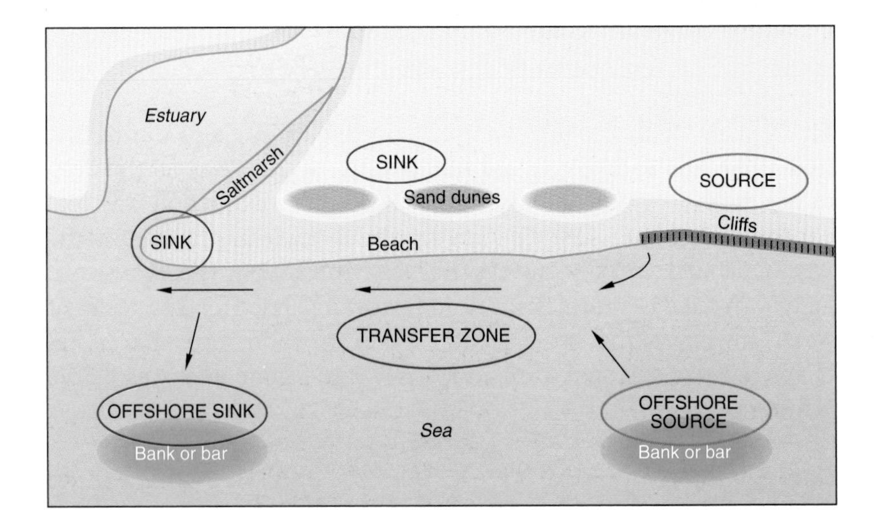

▶ **Figure 2.29** Sediment circulation around a coastline. This diagram shows the relationship between sediment sources from cliff erosion and transfer along the coast by longshore drift to sinks, or points where sand is deposited. If sand extraction continues, the beach becomes depleted, water depth and wave energy increase, and the beach suffers accelerated erosion.

Sand dune formation

Three requisites have been identified for sand dune formation:

- a large supply of readily available sand, dry enough to be transportable by wind
- a reasonably flat surface on which the dune can develop
- a stabilizing agent to control the form of the dune once it has begun to develop, usually in the form of vegetation.

Initially, sand ripples develop as a result of friction between air layers close to the beach. The reduced velocity causes deposition, which in turn creates an increase in deposition, thus increasing the ripple effect. Inland from the high water mark, colonization of sand by salt-tolerant grasses such as marram stabilize sand so that it begins to create an obstacle around which wind must pass, and which creates air turbulence. The turbulence generated causes further deposition on the side away from the wind. Marram is capable of maintaining its growth to suit the increased dune height; its roots can be very long. For further information on dune ecology, turn to Chapter 7, page 155 onwards.

Nuclear energy

Since the early 1980s, Druridge Bay has been the subject of proposals to build a nuclear power station, in the southern part of the bay close to Widdrington. This proposal continues to be the subject of debate. National policies concerning nuclear energy have become somewhat fluid since the Chernobyl disaster in 1986, and will probably depend on whether or not the nuclear industry is privatized. Following privatization, it is highly likely that the new company would seek expansion. Druridge Bay has remained a potential site since its characteristics – on a low flat coast with access to a major conurbation – were first identified.

Conflict of interest and conflict matrices

Figure 2.30 is an outline for a conflict matrix. A conflict matrix identifies uses of land which may conflict with each other. For instance, you may decide that there is a conflict of interest between nature reserves and the extraction of sand. In this case, you would shade in the box where these two areas intersect. Some conflicts are complex, however. Recreational use may conflict with extraction of sand, but only at certain times of the year. In this case, you might use a different colour to shade in the box. You can decide to use different shades of colour depending on how serious the conflict is. For example, light blue might mean a slight conflict, and dark blue a serious conflict.

If you complete a matrix for Druridge Bay, this will help you to analyse the situation. You may identify certain land use types which are not compatible with others. You may decide that some uses should be stopped or prevented. It will also help you to develop ideas on how to manage Druridge Bay, and to identify the features that require most support and attention.

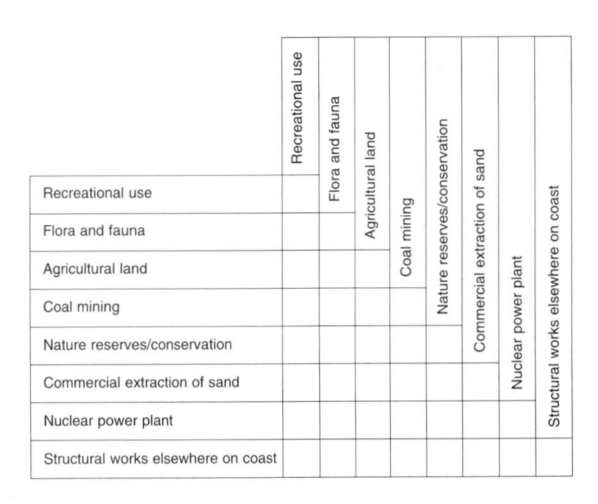

	Recreational use	Flora and fauna	Agricultural land	Coal mining	Nature reserves/conservation	Commercial extraction of sand	Nuclear power plant	Structural works elsewhere on coast
Recreational use								
Flora and fauna								
Agricultural land								
Coal mining								
Nature reserves/conservation								
Commercial extraction of sand								
Nuclear power plant								
Structural works elsewhere on coast								

▲ **Figure 2.30** A conflict matrix for Druridge Bay

- Northumberland County Council
- Alnwick District Council
- Castle Morpeth Borough Council
- National Trust
- Northumberland Wildlife Trust
- British Coal
- English Nature
- Royal Society for the Protection of Birds
- Countryside Commission
- Druridge Bay Campaign
- Friends of the Earth
- Local parish councils
- Alcan Farms

▲ **Figure 2.31** Interest groups in Druridge Bay.

Interest groups in Druridge Bay

Can different groups have the same interests at heart? Consider the list of groups in Figure 2.31. Are they likely to see a conflict matrix in the same way? This exercise will help you to see how large pressure groups develop out of smaller organizations.

1 Individually, select one group from the list in Figure 2.31. On an OHP acetate sheet, complete a conflict matrix like the one in Figure 2.30, as you think that group would perceive it.
2 Compare your matrix with that completed for another group which you feel would agree with yours. For example, the RSPB might agree with the Northumberland Wildlife Trust. Place your acetate sheets together to see where you agree or disagree.
3 Repeat the exercise until you have established groups of like-minded people.
4 Has a conflict situation emerged in your class? Which groups have emerged? What do they have in common? Do they all believe in the same future for Druridge Bay?
5 In groups, decide on a plan that answers the question: 'Should Druridge Bay become part of the English Heritage Coast, or should it be left for development?'

Managing tropical coasts

Coral reefs are one of the most biologically productive and diverse of all natural ecosystems. They are found in shallow waters of less than 30m depth, between 30° north and 30° south of the equator. Corals are polyps that collectively secrete calcium carbonate to build colonies. There are two basic types of coral reef: shelf reefs, such as the Australian Barrier Reef, on the continental shelf just off large landmasses; and oceanic reefs which are found close to deep water and are often associated with oceanic islands.

Many of the coral reefs in Thailand and South-East Asia have been damaged and exploited in recent years. The main causes of this have been identified as:

- intensive fishing
- the use of dynamite fishing, in which areas of reef are quite literally subjected to an explosion of dynamite in order to release catches of fish
- sedimentation due to deforestation
- pollution from urban and industrial areas
- tourism in heavily developed areas
- boat anchorages.

In response to this degradation of coral reefs the Association of South East Asian Nations (ASEAN), which includes Thailand, Brunei, Malaysia, the Philippines, and Singapore, has been working on management strategies for coral reef resources in the region. Some experimentation is taking place with marine parks and reserves. The islands of Ban Don Bay in Thailand have been a pilot for coral reef management, and the strategy for this area provides the focus of this study.

The Ban Don Bay area of Thailand comprises three groups of islands: Ko Samui, Ko Phangan, and Mu Ko Ang Thong. Figure 2.32 shows these island groups, and the relative quality of their coral reefs. Coral resources are of value to the fishing industry and to the growing tourist industry. Even though legislation bans the use of blast or dynamite fishing, it continues, although it is largely practised by people from outside the area. It seems that some officials do not enforce the ban in return for a share in the catch.

The tourist boom in the area means that more people now visit these reefs. Siltation and pollution have also increased, mostly from waste water which is a product of development in the coastal areas. Revenue from tourism in the area increased from US $15 000 in 1980 to US $300 000 in 1987, and tourism is now one of the major income-generators for this province of Thailand.

What is the spatial distribution of reef quality?

Use Figure 2.32 to describe the spatial distribution of coral reef quality in Ban Don Bay, and offer an explanation for these.

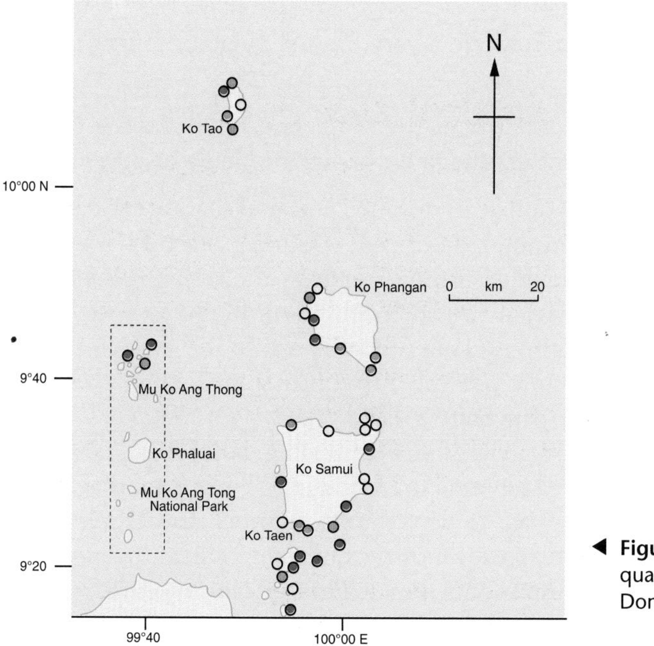

Figure 2.32 Location and relative quality of the coral reefs in Ban Don Bay.

Management of the Ban Don Bay coral reefs

At present there is little positive management, and certainly no co-ordinated management plan for the reefs, although ASEAN is starting to work towards a co-ordinated strategy. There are regulations governing destructive fishing techniques and coral collection. Otherwise, however, management has been left to governors at provincial level, and the quality of management varies enormously.

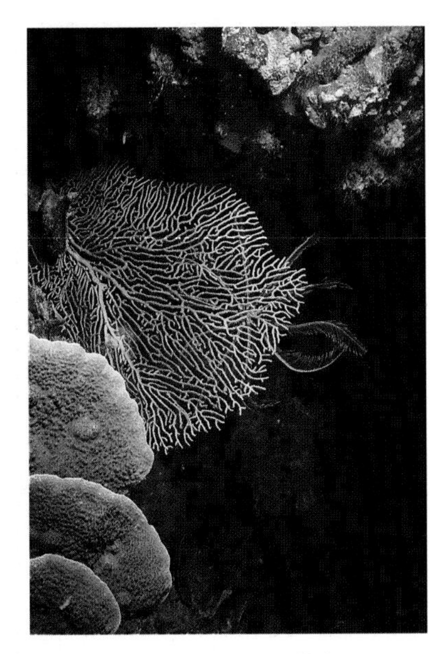

▲ **Figure 2.33** Coral reef off the coast of Thailand.

Degradation of the coral reefs of Ban Don Bay continues in some areas because there is low public appreciation of the value of the coral reefs. There is a lack of co-ordination at different levels, and little knowledge of what management options are possible. There is little money available to police the reefs effectively.

When a management plan was developed for Ban Don Bay, two approaches were adopted. First, it was thought that a 'bottom-up' approach was needed: management could only be successful at the highest level if the local people supported and were involved in the scheme. Secondly, a management plan must allow sustainable use of the coral reefs. Rather than prohibiting use of the reefs, local people could continue to reap the benefits in a way that would sustain both themselves and the reef.

Management programmes come under four headings:

- recreation and tourism
- fisheries management
- water quality maintenance
- public awareness and education.

In order to control the damaging effect that tourism is having on the reefs, without damaging the potential income from tourism, a system of zoning has been proposed. This creates three zones.

1 *Sanctuary zones*, where any destruction of coral is prohibited.

2 *Conservation zones*, where a very limited range of tourist activities is allowed.

3 A *general use zone*, where there are only guidelines for habitat protection and where the range of activities may be greater.

Fishery management seeks to limit and prevent further damage to the reefs by banning destructive fishing practices, upgrading legislation on some other fishing practices, and stopping depletion of the reef caused by coral and shell collection. It also seeks to preserve the traditional culture. Water quality is to be monitored, and legislation drawn up to prohibit any construction activities, especially in areas adjacent to sanctuary zones.

Raising public awareness and educating both local people and visitors is considered essential to the success of any management plan. Information centres are to be set up for visitors, and a trained team will visit villages on an educational campaign. In time, the importance of coral reefs will also be incorporated in the school curriculum.

Why are coral reefs important?

Although coral reefs only cover 0.2 per cent of the world's ocean floor, they are particularly important. The World Conservation Strategy, produced in 1980, identifies coral reefs as one of the 'essential life-support systems' necessary for food production, health, and other aspects of human survival and sustainable development.

Reefs protect coastlines against waves and storm surge, they prevent erosion, contribute to the formation of sandy beaches and sheltered harbours, and they also serve as recyclers of carbon dioxide from sea water and the atmosphere. They are a source of raw materials – corals, coral sands, and shells – for building, for jewellery, and for ornamental objects. Increasing numbers of reef species are being found to contain compounds with medical properties. Figure 2.33 shows a part of the coral reef in Thailand.

What are the threats to coral reefs?

A coral reef is a complex system of interdependent components. It has very specific requirements of light, temperature, water clarity, salinity, and oxygen. The complex nature of the relationships within the reef means that damage even to one component can trigger a negative chain reaction affecting many different organisms in the system (see the theory box on equilibrium, page 17). Figure 2.34 presents some guidelines for tourists in the Red Sea to help protect the reefs.

There are a number of impacts which can have an adverse effect on a coral reef system. Natural disturbances, such as hurricanes, storms, and sea-level fluctuations, have always imposed a temporary pressure on coral reef systems, but there has been a recent increase in the effect of people on reefs. This impact is more significant when much of the damage is chronic (long-term), rather than temporary.

The increase in coastal populations and the development of ports and other coastal urban areas, have led to pollution of waters through sewage and industrial emissions, oil spillages, and the transfer of pesticides into coastal waters from rivers. These pollutants kill the coral. Pollution of reefs is not always terminal; once pollutants are removed, the reef recovers relatively rapidly. But other pressures have a profound effect on coral reefs. Rainforest removal and deforestation in tropical areas have resulted in a huge increase in the amount of sediment reaching coastal areas. The sediment destroys the balance of the coral reef ecosystem. The growth of tourism in tropical areas has certainly had a major impact on the coral reefs. The sheer number of visitors is having an effect, despite careful management and restrictions in some areas.

How can coral reefs be managed?

Coral reefs require careful management if their effective use is to be sustained. One increasingly popular approach is to protect certain areas and to develop marine parks, where fees are charged for entrance, for souvenirs, and for boat permits. Income derived from fees allows both the development of reefs as a resource, and a form of management that recognises the damage that can be done to reefs. In some places lack of funding is a problem and as a result effective management is proving difficult.

ASEAN is seeking to establish a network of heritage parks and reserves. International collaboration is important in developing management strategies for coral reefs. They do not follow political

- Make sure when you enter the water that you are wearing sneakers or some other form of footgear, in order to protect your feet if you step on sharp coral, sea urchins, and so on.
- In the water, make sure you swim, not walk, in order to avoid damaging the coral. When swimming, make sure you do not damage the coral in any way, by touching it or knocking it with a flipper. Remember that such damage remains visible for years.
- When making for the open sea, cross the reef only by the route marked by floats and ropes. Do not cross the reef anywhere else.
- It is strictly forbidden to stand or walk on the reef, which causes major damage to it.
- All rocks and animals in the reserve – both in the water and on the shore – are protected natural assets. Do not touch them and on no account remove them from the water.

By following these rules and keeping the Reserve clean, you will have an enjoyable visit, as well as preserving its unique beauty for other visitors to enjoy too.

Your co-operation is appreciated.

▲ **Figure 2.34** Safety measures designed to protect coral reefs in the Red Sea.

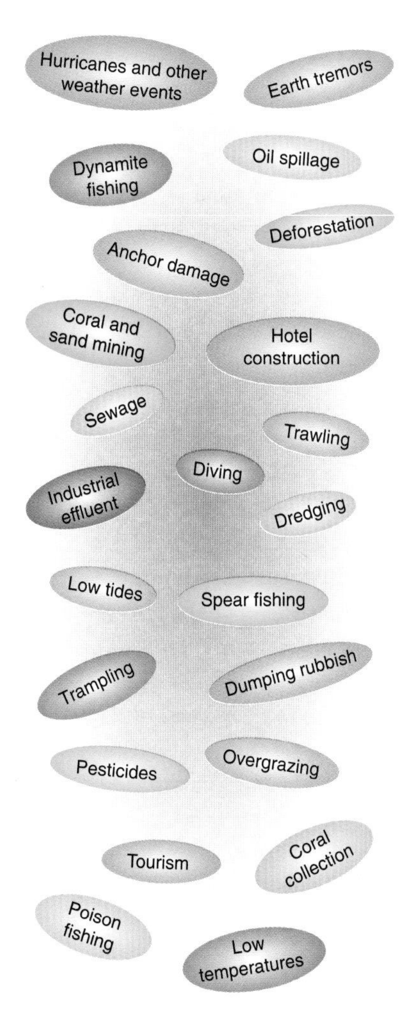

▲ **Figure 2.35** Pressures on coral reefs.

boundaries, and are a collective responsibility. However, conflict may arise between the needs of one country and the needs of another, or between economic development through industry or deforestation on the one hand, and ecotourism and conservation on the other.

Management focuses on three broad categories.

1 *Zoning* Reef management needs to be considered in terms of the whole of the coastal zone, and includes river catchment areas. Coastal zone management involves putting sewage outlets 'down-flow' from coral reefs, discharging thermal effluent into deep water, and generally taking the reefs into account when planning land use. Environmental impact assessments (described in Chapter 3) should be made when development may affect the reefs.

2 *Management of reef fisheries* It has been recommended that coral reef fisheries are maintained at levels that permit fishing without depleting fish stocks. The World Conservation Strategy considers this to be a global priority. Most countries have legislation relating to fishing. It is possible to maintain fisheries in a sustainable way by improving the stock, and developing under-used varieties of coral life such as sea cucumbers and molluscs. It is also possible to stop over-exploitation of shells for ornamental purposes by developing breeding grounds specifically for this purpose.

3 *Research and education* Further research into coral reefs is vital if there is to be effective management. Schools have a major role in educating young people to understand the importance of reefs as important resources and as areas in need of protection and careful management.

Classifying pressures on coral reefs

1 Form groups of two or three people.
   a) Decide how the different pressures on the world's coral reefs (Figure 2.35) might be prioritized. You can do this either by ranking or by grouping similar priorities.
   b) Share your findings with other groups. Explain how you decided to organize the pressures.

2 Read through this study about the attempts made to manage coral reefs in Thailand. Check off each management strategy against the priorities you decided upon. Which ones are being well managed in your opinion? Which are not?

3 Decide on actions that are necessary to improve the management of reefs. Put forward a plan of 750–1000 words. Justify your ideas.

4 Imagine that all your proposals are adopted. How would you know, in five years' time, which ideas have been successful and which have not? How would you evaluate the success of your plan?

Ideas for further study

There is a huge variation in coastlines in the world both in terms of scale and physical and human characteristics. In order to reinforce your study of coastlines as a landform, you should choose another stretch of coastline to study in order to complement your study of the south Northumberland coast. This could be from another part of the world, in a less developed country, or a local example that you have studied as part of a fieldwork exercise. The following are some suggestions that you might use as a route of enquiry. You should add questions of your own where they are relevant to you own case study.

1 Where is the stretch of coastline? How large is it?
2 What are its main geomorphological features?
3 What human pressures are there on the area? How do these affect the area?
4 Who is responsible for managing the area?
5 What management strategies are being or have been implemented to combat the different pressures?
6 What implications are there for the future? Who will gain, who will lose?

Summary

- The coastal zone is a complex scene of many different and often interdependent natural processes.
- There are many pressures on the coastal zone. These are the result of an interaction between natural processes and human activities.
- Human occupation and utilization of the resources at the coast can result in changes to the coastal zone.
- There are often conflicts of interest between users of the coastline. Careful management of the coastline is therefore necessary.
- Successful management of the coastal zone requires a complete understanding of the natural processes at work in it.

References and further reading

R. Case, *Coastal Management: A Case Study of Barton-on-Sea*, Longman, 1984.

R. Collard, *The Physical Geography of Landscape*, Unwin Hyman, 1988.

Coral Reef Management Plan for the Islands of Ban Don Bay, Thailand.

C. Hutchings, 'Back to basics', *Geographical Magazine*, March 1994.

B. Knapp, S. Ross and D. McCrae, *Challenge of the Natural Environment*, Longman, 1989.

Northumberland County Council, *Northumberland Coast Management Plan*, October 1993.

A.T. White, *Coral Reef Management in the ASEAN/US Coastal Resources Management Project.*

M. Younger, 'Will the sea always win?' *Geography Review*, May 1990.

3 Estuaries and deltas

Why do we study estuaries?

So far this section has dealt with rivers and coasts. This chapter is devoted to the places where rivers and coasts meet – estuaries and deltas. It begins with an investigation of the issues facing the Tees estuary on the north-east coast of England, and examines the physical processes within it and the human pressures put on it. The second case study area is the Mississippi delta on the Gulf coast of the USA. Finally, a stretch of the Barrier Reef coastline along the east coast of Australia is examined, with reference to the interaction between rivers and coasts.

The Nature Conservancy Council defines an estuary as 'a partially enclosed area of water and soft tidal shore and its surroundings, open to saline water from the sea and receiving fresh water from rivers, land runoff or seepage'. Particles of fine silt and clay are carried down here from the entire river basin, and form mudflats and mudbanks. This material, together with the plant communities that develop, creates a habitat which is abundant in food resources for animal, plant, and bird life.

Estuaries play a key role in human activity, even though their percentage area is small in terms of the total surface of the world. Long valued as safe anchorages, for their resources of fish and shellfish, and for the grazing provided by local marshes, in Britain they came to serve as focal points linking inland settlements with the coast, the continent, and the rest of the world. Today, one-third of Britain's population – some 18 million people – live around estuaries, and major parts of Britain's industrial and transport infrastructure are located there. Here lies the chief issue facing planners and those interested in managing estuaries. Figure 3.1 shows part of the most industrialized area of the Tees estuary. By contrast, parts of the Tees estuary remain unaffected by industry and building development. Can the animal and plant communities continue to exist with the pressures for development that estuaries are now facing?

▲ **Figure 3.1** Industrial landscape along the Tees estuary.

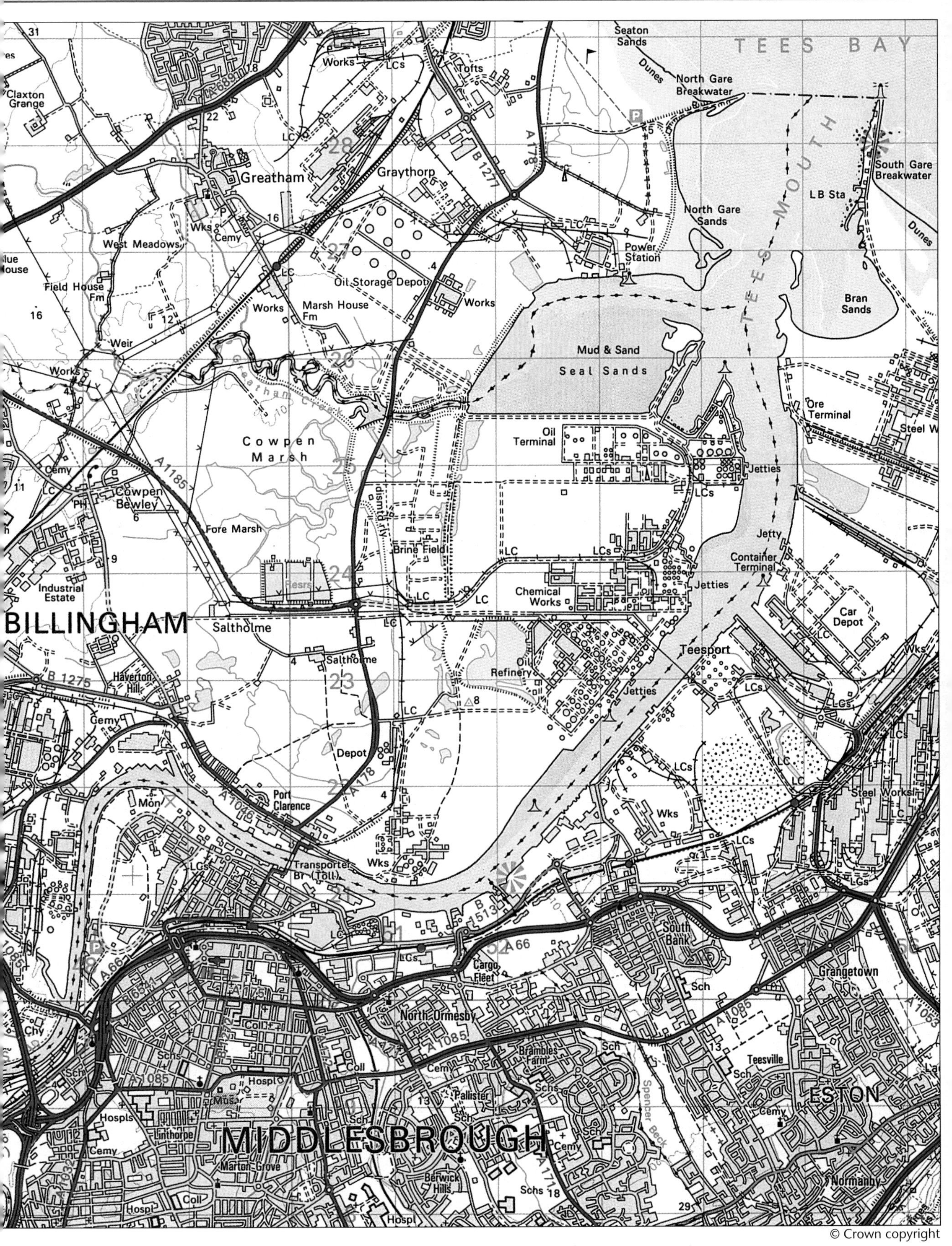

▲ **Figure 3.2** The Tees estuary at a scale of 1: 50 000.

The Tees estuary

The extent of the Tees estuary was determined by the length of the river that was affected by the tides of the North Sea. It is now determined by the Tees barrage. The full tidal limit stretches inland for 40km from the mouth of the Tees to Low Moor.

1 What are the various possible uses of an estuary? Use Figures 3.1 and 3.2 to help you answer this question. Also consider other estuaries, in the UK and in other parts of the world.

2 How might these uses conflict with each other?

3 Using evidence from Figure 3.2, explain why estuaries make good industrial sites.

What processes are involved in the development of the Tees estuary?

Estuaries are constantly changing, on a daily and seasonal basis, as a result of tides. Tidal processes are the main causes of transportation and deposition of sediment in estuarine areas. **Tides** are rising and falling water-levels, caused by the gravitational pull of the Sun and the Moon. As the Moon rotates around the Earth, it exerts a mild gravitational force on the Earth, to which water responds. This creates a tidal bulge. A similar bulge is produced on the opposite side of the Earth. As the Earth rotates during a 24-hour period, this means that two bulges affect each part of the Earth every day. Twice each day the tide flows in to produce a high tide, and then ebbs (flows out) to produce a low tide. The difference in water-level between high and low tides is the **tidal range**. The two high tides are usually slightly more than twelve hours apart. Tides produce changes in water-level in areas of shallow coastline such as estuaries.

This daily cycle, which occurs as the Earth rotates, is part of a broader 28-day cycle, caused by the time it takes for a single lunar orbit of the Moon around the Earth. The Sun also exerts a gravitational pull on the Earth, although the pull of the Moon is greater. When both pull in the same direction, tides are higher than normal and low tides lower than normal; that is, the tidal range increases. Such tides are known as **spring tides** (Figure 3.3a). Between spring tides, the rotations of the Sun and the Moon move out of sequence, and pull in opposite directions. This reduces the tidal bulge. These tides are called **neap tides** (Figure 3.3b), during which time water does not travel so far inland at high tide, nor move as far out to sea at low tide. Thus the tidal range decreases.

▼ **Figure 3.3** The tidal process.

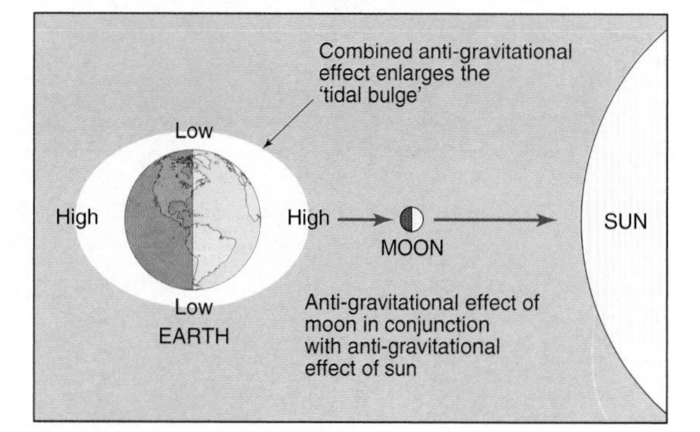

(a) Spring tide.

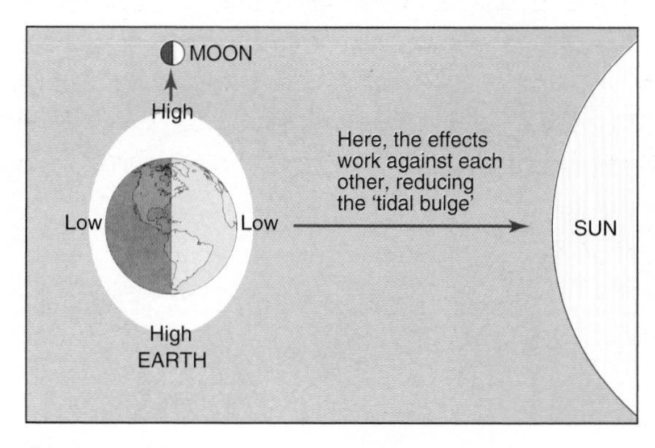

(b) Neap tide.

The tidal regime of the Tees estuary

The Tees estuary is a narrow, partly canalized channel extending from Tees Bay to Low Moor Weir, some 40km upstream. The estuary has a strong tidal regime, with an average tidal range of 4.5m at spring tide and 2.3m at neap tide. The cycle of neap and spring tides repeats itself every fifteen days. The average high-water spring tide is equivalent to +2.65m above normal sea-level (known as AOD – Above Ordnance Datum). At times, therefore, high tides can threaten low-lying land around the estuary.

Movements of, and interactions between, tidal marine and fresh river water lead to typical patterns of sediment in the lower parts of the estuary where the circulation of water is greatest. The circulation pattern is shown in Figure 3.4, and takes place when water flows and ebbs during tidal movement. It results in a net seaward flow of water in the upper one-third of water and a net landward drift of water in the lower two-thirds. Part of the surface layer ebb flow is carried back on subsequent flood tides, while the remainder is removed by diffusion and transfer to the sea. This produces different patterns of sediment at different points in the estuary.

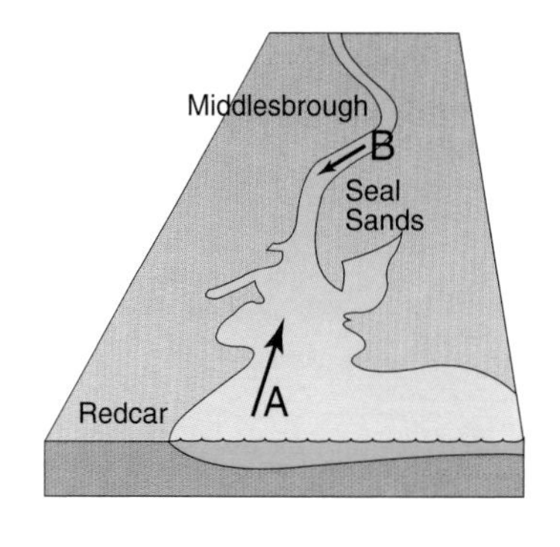

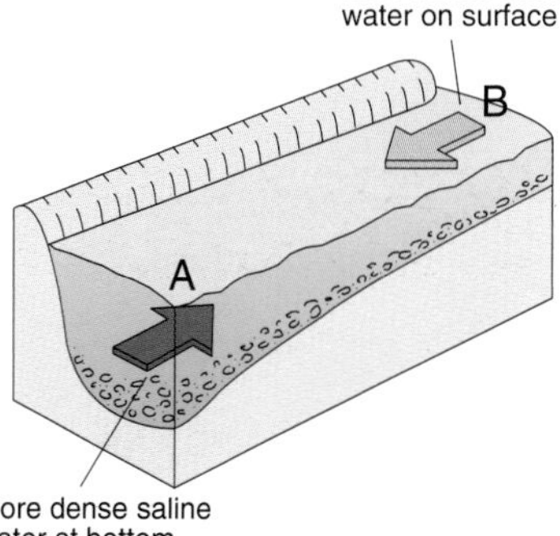

▲ **Figure 3.4** The dynamics of the Tees estuary.

Distance upstream of mouth (km)	**Sediment type**	**Size of particles**
Tees Bay	Coarse silt	20–65 microns
0–2	Sand	50% 0.15mm
2–13	Soft mud	90% under 60 microns
Upstream of 13km	Fine gravel	Over 5mm

▲ **Figure 3.5** Sediment types within the Tees estuary.

Sediment deposited in the estuary has been carried there in suspension either by rivers or by tidal currents. The majority of sediment deposited in the lower estuary is of marine origin. The range of sediment types and their location within the Tees estuary is given in Figure 3.5.

In time, tidal sediments fill estuaries to produce mudflats, which are expanses of silt and clay exposed at low tide but covered at high tide. Salt-tolerant plants may then colonize the mudflat. The stems of these plants trap further sediment and the flat builds up to a level about that of the high tide, becoming a saltmarsh.

1 Explain why the particle size of sediment gets smaller as it reaches the mouth of the river.

2 What are the implications of the sedimentation process for port authorities? How can these be managed?

On occasion, a combination of high tides and severe weather can result in coastal flooding. Flooding from North Sea tidal surges is a feature of the lower Tees estuary. The most severe flood in the 20th century was in 1953. This flood affected the entire east coast of Scotland and England: 300 people drowned, 65 000ha of land were flooded, and 24 000 houses were damaged. It occurred as a result of a combination of one of the worst northerly gales on record (winds of Force 10 and 11 were recorded), and a spring tide. This flood serves as a benchmark for land planning and development in the coastal zone. The National Rivers Authority (NRA), in conjunction with the Meteorological Office Storm Tide Warning Service, now provides warnings in advance of surges combined with high spring tides.

How can the interaction of river and coastal systems change the form of an estuary over time?

Study the aerial photograph in Figure 3.6 of the Tees estuary in 1994. Describe the land use in this section of the estuary. The amount of building and development has altered the landscape significantly. However, coastal and river processes have together contributed to some of the changes.

▶ **Figure 3.6** The Tees estuary in 1994.

Coastal change

Sediment accumulation or removal along a coast depends on a combination of coastal and river currents. Coastal currents depend on prevailing winds. In the case of the Tees estuary, these are seasonal; movement of water tends to be from north-north-west to south-south-east during winter, and the reverse in summer. As a result, sediment builds up on either side of the Tees estuary. It is prevented from extending further across the mouth of the estuary by breakwaters. The north side of the estuary is sheltered by the North Gare breakwater, and the south side by the South Gare, as shown in Figure 3.6.

However, the movement of river water also affects the estuary. The inward and outward flow of water twice daily makes sediment transport and deposition a constant process, and sediment is constantly being re-worked. As a result, channel changes occur within the estuary, modified only by the process of dredging.

What are the human demands on the Tees estuary?

There is also evidence in Figure 3.6 of considerable human change. Over time there has been an expansion in human activity on the land fringeing the estuary. This has largely been due to the growth in importance of North Sea oil and associated industries. The estuary has been transformed from an essentially 'natural' environment to one that is dominated by human activity.

The changes within the estuary are largely due to:

- industrial development
- land reclamation
- areas designated as Sites of Special Scientific Interest (SSSI)
- navigation and dredging
- coastal management against flooding
- urbanization.

Industrial development

The area around the estuary of the Tees is the heart of industrial Teesside, home of the second largest concentration of chemical and petrochemical works in Europe, and one of the UK's largest ports. The Stockton and Darlington railway brought coal to be shipped from the port. Iron and steel works were built on the reclaimed mudflats, and shipbuilders followed. In 1926, ICI built its first plant at Billingham to produce ammonia and other nitrogen-based products from locally extracted coal and anhydrite. Now ICI has various sites along the estuary. Other companies located here are British Steel, Amoco, Enron, Phillips Petroleum, BASF, and Tioxide. While these industries argue that they bring many benefits to the area, most have also created serious pollution threats at different points. Discharges from industry are shown in Figure 3.7.

Victoria Bridge to Cargo Fleet		
	BOD	Ammonia
1970	102	73
1979	56	64
1985	34	39
1988	49	25

Cargo Fleet to Teesmouth		
	BOD	Ammonia
1970	402	15
1979	122	20
1985	126	10
1988	118	6

▲ **Figure 3.8** Pollution loads in the Tees estuary (tonnes per day).

1 Using Figures 3.8–3.10, describe how pollution levels in the Tees estuary have changed between 1970 and 1990. Select different ways of presenting this data.

2 The National Rivers Authority was established in 1989 as a public body to safeguard and improve the water environment in England and Wales. Find out what powers it has to control pollution, and what action it can take against polluters. Do you think its powers make it effective in fighting pollution?

	Km	%
Class A (good)	2.1	4.8
Class B (fair)	21.9	49.8
Class C (poor)	6.5	14.8
Class D (bad)	13.5	30.7

▲ **Figure 3.9** Water quality survey of the Tees, 1992.

▼ **Figure 3.7** Major points of industrial discharge along the Tees estuary.

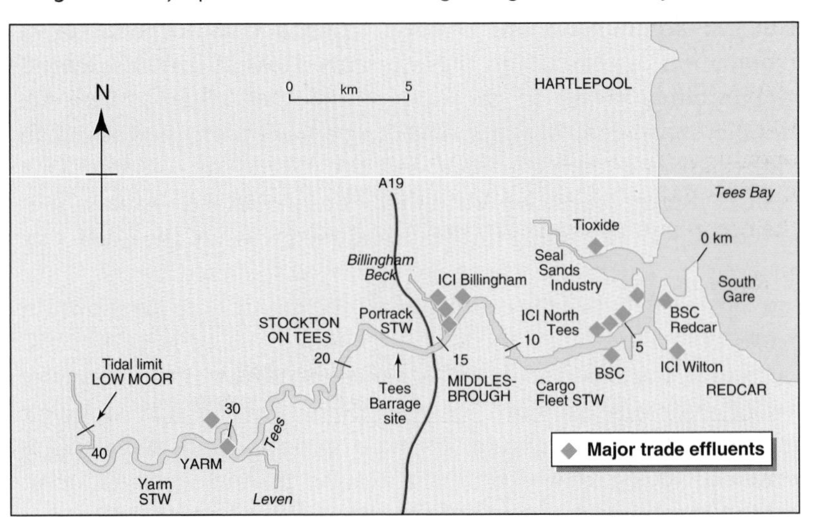

Perhaps the most serious form of pollution in estuaries is that which results in a decrease in oxygen content of the water. The extent to which a given discharge will deplete oxygen in the receiving area is measured as a '5-day BOD' (biochemical oxygen demand), which is the quantity of oxygen used in five days for the partial oxidation of a sample of effluent under standard conditions. By 1970 the estuary was one of the most heavily polluted in the UK, with over 500 tonnes of BOD load discharged daily from chemical, petrochemical, and steel-making industries, and from untreated domestic sewage. The variety of pollutants is shown in Figures 3.8–3.10.

▼ **Figure 3.10** Pollution in the Tees estuary 1970 and 1990.

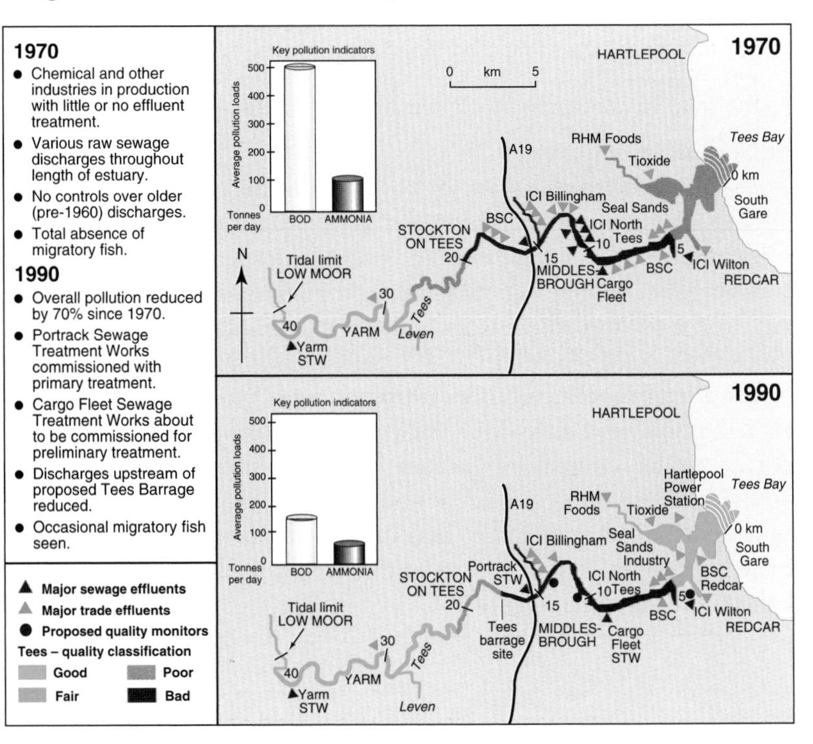

The Tees estuary, like others on the east coast such as the Forth, Tyne, Humber, and Thames, is a pathway for pollutants collected from its inland catchment which are discharged into the North Sea. Figure 3.12 shows the North Sea from a different perspective. Here its true form is revealed: it is a bay on the north-west coast of Europe, opening into the Atlantic Ocean in the north, with a narrow channel into the ocean in the south. Pollutants are not rapidly flushed out to the ocean but linger in a sea that also meets many other demands.

▼ **Figure 3.11** River pollution of the North Sea.

River	Mercury (tonnes/ year)	Cadmium (tonnes/ year)	Nitrogen ('000 tonnes/ year)	Phosphorus ('000 tonnes/ year)
Forth	0.1	2.0	1	–
Tyne	1.4	1.3	1	0.2
Tees	0.6	0.6	2	0.2
Humber	0.7	3.5	41	0.6
Thames	1.1	1.5	31	0.1
Scheldt	1.0	7.4	62	7.0
Rhine	3.9	13.8	420	37.0
Ems	0.4	0.7	22	0.7
Weser	1.1	2.9	87	3.8
Elbe	7.3	8.4	150	12.0

▼ **Figure 3.12** The North Sea and surrounding industrial nations.

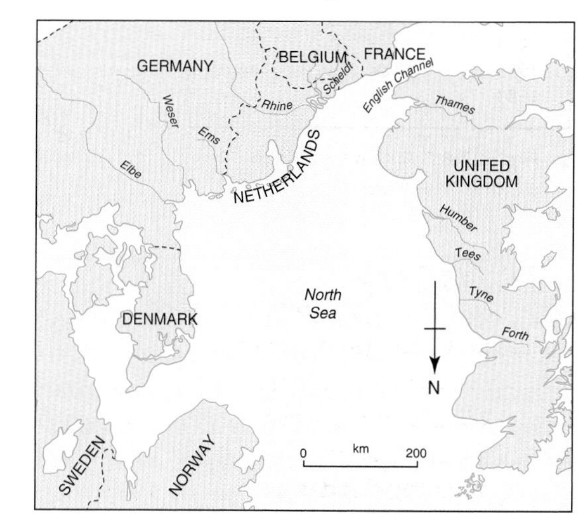

Proportional circles

Proportional circles can be used to represent variations in the distribution and quantity of data. For instance, city populations can be shown in this way. The area of the circle is in direct proportion to the magnitude of the data. To draw proportional circles, follow these steps:

1 Decide on the maximum size of circle that can be used to show the largest quantity on the map you are using. For this example, imagine that this is 2cm diameter or 1cm radius. This will always depend on the scale of the map and how many circles you have to draw. Make a note of the radius you need to draw a circle of this size for step (3).

2 Calculate the square root of the largest figure you want to represent. Imagine the value is 10 000; so the square root is 100.

3 Your largest radius in step (1) – e.g. 1cm – is therefore equivalent to 100 because that is the largest figure you have to show. This means that 1mm is equivalent to 10.

4 A second value of 6500 would therefore have a square root of 80.62. The circle to represent this will be 8mm radius. You can calculate the size of all the circles you need on this basis.

5 Construct a scale beside your map like the one shown below.

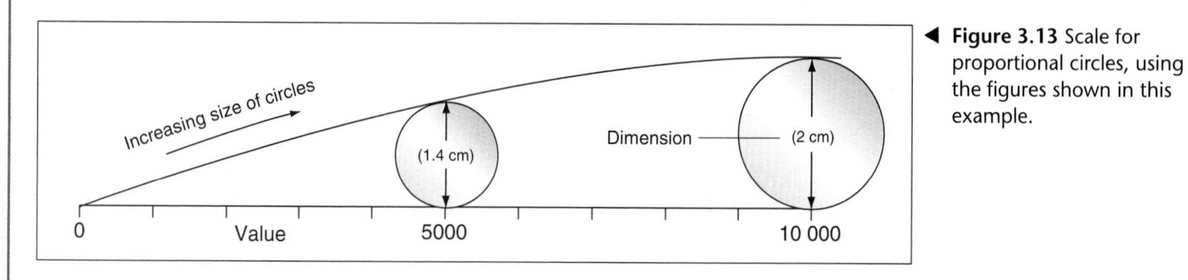

◀ **Figure 3.13** Scale for proportional circles, using the figures shown in this example.

1 Make a copy of Figure 3.12. On it, mark and label Europe's key industrial regions, such as Teesside and the Ruhr.

2 Using the data in Figure 3.11, construct proportional circles for the mercury pollution from each river that discharges into the North Sea.

3 Who, if anyone, should manage such pollution issues? Discuss how effective you think each of the following would be, or could be:
   a) private companies
   b) Teesside local borough councils
   c) the UK government
   d) the European Assembly
   e) some other suitable body
   f) nobody.

4 Should responsibility for pollution in the North Sea be shared amongst the bordering countries?

5 How might different individuals and organizations use the pollution issue
   a) to argue that European countries need closer links
   b) to argue that European countries need fewer or no links?

Land reclamation

Much of the original marshland that existed along the Tees estuary has been purchased, drained, and protected to allow the expansion of industry. This process is known as **reclamation**. The extent of this reclamation is shown in more detail in Figure 3.14. The main effects are felt by the migratory estuarine birds, seals, and shellfish, as it is their habitats that are lost.

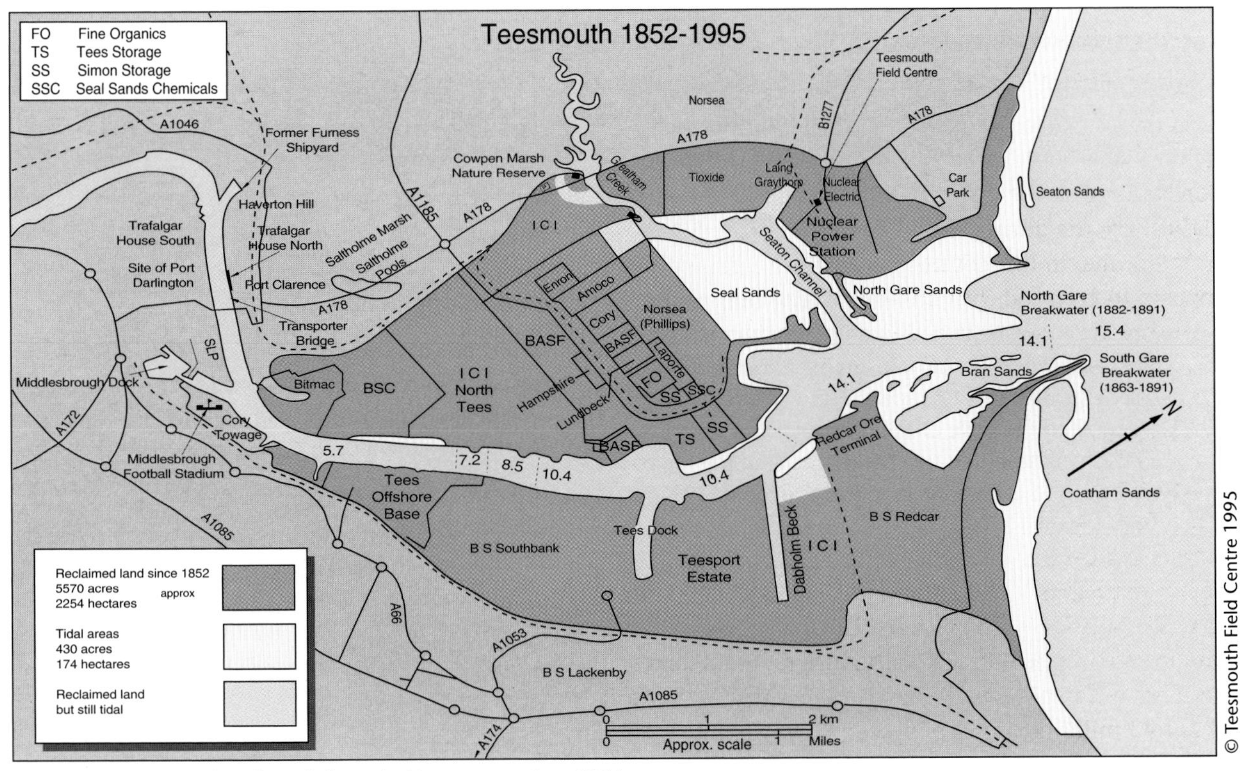

▲ **Figure 3.14** Land reclaimed along the Tees estuary since 1856.

Areas designated as Sites of Special Scientific Interest

Despite being heavily industrialized and densely populated, the Tees estuary is a very important area for wildlife and birds. Here, four Sites of Special Scientific Interest (SSSI) protect what is left of the original inter-tidal habitats. These SSSI are collectively known as the 'Teesmouth Flats and Marshes'. It is the only extensive area of inter-tidal mudflats and marshland left on the east coast of the UK between Fenham Flats, Northumberland, and the Humber estuary. It is an invaluable break in the journey for many migrating waders; it also provides a winter refuge for birds from further north, and summer breeding grounds for others from further south.

The four SSSI are as follows.

1 South Gare and Coatham Sands – these include large areas of sand dunes, saltmarsh, and freshwater marsh. Areas of mudflat on Bran Sands also provide important winter feeding grounds for many birds such as the curlew, redshank, and grey plover.

2 Cowpen Marsh includes the largest area of saltmarsh between Lindisfarne and the Humber estuary. This saltmarsh has a very rich flora, which provides an important roosting and feeding ground for migratory waders and wildfowl. Part of Cowpen Marsh is managed as a nature reserve.

3 Seaton Dunes and Common on the north bank at the mouth of the Tees estuary has a wide range of habitats which support a rich flora, invertebrates, and bird life.

4 Seal Sands consists of mudflats of great international ornithological importance for large numbers of shelduck, knot, and redshank. Many other species of wildfowl and waders feed on Seal Sands and use adjacent areas of reclaimed land as roosting sites.

Common seals are present consistently throughout the year, with highest counts during the moulting season of August–September. In 1992/93, maximum counts were 34. The general health of the seals appears to be good, but there was a total breeding failure over the monitoring period from 1989. Only three pups were born and all three died within a few days of birth. The cause is uncertain.

Navigation and dredging

Navigation

The Tees estuary is a major commercial port serving the petrochemical industry, offshore oil and gas installations, and other heavy industries. Currently the Tees and Hartlepool Port Authority Ltd (THPAL) has control over the tidal reaches of the Tees as far inland as the Tees barrage. THPAL is the third largest port in the UK, both in terms of tonnage and profitability, handling a volume of about 43 million tonnes per year.

Transporting large volumes of oil into and out of estuaries can cause problems. In 1983, 6000 tonnes of crude oil were spilt into the Humber estuary when the tanker *Sivand* collided with the terminal at Immingham. Oil was rapidly distributed throughout the estuary, affecting the fringeing reedbeds, oiling hundreds of birds, and destroying ragworm, an important food for migrating birds. No similar incidents have occurred on Teesside, although illegal discharges from tankers flushing out between cargos over a period has caused significant accumulations of oil-based pollutants.

▲ **Figure 3.15** Tees and Hartlepool ports.

Dredging

Approximately 1.55 million cubic metres of sand and silt are removed each year by dredging from the tidal stretch as far upstream as Billingham oil jetty, 13km from Teesmouth (see Figure 3.2). This is necessary to prevent the silting-up of both the entrance to the estuary and the river itself. Without dredging, siltation rates of 1m depth per year would lead to problems for large ships. With large crude oil and petrochemical tankers, it is crucial to keep waterways clear: the loss of one extra metre's depth in a single year would prevent some ships from using Tees oil and petrochemical terminals. Around 90 per cent of sediment entering the estuary comes from the sea and, once it has been dredged, it is returned there.

Coastal management against flooding

During 1993/94, tidal defences were completed at Greatham Creek and on the Greenabella sea wall. These provide additional tidal protection to the low-lying industrial zones on the north of the estuary.

Urbanization

The Tees estuary is surrounded by the urban areas of Middlesbrough, Stockton, Hartlepool, and Redcar. As well as pressure for space, which has tended to push the port further north-east and to seek sites that can be drained and filled in, the main effect of these urban areas is the sewage discharged into the estuary. The main sewage disposal points are shown on Figure 3.10.

The Tees barrage scheme

A barrage has been constructed across the river Tees at Stockton, 16km from the sea. The main purposes of the barrage are:

- to raise water-levels in the upper part of the tidal reach so as to improve development opportunities
- to enhance the amenity in the surrounding area by covering what some believe to be 'unsightly' mudflats that appear at low tide
- to install a fish-pass to enable the passage of migratory fish
- to create a navigation lock and a canoe slalom for recreation.

The barrage has been sponsored by the Teesside Development Corporation (TDC) and has provoked comment from various other organizations such as English Nature (formerly the Nature Conservancy Council for England), the Royal Society for the Protection of Birds, and the Tees and Hartlepool Port Authority Ltd. The TDC is a government-appointed **quango** – that is, an appointed, unelected group of people, responsible to central government for its funding. Its purpose is to promote economic development on Teesside.

▲ **Figure 3.16** The Tees barrage.

1 a) Should people living around the Tees estuary be concerned about its condition? Who might be concerned? Why? Who might not? Why not?
  b) Should people elsewhere in the UK be concerned? Why?

2 What problems are being caused by the various demands on the Tees estuary? By whom are these problems being created?

3 For each of the problems you have listed in (2), suggest ways to reduce its impact. Who, if anyone, should be responsible for managing the estuary? What powers should they have, and over whom?

4 a) Work in groups of three or four. Devise a five-year plan for the Tees estuary. Outline what you want to achieve, and identify people whom you recommend should be responsible for achieving it.
  b) Would your plans raise objections from other bodies? If so, from whom? How would you deal with these?

5 There is much current interest regarding a possible future rise in sea-level. Predict how this might affect estuaries like the Tees. You can find out more about this in Chapters 4 and 6.

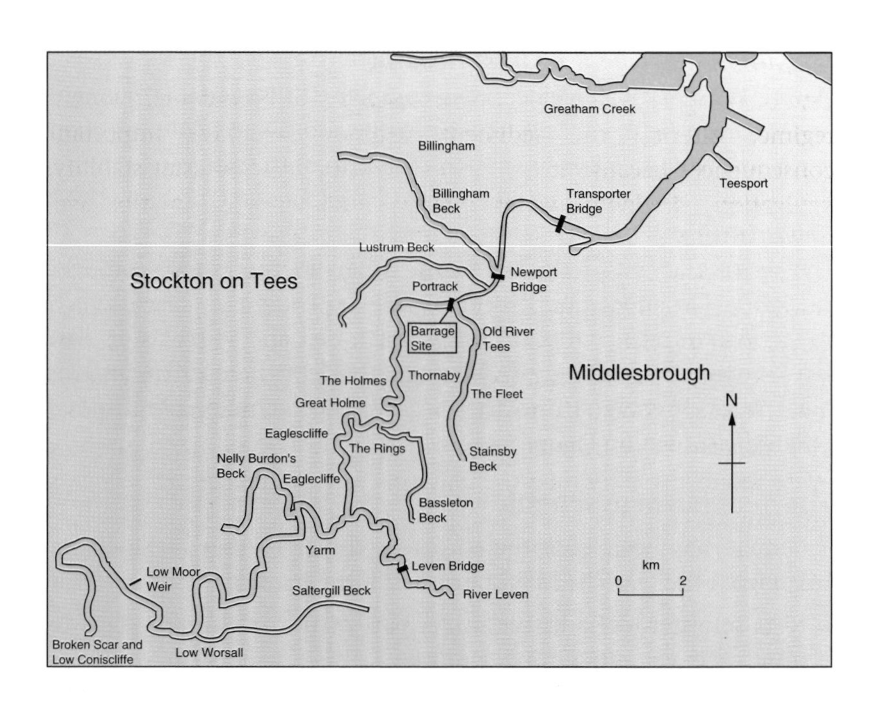

▶ **Figure 3.17** Location of the Tees barrage.

1 Which group(s) of people might disagree that the mudflats are unsightly? Why?

2 Study Figure 3.17. Suggest reasons for the location of the barrage at this point.

The impact of the barrage

The barrage will affect the movement of water, the sediment regime, flooding, salinity, water quality, land drainage, the ecology of the estuary, and the visual landscape. The effects outlined below are mainly the views of the planning consultancy that the Teesside Development Corporation commissioned to investigate the likely impacts of the scheme.

Movement of water

Movement of water will be affected both above and below the barrage. The following effects will be felt.

- Above the barrage, tidal range will be reduced, as will tidal current velocities. The rise in mean water-level will cause many existing tidal mudflats to be covered permanently.
- The tidal regime will also be affected. Computer modelling shows that under average flow conditions, tidal volumes in the Tees estuary may be reduced by 3 million cubic metres of water. The barrage will remove most of the shallow part of the existing estuary, where tidal energy is dissipated or spread over a wider area, and where most of the vertical mixing occurs.
- The barrage will prevent the upstream movement of tidal water. The landward extent of the estuary will therefore be reduced and the tidal regime downstream of the barrage altered. This will increase circulation downstream of the barrage.
- Velocities upstream of the barrage will fall, especially during normal river flows, whereas velocities immediately downstream of the barrage may increase, particularly in surface waters.

Together, these effects will create a major environmental change in the estuary caused by alterations to the surface-water regime of the Tees.

Sediment regime

One of the greatest impacts of the barrage will be on the sediment regime. Altering the sediment regime can have important consequences because it influences land drainage, coastal stability, navigation, dredging, water quality, and wildlife. In the area upstream of the barrage, the quantity of fine sediment carried by spring tides will be reduced, because the barrage will limit the amount of sediment brought in by the tide. On the other hand, the barrage itself will act as a lake for sediment brought downstream by the Tees river, and there will therefore be a slight increase in sedimentation upstream of the barrage. In contrast, there are indications that sedimentation downstream of the barrier may show some reduction. This may benefit the Port Authority by reducing dredging costs.

Flooding

There is an increased likelihood of flooding of land upstream of the barrage site as a result of higher water-levels. Flood defences are already inadequate in some parts of the Tees estuary, especially close to Yarm, and these will have to be raised slightly. Raising bank defences should prevent any potential problem.

Salinity

Operation of the barrage will shorten the estuarine section of the river, so the salinity of the stretch of river from Preston Park to Blue House Point will fall until it can be characterized as fresh water. This will cause a change in plant species along this stretch of the river.

Water quality

This is explored in Figure 3.18, which sets out the view of the Teesside business community. Like any viewpoint, it should not be read and accepted without question. The business community aims to promote development on Teesside by encouraging companies to locate there. This is not to say that the improvements they predict will not happen, but they are likely to seek ways of promoting Teesside, and water quality may be seen as a selling point. You may find that other extracts in this chapter disagree with that view. English Nature believe that the barrage will concentrate pollutants below the barrage, which will have an adverse effect on the mudflats. This, they say, will have a detrimental knock-on effect on the bird populations that feed in this area.

Land drainage

Barrage operation will increase river levels upstream of the site. This will cause a localized rise in the water table in the river corridor, which is a change to the existing groundwater regime. Land drainage will be moderately affected, though it will vary and will be worse in some localities than in others. In the area of Stockton race course, a cut-off ditch will intercept and tap pressure increases in groundwater.

'The barrage will possibly reduce the extent to which the lower reaches of the Tees dilute and disperse effluent. Upstream of the barrage the quality of water is predicted to improve to some extent, resulting from the prevention of tidal inflow. Downstream of the barrage, the new hydrological regime will result in a deterioration of the quality of surface waters at Teesmouth, made worse under high river flow and ebb tide conditions. Although this may be substantial at times, relatively clean sea water occupies the bed of the estuary, maintaining a reasonable habitat, and overall the impact is considered to be minimal.'

▲ **Figure 3.18** Opinion of the Teesside business community.

The ecology of the estuary

Changes to the hydrological regime of the estuary, as well as changes to salinity, sedimentation, water quality, and groundwater levels, will all have an impact on the ecology of different parts of the estuary in different ways. Locally, these impacts may be substantial, such as on plant communities along the inter-tidal area, resident fish populations upstream of Blue House Point, and movement of some species of migratory fish. However, specific impacts can be averted, either by moving animals or plants elsewhere, or by adjusting the ways in which the barrage is operated daily, such as the operation of the fish pass. Some effects can be predicted with some certainty and action taken to avert them; for example, the raised river level upstream of Blue House Point may cause industrial pollutants from contaminated land to leach out into the raised water table.

The visual impact

The visual impact is significant. Again, viewpoints differ and depend a great deal on individual attitudes and feelings.

Environmental impact assessment

It is now a requirement by law that an environmental impact assessment (EIA) is carried out before any major development project is approved. An EIA is a means of estimating possible change to the environment as a result of a project, and a weighing-up of whether apparent problems outweigh apparent benefits. It usually involves the following.

1 An assessment is made of the existing environment.
2 The proposed development is described.
3 The probable environmental impacts of the development are assessed in relation to:
   a) the impact on the natural environment: its ecology and habitats, and the impact that any change or pollution might have
   b) the impact upon the human environment, such as recreation, health, aesthetics, well-being of people, and local employment (increased or reduced opportunities).
4 These environmental impacts are scored on a grid like the one in Figure 3.19.
5 Any modifications which could be made to minimize adverse impacts are considered, and their impact re-assessed.
6 A decision is made.

−3	−2	−1	0	+1	+2	+3
strong negative impact	negative impact	slight negative impact	no impact	slight positive impact	positive impact	strong positive impact

SCORE / Environmental Factor	−3	−2	−1	0	+1	+2	+3
Tidal regime							
Sedimentation							
Flooding							
Drainage							
Water quality							
Ecology							
Visual impact							
Development							
Recreation							
Wildlife habitats							
Employment							
Pollution							

▲ **Figure 3.19** An environmental impact assessment

This method is useful because it focuses on environmental consequences of projects, and takes a wider view than cost–benefit analysis, described in Chapter 1. However, more information is needed for an EIA. Although both are useful for measuring impact, the final decision may be based on economic or political considerations rather than environmental ones. In a number of cases recommendations have been ignored or overturned.

The Tees barrage: carrying out an environmental impact assessment

Work in groups of two or three. Each group should complete the following exercise in *one* of the following roles:

- a planning executive for a local Teesside company
- a Port Conservancy Engineer for Tees and Hartlepool Port Authority Ltd
- a member of English Nature.

In your role, you have been asked to contribute to an EIA of the barrage scheme on the river Tees. Your final report should consist of the following.

1 An outline of the present situation.
2 A brief description of the proposed scheme.
3 An environmental impact assessment (EIA), carried out as shown on page 68 and completed using the scoring system in Figure 3.19.
4 An outline of each of the impacts you have scored, justifying your decision.

The method of assessing environmental impact shown here assumes that each factor has an equal weighting in terms of importance. In reality, certain impacts tend to be more important than others. For instance, to a member of English Nature, wildlife habitats are more important than recreation. To a member of the Teesside Development Corporation, recreation may be an important factor in promoting Teesside as a place to establish a company. Still in role, rank the factors you have scored, in order of importance.Once you have done this, award the factor you have ranked most important 12 points, the second factor 11 points, and so on. Multiply each factor by the score you have given, for example:

Ecology
–3 (score) x 12 (most important) = –30

Employment
+2 x 11 (second most important) = +18

1 Compare the rank orders arrived at by members of the three different roles. How are they different? Why should this be so?

2 Compare your final scores for each factor. How are they different, and why?

3 Can the three groups reach an agreed EIA between them? If so, a final report should be written after agreeing on all aspects of the EIA. If not, a majority may write a report, and leave the minority to decide how they wish to have their opinions known.

River deltas

The point where a river meets the sea can be an estuary but it can also be a delta – see Figures 3.21–3.23.

A delta is a low, almost flat, area of land at the mouth of a river where it flows into a quiet body of water, such as an ocean, sea, or lake. At the point of entry, the velocity of the river is reduced and its load deposited. Estuaries and deltas are different: deltas carry greater supplies of sediment and therefore deposit far greater volumes than estuarine rivers, where sediment volumes are less and tidal currents and river discharges keep channels clear. Figure 3.20 shows the Nile delta where the river Nile enters the Mediterranean Sea. Notice the shape of the delta – it appears to 'fan out' from a confined river channel into a broadening area which extends into the sea. On either side of the delta, lighter

▲ **Figure 3.20** Satellite view of the Nile delta.

colours show the Arabia and Sahara deserts. There is an abrupt colour change to red, where irrigation from the Nile river has allowed land to be cultivated.

The formation of deltas

The formation of a delta is relatively straightforward. At the point of entry to a sea or lake, river velocity slows and halts, and as a result deposits much of its load, called **alluvium**. This alluvium partially blocks the river channel, and forces the river to seek another route. This route in turn becomes clogged and the process is repeated as new channels form. These form a maze of semi-parallel channels known as **distributaries**, through which the river flows towards the sea. As a result of the distributaries, a delta grows laterally, or sideways, 'fanning out' as it reaches further into the sea or lake (see Figures 3.20 and 3.21).

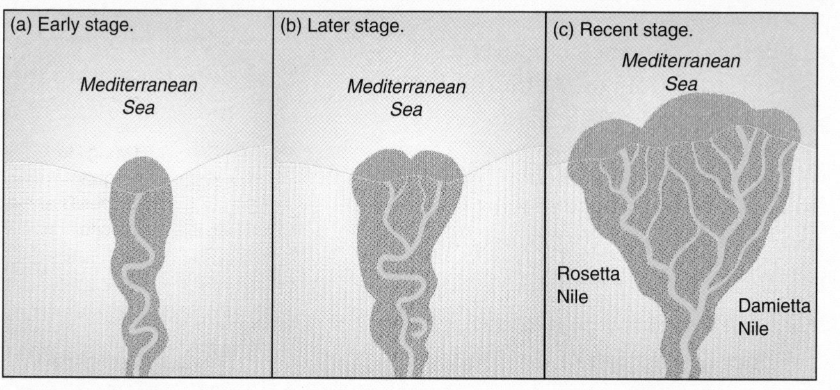

▲ **Figure 3.21** Stages in the formation of the Nile delta.

Figure 3.22 is a cross-section through a delta, and shows how a delta develops. It grows seawards as further layers of sediment are added. At the point where sediment is deposited on the edge of a delta, there is a steep drop to the ocean or lake floor and rapid subsurface currents form. These are known as **turbidity currents**. The surface of the delta is built up by continued deposition so that it is partially exposed above sea-level, especially at low tide. As the alluvium is rich in nutrients, vegetation growth is prolific. Roots bind the alluvium together and this provides a more secure base for further expansion of the delta. Later in this chapter we will see how, in tropical areas, particular plant communities develop along river deltas and estuaries.

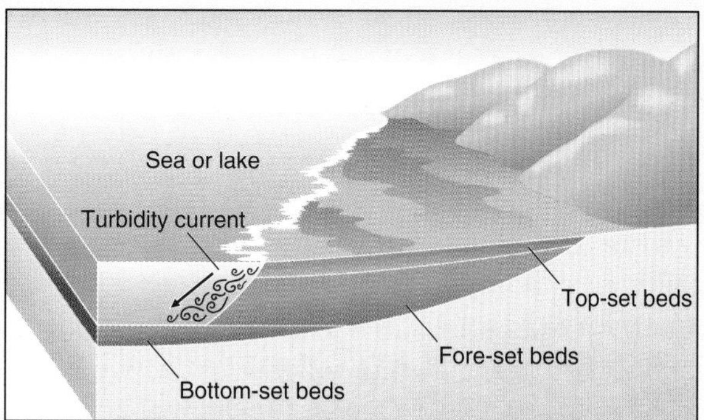

▲ **Figure 3.22** The formation of a delta in cross-section.

The size and shape of a delta depends on the load of sediment, the rate of flow of the river, and the wave power and tidal range of the ocean. In some deltas, shape is modified as a result of the balance between river deposition and the removal of sediments by ocean waves and currents. The Mississippi delta, for example, exhibits the 'bird's-foot' shape which results from large quantities of sediment carried into quiet water, and the volume of discharge from the Mississippi river creating permanent distributaries. Tides and currents cannot remove all the sediment carried, and so the delta grows out to sea.

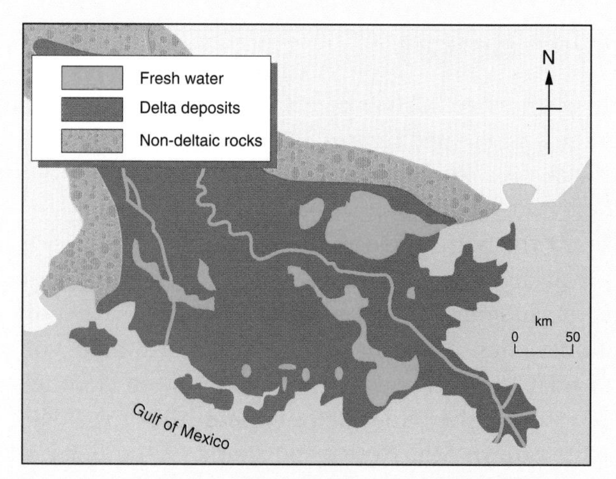

▲ **Figure 3.23** The Mississippi 'bird's-foot' delta.

The Niger delta is different from that of the Mississippi because it has no seaward fingers. While currents and waves along the coast by the Niger delta cannot prevent the formation of a delta, they do sweep sediment along the coast rather than allow the formation of a bird's-foot shape. Also, further offshore, sea currents prevent any possible extension of sediment out to sea.

Some rivers, such as the Amazon, have no deltas at all as the removal of sediment by ocean waves and currents is so vigorous.

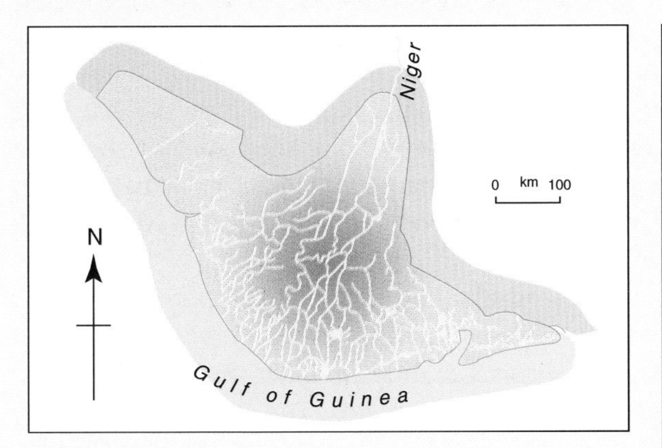

▲ **Figure 3.24** The Niger delta. Note how the smooth shape of this delta contrasts with the 'bird's-foot' of the Mississippi.

Rank	River	Country	Area ('000 km²)
1	Indus	Pakistan	163.0
2	Nile	Egypt	160.0
3	Hwang Ho	China	127.0
4	Yangtze	China	124.0
5	Ganges/Brahmaputra	Bangladesh	91.0
6	Orinoco	Venezuela	57.0
7	Yukon	USA	54.0
8	Mekong	Vietnam	52.0
9	Irrawaddy	Myanmar	31.0
10	Lena	CIS	28.5
11	Mississippi	USA	28.0
12	Chao Phraya	Thailand	24.6
13	Rhine	Netherlands	22.0
14	Colorado	Mexico	19.8
15	Niger	Nigeria	19.4

▲ **Figure 3.25** The world's largest deltas in terms of area.

▼ **Figure 3.26** The world's largest deltas.

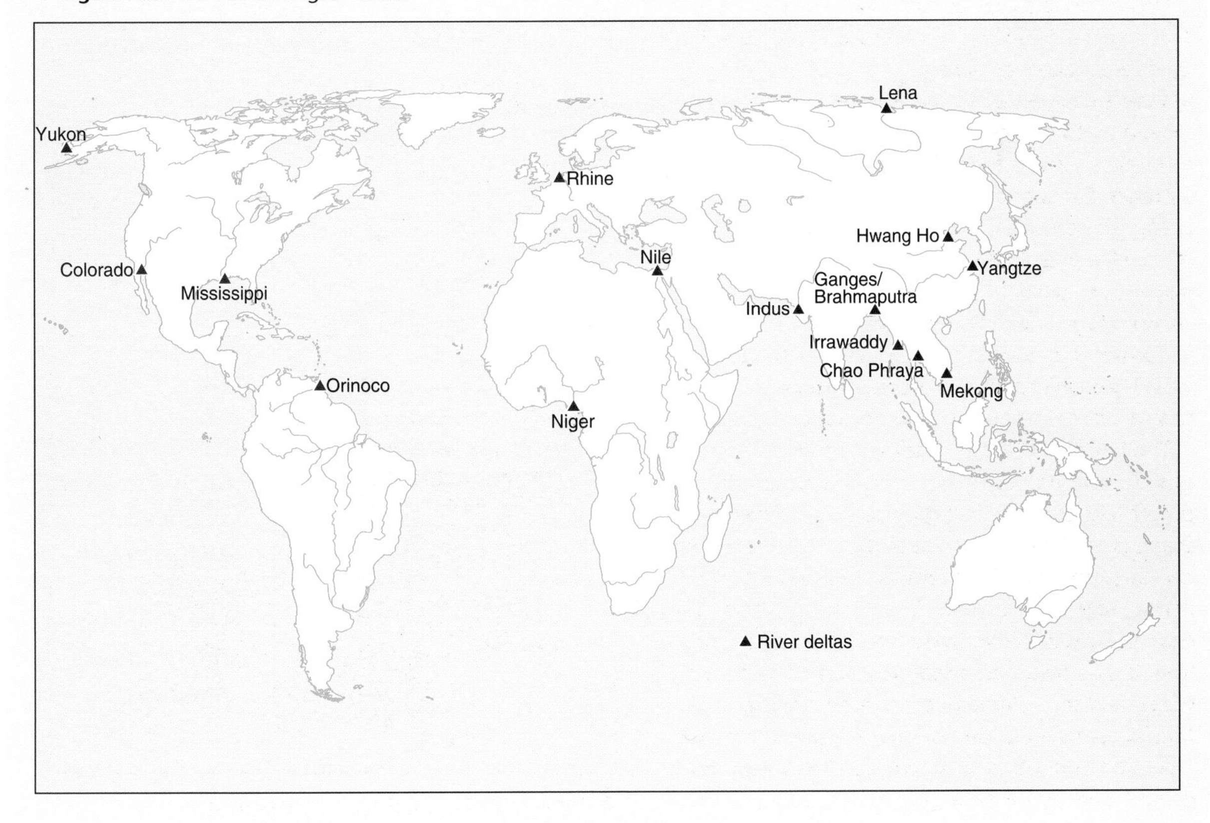

At least seven major deltas have formed as the Mississippi has changed channels over the past 9000 years.

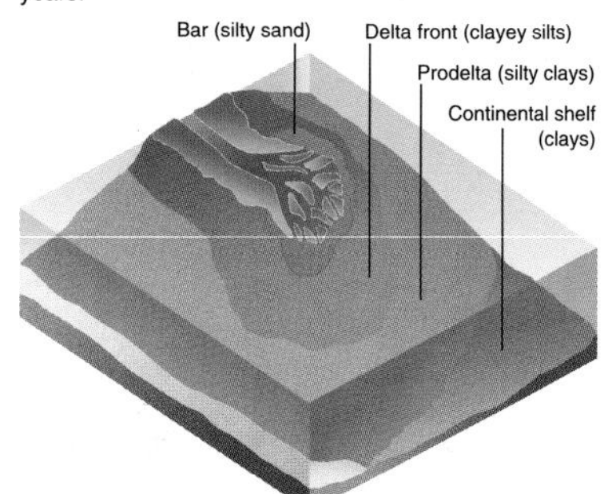

15th century
After a course change, the river reaches the sea and slows, dropping first heavier then lighter sediments. Natural levees build as floods drop sediments on the banks.

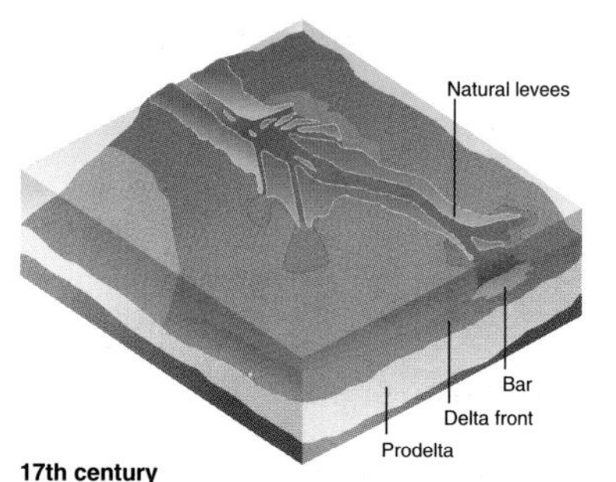

17th century
Delta grows rapidly seawards. Flooding expands land area. Freshwater marsh plants are nourished by the river.

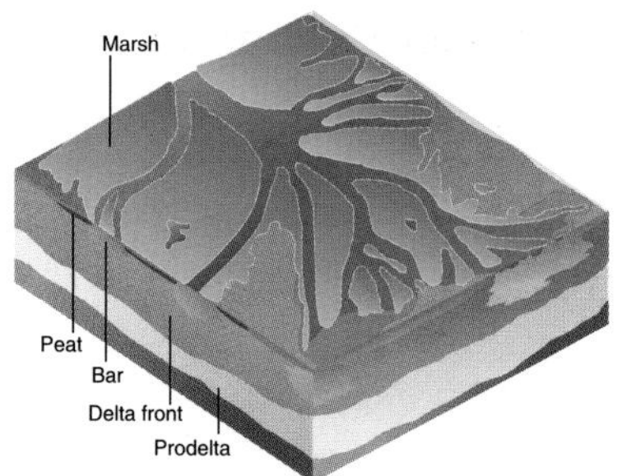

1956
Although now less than the peak size reached in the 1800s, wide marshes still extend beyond the levees. The levees have grown naturally, while the river's channels have been deepened and the marshes have been cut off, leaving them susceptible to erosion.

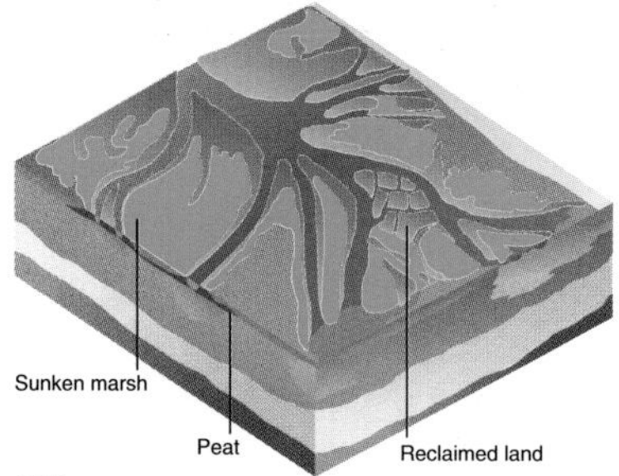

1978
Ocean intrudes as marshes are eroded, because they are no longer supplied with so much river sediment. Although the river still runs through channels, canalisation of the river has altered velocity levels. These have increased and eroded parts of the delta.

Case study: the Mississippi delta

The Mississippi delta is one of the biggest in the world. The river carries so much silt that it was known as 'The Big Muddy' to European settlers. For thousands of years the river overflowed its banks each spring, sprawling over lowlands and depositing sediments, which gradually extended the delta southwards into the shallow waters of the Gulf of Mexico, creating a few square kilometres of new land each year. At least seven major deltas have formed as the Mississippi has changed channels over the last 9000 years. The formation of today's active delta, the Balize, is shown in Figure 3.27. This shows that the pattern of delta growth has been reversed – Louisiana now loses about 100km^2 of its wetlands each year. Why has this happened?

The Mississippi carries mostly fine silts and clay which compact much more easily than larger sediment particles such as sand. When clay settles, its volume consists mainly of water, which is squeezed out as new sediment is deposited on top. The land surface of the delta sinks as a result of the weight of sediment on the Earth's crust. As long as the river continues to add sediment to the delta, this offsets the settling of earlier deposits.

However, the deposition of sediment in the delta has been severely affected by human activity. Embankments or levees were built throughout the delta to control the annual spring floods, and channels were dug to make navigation easier. These two measures deny marshes new layers of silt and fresh water, while the land continues to sink. A further threat to the wetlands is canalization.

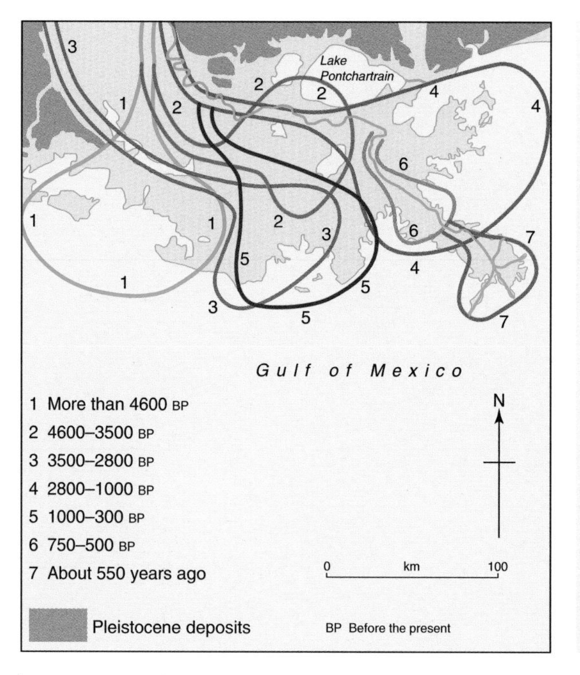

Using Figure 3.27, construct a flow diagram to show the sequence of events which have produced the present shape of the Mississippi delta. Highlight natural or physical factors in one colour, and the effects of human interference in another.

◄ ▲ **Figure 3.27** The dynamic delta. The changing shapes and direction of growth of the delta reflect changing volumes of sediment and sea currents, the two factors on which delta growth and shape depend (*after Morgan, 1970*).

The delta area is rich in oil and gas, and has more than 22 000 wells. Drillers have dug canals for access during the construction and maintenance of pipelines and rigs. It is estimated that there are 16 000km of canals throughout the delta.

Canals provide easy routes for salt water to penetrate inland with the rise and fall of the tide. Before canals were dug, fresh water trickled steadily across the flat surface of the delta, limiting the width of the saltmarshes. Now, with salt water intruding, it literally tears the marsh apart. No longer anchored by freshwater vegetation, banks crumble away before saltwater plants can take root. Better soil conservation upstream has also reduced the amount of sediment carried by the river.

The problem is that the flood control and navigation projects that have starved the delta of sediment are vital to the local economy, as is the drilling and exploration for petroleum in the delta.

Use Figure 3.28 to answer the following questions.

1 What evidence is there that the delta area is rich in natural resources?

2 Make a simple tracing of Figure 3.28 and show on it the extent to which sea water has encroached into the delta forming saltmarsh. Why is this detrimental?

▼ **Figure 3.28** The Mississippi delta today.

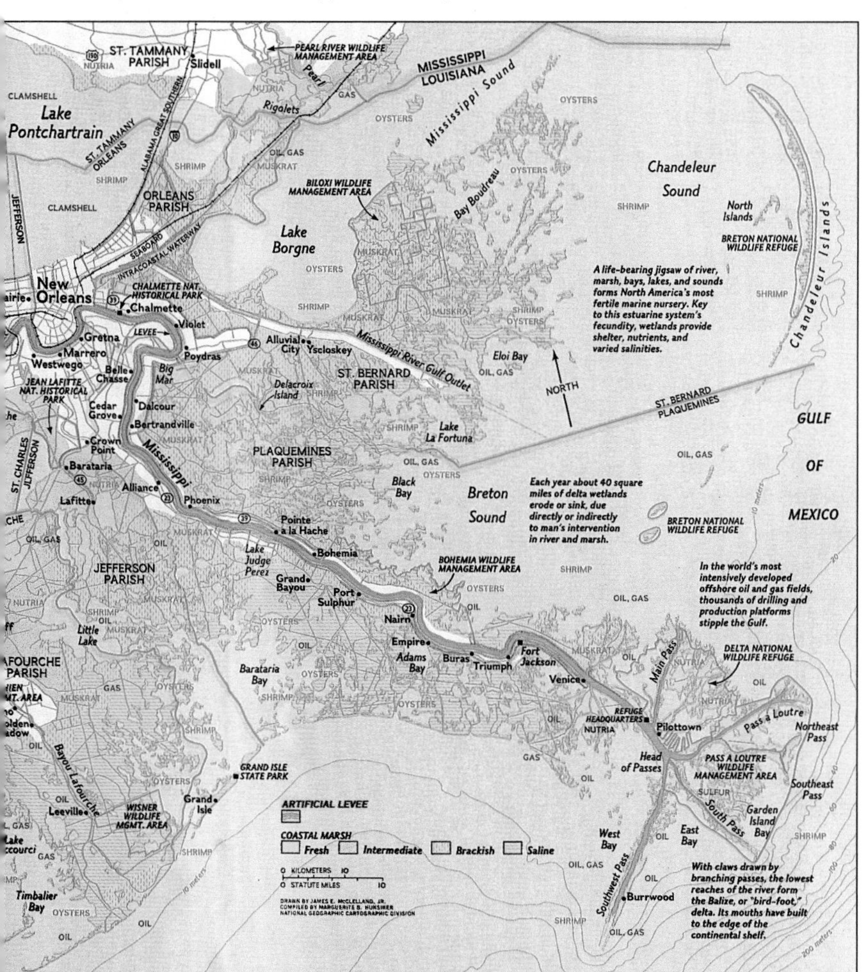

Identifying concerns about deltas – a point for debate

1 Form two groups – A and B. Prepare a 30-minute debate in which Group A will argue that 'Drilling, exploration, and navigation interests are the most important factors to consider in managing the future of deltas'. Group B will oppose this, arguing that 'In order to save the delta, these activities should not be allowed to continue'. Each group must justify its position, and listen to the argument of the opposing group.

2 Does your personal opinion alter as you listen to others?

The Fitzroy estuary and the Barrier Reef coast

Processes within estuaries have implications that extend beyond the estuary itself. For example, we have seen how the condition of the water being discharged from the estuaries around the North Sea directly affects its condition. This is also true of the eastern coast of Australia, where the Great Barrier Reef (Figure 3.29) stretches for 1500km along the coast of Queensland. With a population of 18 million, Australia's apparent surplus space is deceptive. In fact, many issues affect and in some cases threaten the Great Barrier Reef. One of these issues affects the estuary of the Fitzroy river, and illustrates how coastal and river management are inter-connected.

The Great Barrier Reef, like all coral reefs, consists of billions of coral polyps growing from a calcium carbonate base. The reef is a vast community of corals, of which there are thousands of types. Chief requirements are consistent warm temperatures and clear, relatively shallow water – corals cannot function without sunlight.

There is concern that agricultural fertilizers and other nutrients, such as those from urban sewage, which are discharged along the Great Barrier Reef coast from the mainland, are affecting coral reef communities. If river-borne nutrient and sediment discharges and dredging activities increase, coral reefs may be exposed to unacceptable levels of stress.

Figure 3.30 shows the location of the Fitzroy river basin. It covers an area of 140 000km^2 on the north-east coast of Australia, and the river flows into the sea 40km south-east of Rockhampton. Sediment carried to the estuary from the Fitzroy basin is transported out to sea and threatens the fringeing reefs. Sediment has a considerable influence on coral reef communities, because if the water is cloudy, sunlight cannot penetrate and this reduces the productivity of the zooxanthellae. These organisms have a **symbiotic** relationship (an association between two different organisms through which both benefit) with coral, and contribute sugar and other nutrients to the coral cells by photosynthesis. In return the coral provides the algae with a protected environment, and carbon dioxide. Too much sediment can also be a problem because in storm conditions it can erode the reef structure.

▲ **Figure 3.29** The Great Barrier Reef.

1 Using an atlas, locate the Great Barrier Reef and measure its length.

2 Is it one continuous reef? Describe the reef.

3 What impact would the degradation of the Great Barrier Reef have on:
   a) waves breaking on the shore
   b) the variety of sea life on the east coast
   c) Australia's world reputation as a tourist attraction.

Inputs of fresh water close to the coast can mean that salinity levels are lower than in more open waters. Coral growth is better in areas of high salinity. A sudden input of nutrients from the land can favour the growth of other species which affect the equilibrium of the reef communities. For example, an increase in nutrients is often followed by a population explosion of the crown-of-thorns starfish, which feed on coral. River-borne phosphorus levels are often particularly high after a major storm, and this has an adverse effect on the coral.

The effects of flooding on the coral reefs of the Keppel Islands

In late December 1990 and early January 1991, there was heavy rainfall in the Rockhampton region of Queensland, Australia. A tropical cyclone (Joy) crossed the Queensland coast near Ayr – a typical event during the Queensland summer. This led to considerable flooding of the Fitzroy catchment. Over 18.5 million litres of floodwaters flowed down the Fitzroy river and into Keppel Bay between 28 December 1990 and 21 January 1991.

Keppel Bay contains fifteen islands, a number of which have extensive fringeing reefs. The effects of the freshwater flood from the Fitzroy river were evident on most of the reefs in Keppel Bay. Of those reefs affected, the extensive inshore fringeing reef communities on the southern and western sides of the islands were the most devastated.

The cumulative effects of increasing silt deposition, combined with massive additions of fresh water, reduced the viability and potential for the re-establishment of the coral communities in parts of Keppel Bay.

Sea water and fresh water have different densities, the salinity of sea water making it heavier. A boundary forms between the two, and the fresh water overlies the sea water, remaining essentially a separate body of water. In Keppel Bay there is little mixing with the salt water underneath. The fresh water contains suspended clay particles, and the upper layers of water form a cloud around the coast. This is known as a 'freshwater plume' – see Figure 3.31.

▼ **Figure 3.30** The Fitzroy river and Keppel Bay.

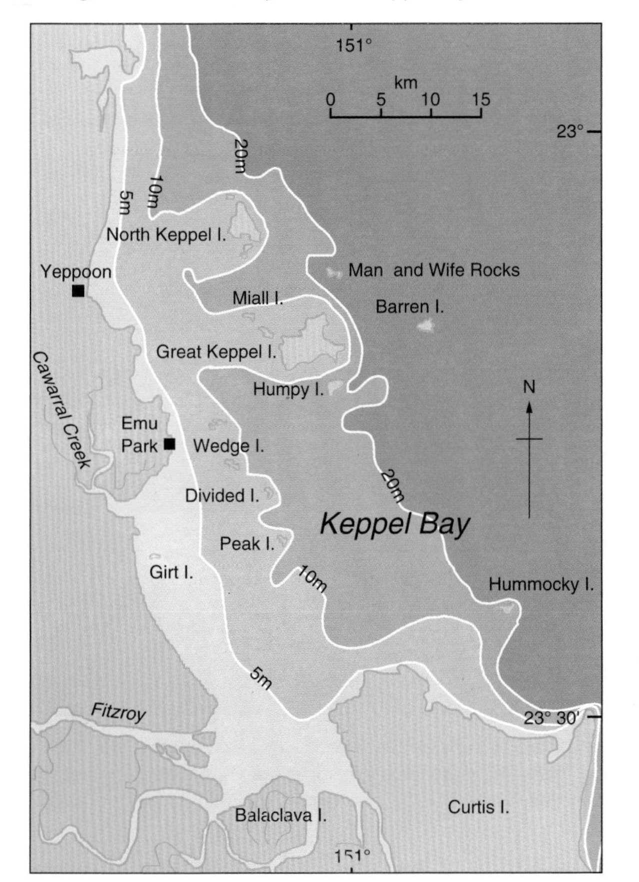

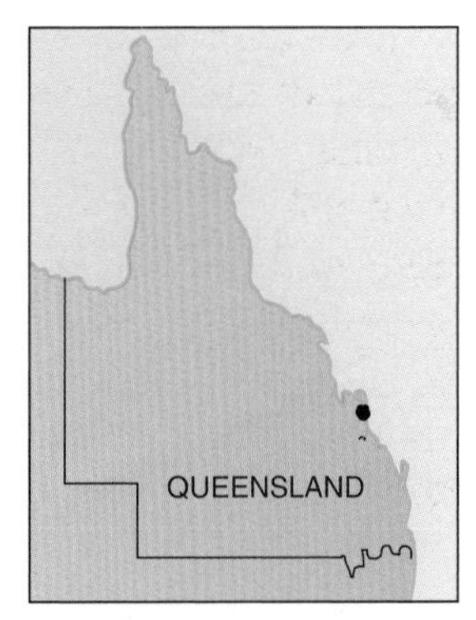

Why would human actions such as land clearance affect the impact of storms such as those described here?

Managing landform systems: How do the Keppel Islands and the Fitzroy river illustrate the need to manage landform systems?

In groups, consider the following aspects of the Keppel Island freshwater plume.

- Sedimentation by the Fitzroy river (some of which is the result of land clearance in the river basin).
- The threat posed to the Barrier Reef.
- The possibility of establishing mangroves along the margins of the estuary, and protecting them.

1 Should the system be managed as a whole, or in sections? Should it be managed at all?

2 If someone should manage it, who should that be?

3 How should this management be funded?

4 What do you see as the potential problems:
   a) in doing something
   b) in doing nothing?

Write your findings as an essay of 1000 words.

The role of mangroves

A partial solution to the problem could be to create a mangrove barrier along the coast. Mangroves are trees and shrubs which produce stilt roots and are able to survive in swampy, unstable clays and silts which form around deltas and estuaries (Figure 3.32). Once established, the mangroves enable sediments to be deposited at an accelerated rate. Mangrove trees and soils around them have an ability to absorb fairly substantial inputs of inorganic nutrients and can act as interceptors of fresh water and nutrients before they are flushed out onto the Barrier Reef. However, the ability of mangroves to absorb nutrient inputs is heavily dependent on the placement, timing, quantity, and nature of the effluent, and in some cases eutrophication may result. For discussion of eutrophication, its causes and effects, turn to pages 241–42 in Chapter 10.

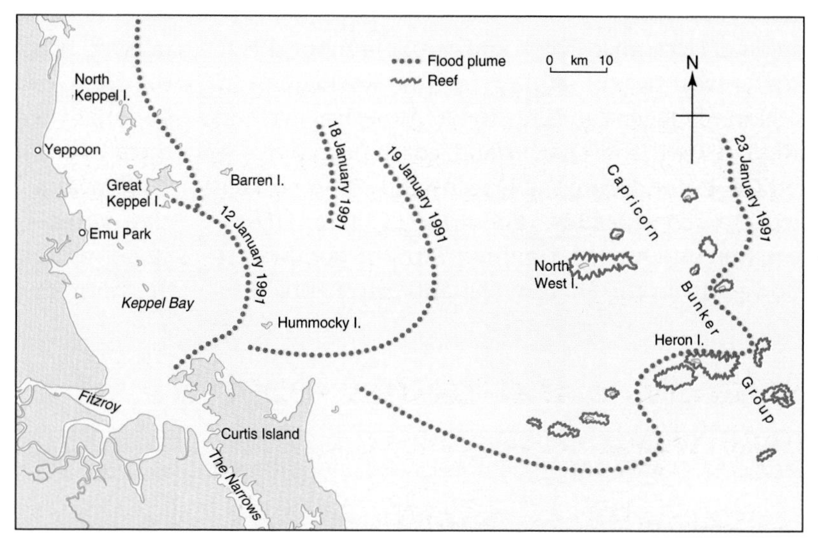

▲ **Figure 3.31** Movement of the eastern edge of the flood plume, January 1991.

▲ **Figure 3.32** Typical mangrove scenery on the Queensland coast. Notice the 'stilt' roots at the base of the trees. These are exposed at low tide; high tide brings in silt-laden water. The roots reduce water velocity and enable sediment to be deposited.

Ideas for further study

1 The National Rivers Authority was established on 1 September 1989. It was created as a public body responsible for safeguarding and improving the water environment in England and Wales. Investigate the role played in pollution control by the NRA. Find out what action the Authority takes against polluters. Do you think this is effective?
2 Investigate the physical, human and environmental threats to other estuaries and deltas. Possible areas for study include the Niger delta, and Chesapeake Bay on the eastern seaboard of the USA.
3 Investigate the impact of the disappearance of wetlands in Louisiana on different cultures, such as Cajun.

Summary

- Estuaries and deltas are formed through the interaction of river and coastal systems over time. They differ in the relationship between sedimentation rates and water currents and velocities.
- Many estuaries and deltas have been modified by the various demands put upon them by human activities. These environments need to be carefully managed to avoid any conflicts that may arise.
- Estuaries and deltas need to be seen as part of a whole system which combines river and coastal management.

References and further reading

R. Arnett, 'Estuarine pollution: a case study of the Humber', *Geography*, January 1991.

A. Bingham, 'Muddy waters', *Geographical Magazine*, August 1992.

English Nature, *Caring for England's Estuaries: An Agenda for Action*, 1992.

J. Herscht, 'The incredible shrinking Mississippi Delta', *New Scientist*, 14 April 1990.

D. Lee, 'Mississippi Delta', *National Geographic*, August 1983.

D. S. McClusky, *The Estuarine Ecosystem* (especially Chapters 5 and 6), Blackie, 1989.

National Rivers Authority – Northumbria and Yorkshire Region, *River Tees Catchment Management Plan Consultation Report*, 1994.

A. H. Strahler and A. N. Strahler, *Modern Physical Geography*, John Wiley and Sons Inc., 1992.

Managing landform systems: Summary

Key ideas	Explanation	Examples
1 Landform formation	Landforms result from complex interactions between the materials and structures of the Earth's crust and the natural geomorphological processes acting on them over time.	• The hydrological cycle in Keswick and other river basins • Causes of flooding • Processes of erosion, deposition, and cliff movement at Collywell Bay; dune formation at Druridge Bay; longshore drift processes along coastlines • Estuarine processes caused by tidal movement of water • The formation of deltas
2 Natural processes change landforms	Natural processes operating on landforms may cause changes to occur. These may be apparent within human time-scales in terms of changing forms and features.	• Changes in the Keswick area to manage flooding • Effects of water management on the Colorado basin • Changes along the Northumberland coast • Processes acting on the Tees estuary and the Mississippi delta • Effects of flood plumes on the Barrier Reef
3 Landform systems have an impact on people's daily lives and activities	Changes in landforms may give rise to the need to adapt the human use and activity of the area.	• Effects of flooding on Keswick • Changes to the Northumberland coast, e.g. the construction of sea walls • Changing form of the Tees estuary and the Mississippi delta
4 Human activities modify landform systems both in the short term and the long term, often with adverse effects	Landform systems may be consciously or unconsciously influenced and modified by often conflicting human activities. Such modifications may result in adverse consequences for the future use of natural environments.	• Changes brought by flood management to Keswick • Changing water management and dam construction in the Colorado and the Narmada basins • Impact of changes to the Northumberland coast, e.g. sea walls, sand dredging • Industrial change in the Tees estuary
5 Management of landform systems poses a continuing challenge for people	Successful management requires an understanding of landform systems and processes.	• Links between river basin management and hydrological processes • Links between coastal processes and coastal management
	People attempt to adjust their use of land areas to optimize their relationship with landforms and to minimize the problems and risks associated with such relationships.	• Different management options, such as flood alleviation in Keswick and coastal protection in Northumberland
	Management can be seen as a three-way process of policy, planning, and practice.	• Contrast between the attempts to manage coasts in Northumberland and Thailand • River basins in the Lake District, the western USA and India; estuaries in the UK, the USA, and Austrailia
	Catchment and channel management plans and coastal management plans can be evaluated using such techniques as cost–benefit analysis and environmental impact studies.	• Cost–benefit analysis in Keswick • Environmental Impact Assessment (EIA) in the Tees estuary and the Tees Barrage Project

2

People, weather, and climate

How are people influenced by weather events?

Weather events make startling news. On Saturday 27 August 1994, tornadoes hit the state of Wisconsin in the USA. This event featured in the national news on both television and radio, and was a major news item in national newspapers (Figure 1). Why? What is it about events in the weather that makes exciting news?

Disruptive weather

Careful study of Figures 1 and 2 shows that the effects on people are clearly of interest. People going about their daily lives are suddenly disrupted by weather events. Injuries or even deaths may occur. Other considerations are also important. Notice that the Wisconsin tornadoes (Figure 2) had a financial cost of US$4.5 million, to cover repairs, insurance, and loss of income from damaged crops. Such events can cost time and money, as was the case in the flooding of the London Underground during heavy rain on 11 August 1994 (Figure 3).

Four are killed in Wisconsin as tornadoes cut across State

BIG FLATS, Wis., Aug 28 (AP) – Four people were killed as tornadoes tore across Wisconsin on Saturday night, ripping up small towns and farms. One tornado cut a 13-mile swathe through central Wisconsin and left this small town's main street a tangle of metal, lumber and trees.

'All I could think about was all this stuff was going to come down on my head and it was going to hurt,' said Shirley Warner, 57, who was staying at a friend's mobile home in Big Flats with her 6-year-old grandson, Nicholas Forslund. 'I tried to pull the mattress over my head, but I couldn't get it off the bed. And then all of a sudden, wooooof. It was over.'

The walls were torn away and most of the furniture blown out.

Damage from the tornado in Adams Country, where Big Flats is located, was estimated at $4.5 million, Sheriff Robert Farber said.

▲ **Figure 1** From *The New York Times*, 29 August 1994.

Dangerous weather

Is there a more serious reason for our fascination with the weather? Storms in Hong Kong in 1994 were responsible for three deaths, the evacuation of 2500 people from their homes, and damage to property.

A matter of increasing concern to people who enjoy sunbathing is the very real threat of skin cancer. Is one of our concerns the danger of the weather? Do we feel that we should be able to control it a little more?

▶ **Figure 2** Damage caused by a tornado in Alabama, USA, 24 March 1994. The tornado destroyed the Methodist Church during a Palm Sunday service. Twenty people died, and more than 80 people were injured.

LONDON today suffered one of its worst storms for years as a deluge flooded hundreds of homes and offices, disrupted journeys for 50 000 Tube passengers and stretched Fire Brigade resources to breaking point.

The Underground network was thrown into chaos as the huge storm which struck the capital last night returned at about 6am, closing 24 Tube stations, starting fires and sending both BBC television channels off the air.

Services on the Central line were halted in both directions between Marble Arch and White City, and the Richmond branch of the District line was shut down completely from Turnham Green. Circle, District, and Metropolitan services were halted between Baker Street and Aldgate.

The Liverpool Street area, where Central, Metropolitan, Circle, and District lines meet, was said to be 'very bad'.

▲ **Figure 3** From the *Evening Standard*, 11 August 1994.

The big dry that reached out from beyond the Bush

The unthinkable is about to happen. Australia, one of the world's great breadbaskets, could have to import grain as one of the worst droughts on record drags on and on. The federal government says there is no need to panic but the National Farmers' Federation says there is only enough grain to satisfy domestic demand until Christmas.

This is the drought which has reached out from beyond the Bush to affect almost everything and everyone. Forget sport, Australians' usual talking point: the distinct lack of rain is the number one topic of conversation from Budgee Budgee to Bondi.

Food prices are shooting up, there are water restrictions for those lucky enough to have water, and even the fish are deserting coastal areas as the dried-up rivers fail to flush out the nutrients they feed on. Eastern Australia is tinder-dry. It is still spring and the bushfires have already begun: fire brigades warn that there is even worse to come. On the world's driest continent huge dust storms are simply blowing millions of tonnes of topsoil away.

A fleet of old milk and oil wagons is shipping water out west by rail to places where there has not been a decent shower for four years. There's even talk of building emergency pipelines just to keep some towns alive.

Drought is a harsh fact of life in the sunburned country with nine major 'drys' since white settlement. But few droughts have been so widespread, making it impossible to move the stock to fodder and increasingly expensive to bring fodder in. Many farmers have already spent more feeding their animals than they are worth.

▲ **Figure 4** From *The Guardian*, 22 October 1994.

Impacts of weather

Figure 4 shows very clearly the economic and human impacts of prolonged weather events. The droughts that affected Australia during the early to mid-1990s have had a marked effect on people, and in particular on those who rely on the weather for their livelihood. Food shortages were unheard of in Australia from the end of the Second World War, yet were a very real threat by October 1994. The bush fires in New South Wales in early 1994 made the world aware of the dangers of fire in a state that was 'tinder dry' at the time, and where prolonged drought threatened another series of fires in the summer of 1994/95. It seems that Australia faces the economic and political effects of a continued drought.

Threats to the atmosphere

One of the challenges to people around the world is to maintain a sense of control over weather events before they control us. The news events on these pages make it very clear that we are far from that stage. It is all the more alarming that the very atmosphere on which we rely and which generates all the events shown here is under threat from the impact of human activities. Pollution in the former Soviet Union is at dangerously high levels. Will economic pressures create resistance to the economic costs of 'cleaning up'? If they do, will the world be able to tolerate the consequences of increasing pollution, added as it is to the smogs already being created by vehicle emissions?

Controlling the weather

Human societies clearly do not have control over weather and climate. This section is designed to help you understand some of the processes that operate to produce weather events, and to gain insight into longer-term issues about climate and climatic change. This will help you to understand some of the issues behind the news stories, and to realize that both individual and international actions may be necessary to control the extent to which societies affect the world's weather and climate.

1 Study Figures 1–4 and the events mentioned in the text. With a partner, summarize:
   a) the nature of the weather issues
   b) the effects that the weather is having on human activity
   c) why these issues are relevant for geographical study.

2 Work together as a group.
   a) On the basis of the extracts presented here, discuss whether you agree with the statement: 'People are at the mercy of the weather.'
   b) Brainstorm examples of how people can – or might – change the weather and climate. Does your list suggest any categories, for example global or local, short-term or long-term? Is it desirable to be able to change weather and climate artificially?

Why do geographers study weather and climate?

Weather and climate – what's the difference?

Weather events such as thunderstorms, tornadoes, and blizzards occur at different times and places, having a major impact on people's lives. To protect ourselves from these events, we need to forecast when they may happen. To do this we must understand the changes in the Earth's atmosphere that produce such weather, and the resulting climate patterns.

Short-term daily variations in temperature, air movement, and moisture in the Earth's atmosphere are known as weather. These may be experienced locally, like the tornado in Figure 1. They may also be experienced at the wider scale, such as the depressions that bring wind and rain to all parts of the British Isles.

When rainfall, temperature, wind speed, and sunshine data are collected for periods of 40 years or more, distinct trends emerge. The averaging of the data produces patterns known as **climate**. These patterns suggest that:

- there are seasons, characterized by combinations of temperature and rainfall
- particular weather hazards such as hurricanes follow regular seasonal patterns.

Climate data enable weather scientists, or climatologists, to draw graphs and maps of average monthly or yearly temperatures, average yearly rainfall, and seasonal winds.

These are used to divide up the world into zones of similar climatic conditions. This is known as climatic classification. Figure 5 is an example of climate classification of Australia. Climate data has been used to divide Australia into different climate zones.

Weather, climate, and human activity

Human activity is closely linked to these yearly patterns of climate, especially in countries or societies where many people make their living from farming. People in climates where weather extremes make life difficult, such as the desert margins where rainfall varies greatly from year to year, also need to adapt to the climate. Scientists have looked into the effects of climate on human health and well-being, and have produced classifications of discomfort.

Figure 6 shows levels of heat discomfort in Australia, which are significant between November and April. During this time temperatures and humidity are high for long periods around the northern coasts, and it is very hot in the interior. The cooling system of the human body depends on evaporation of moisture. This keeps the body temperature from rising to life-threatening levels as air temperatures rise. High humidity restricts the body from releasing moisture into the air, because the air already contains high levels of moisture. Heat discomfort is therefore closely correlated with both high temperatures and high humidity.

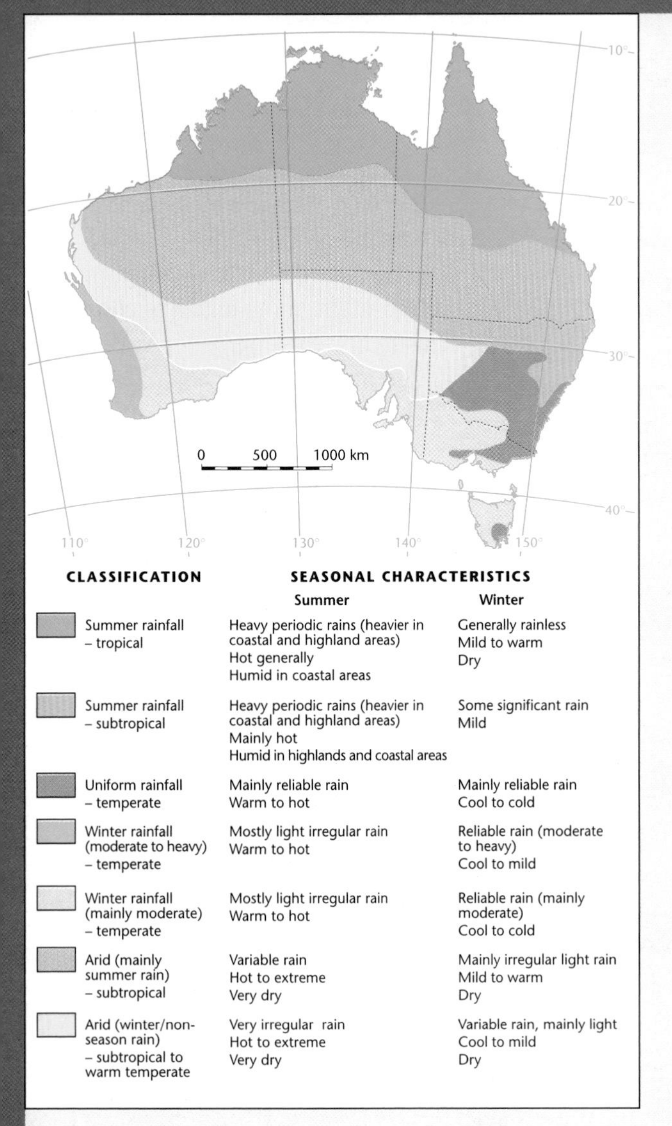

CLASSIFICATION	SEASONAL CHARACTERISTICS Summer	Winter
Summer rainfall – tropical	Heavy periodic rains (heavier in coastal and highland areas) Hot generally Humid in coastal areas	Generally rainless Mild to warm Dry
Summer rainfall – subtropical	Heavy periodic rains (heavier in coastal and highland areas) Mainly hot Humid in highlands and coastal areas	Some significant rain Mild
Uniform rainfall – temperate	Mainly reliable rain Warm to hot	Mainly reliable rain Cool to cold
Winter rainfall (moderate to heavy) – temperate	Mostly light irregular rain Warm to hot	Reliable rain (moderate to heavy) Cool to mild
Winter rainfall (mainly moderate) – temperate	Mostly light irregular rain Warm to hot	Reliable rain (mainly moderate) Cool to cold
Arid (mainly summer rain) – subtropical	Variable rain Hot to extreme Very dry	Mainly irregular light rain Mild to warm Dry
Arid (winter/non-season rain) – subtropical to warm temperate	Very irregular rain Hot to extreme Very dry	Variable rain, mainly light Cool to mild Dry

▲ **Figure 5** A climate classification of Australia.

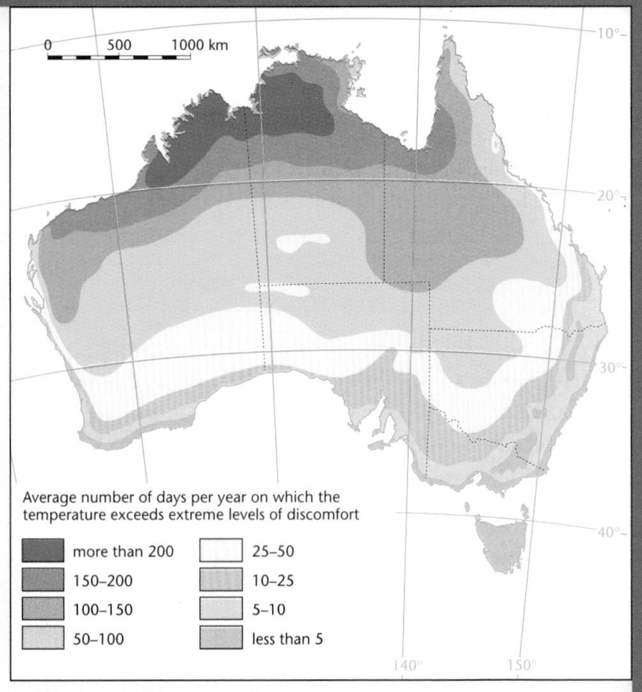

▲ **Figure 6** Heat discomfort in Australia.

Global atmospheric machine

The complexity of the world's climate is illustrated in Figure 9. What we experience and what we record about the atmosphere are based on the interaction of several aspects of the natural world. The global weather record is a relatively short one. Reliable observations began only in the late 20th century. It is, then, only recently that technology has enabled us to model the interactions of this 'climate machine' using sophisticated computers.

The human impact presents an enormous challenge to our understanding of the atmospheric machine, and this is investigated in Chapter 6.

1 a) Study Figures 5 and 6 and suggest some connections between levels of heat discomfort and different climate zones.

b) What other sorts of discomfort might people experience?

2 Look at Figure 5. It has been drawn using average climate data. What problems might arise in using averages like these?

3 a) Which groups and organizations would find these classification maps useful in planning and organizing their activities?

b) What are the limitations of maps such as these?

4 a) With a partner, use Figure 7 to draw a sketch diagram showing the main elements of the climate system.

b) Draw arrows to show how and why some of the different parts may be linked.

c) As a group, discuss your diagrams.

ATMOSPHERE

The atmosphere is a mix of gases which form a blanket around our planet. These vary in thickness according to location and altitude. It is here that changes in day-to-day and season-to-season weather occur. The average, extremes, and variability of weather and its seasonal changes together define the climate of a given region. The dynamic nature of such regional climates is a challenge to any prediction of climate change.

HUMAN AND ANIMAL INPUT

Human occupancy of the planet has changed the nature of every natural component of the climate system. Most significantly, our exploitation of resources, and our settlements, have permanently altered much of the biosphere and geosphere. One hundred and fifty years of increasing gaseous emissions have created an 'enhanced' **greenhouse effect**. The dynamic interactions of the natural components in global climate are so complex, however, that predicting the scale and speed of human-induced climate change is at a very early stage of development.

BIOSPHERE

(*the area of the Earth's crust and the atmosphere where life is found*)

This area provides a major link between the non-living environment of the geosphere below and the atmosphere above. Heat, moisture, and greenhouse gases – carbon dioxide, for example – are passed backwards and forwards between the three as living organisms grow, decay, and die. For example, plants take up carbon dioxide as they grow, but release some of it as they die and decay.

OCEANS

Atmosphere and oceans are very much part of the global weather machine. Winds pass over the surface of the ocean, and create currents. These ocean currents then dictate the patterns of moving water within ocean basins. This has a major effect on the circulation of the atmosphere. The atmosphere responds to sea surface temperatures quickly, often with immediate effect, but ocean circulations in deeper water may take up to a century to create any climate change.

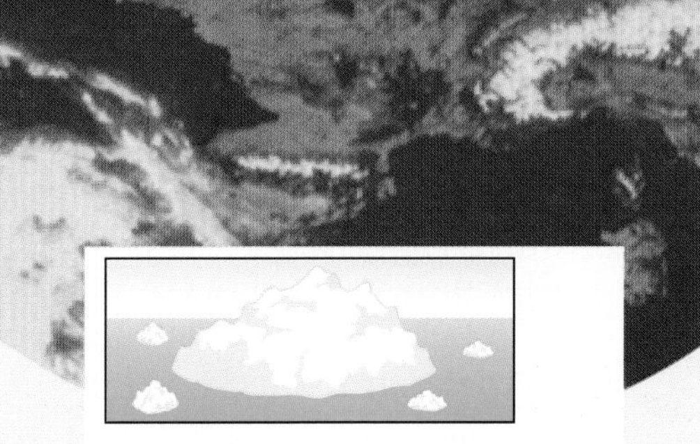

CRYOSPHERE

(*the permanently frozen region of the Earth's surface*)

Seasonal sea-ice and snow cover show a rapid response to changes in the atmosphere. They affect much of the global heat budget, such as the increase or decrease in surface **albedo** (how much sunlight is reflected back to space), the availability of fresh water to plant and animal life, and the growth of vegetation cover.

GEOSPHERE

(*the Earth, excluding the* ***atmosphere, lithosphere, hydrosphere****, and* ***biosphere****)*

This is the non-living section of the Earth, consisting of rock and soil. It affects climate in influencing the amount and rate of transfer of water and heat between biosphere and atmosphere. The reflecting of solar radiation (called albedo) also varies with land surface. For example, sand has high albedo, and dark volcanic soils have low albedo.

▲ **Figure 7** The world climate system.

4 Weather and climate in the British Isles

Talking about the weather

In Britain, the weather is a frequent topic of conversation. Its variability from cold and wet to warm and sunny, from mild and balmy to freezing fog, provides a talking point and also affects people's lives. Journeys, holidays, and everyday activities, are all influenced in some way by the weather. The variability of weather helps to characterize places. Melbourne, Australia, is known to Australians as the place where you are likely to experience 'four seasons in one day'. If you don't like the weather, you are told, just wait for a few minutes and it will change.

Why do the British Isles have such varied weather?

The British Isles lie approximately between latitudes 50° and 61° north of the Equator. At this latitude, patterns of both daily weather and seasonal climate are influenced for most of the year by prevailing westerly winds, which bring unsettled weather. These are winds which blow for much of the year from west to east, and which bring moisture in the form of cloud and rain from the North Atlantic. Unsettled weather is associated with areas of low atmospheric pressure, known as depressions. Depressions bring with them windy, wet, and disturbed weather. Generally, the further north you go in the British Isles, the more unsettled the weather becomes.

Depressions share their place in the British climate with patterns of more settled weather. Sometimes there are long periods when daily weather changes very little. Many of the British summers of the late 1980s through to the mid–1990s were characterized by settled spells of warm summer weather. In July 1990, the highest ever daily temperature of 37 °C was recorded in Britain. Similarly, winter weather is sometimes settled. There may be intense frosts, sometimes accompanied by freezing fog, at other times by crisp, sunny winter days. In December 1981, Britain's record low temperature of –25 °C was recorded at Shawbury, Shropshire, during a long spell of intense winter cold. Settled weather is associated with areas of high atmospheric pressure, known as anticyclones.

The regular change from depressions to anticyclones produces our changeable weather. British weather matches a type known as 'Temperate maritime' in climatic classification. 'Temperate' means that there are few or no extremes of temperature. 'Maritime' means that the weather is influenced by the sea, since all airflows which affect the British Isles have to pass over water.

One of the main influences on temperature is the warm North Atlantic Drift. This is an ocean current which originates in the warm subtropical waters of the Gulf of Mexico, and which tracks north-eastwards to the west coasts of Britain and Scandinavia. The absence of ice in winter in northern Scotland and Norway is largely due to the presence of this current. At Inverewe, in Ross and Cromarty, north-west Scotland, subtropical plants grow in gardens which are nearly 58° north of the Equator. This is the same latitude as the southern parts of Hudson Bay, in Canada, and Siberia in Russia. Each of these regions is intensely cold in winter. Water is slower to cool than land, so the presence of warmer water offshore reduces the occurrence of frost and freezing temperatures, especially in the south and west of Britain in winter.

Location	Average annual rainfall (mm)	Average daily temp. Jan. (°C)	Average daily temp. July (°C)	Average annual sunshine hours
1 Belfast Airport	837	3.5	14.5	1298
2 Douglas	1131	4.8	14.3	1572
3 Ambleside	1865	3.1	15.0	1185
4 Durham	645	2.8	14.7	1300
5 Leeming	617	3.0	15.2	1331
6 Blackpool Airport	847	3.7	15.4	1534
7 Hull	653	3.7	16.1	1380
8 Rhyl	661	4.8	15.6	1475
9 Manchester Airport	806	3.5	15.6	1359
10 Buxton	1289	2.0	13.9	1149
11 Lincoln	593	2.8	15.5	1401
12 Skegness	601	3.4	13.9	1512
13 Shrewsbury	624	3.6	15.9	1349
14 Aberystwyth	959	5.2	15.1	1473
15 Llandrindod Wells	1003	2.8	14.9	1244
16 Stratford-upon-Avon	627	3.2	16.0	1371
17 Cambridge	552	3.5	16.4	1508
18 Cardiff	1064	4.4	16.5	1497
19 Oxford	663	3.8	16.8	1517
20 Heathrow Airport	610	4.0	16.5	1494
21 Clacton-on-Sea	542	3.6	16.7	1635
22 Ilfracombe	1063	6.3	16.2	1631
23 Penzance	1131	6.9	16.1	1738
24 Plymouth	992	5.7	15.9	1687
25 Bournemouth	802	4.6	16.4	1777
26 Shanklin	906	4.8	16.1	1908
27 Eastbourne	811	4.9	16.6	1827
28 Folkestone	727	4.3	16.6	1732

▲ **Figure 4.1** Climate data for selected places in England, Wales, and Northern Ireland. This table shows climate data for selected places in England, Wales, and Northern Ireland as an average between 1951 and 1980.

Describing Britain's climate

1 Read about how to draw an isoline map on page 86. Divide into groups of four. Using Figures 4.1 and 4.2, produce isoline maps to show variations across the British Isles in:

a) average annual rainfall

b) average daily temperatures for summer

c) average daily temperatures for winter

d) annual sunshine hours.

Draw one map each and display your maps as a group.

2 Compare each of the maps you have drawn. Describe the main patterns that emerge. What conclusions can you reach?

3 Find climate and relief maps of the British Isles in your atlas. Using data from the atlas maps, suggest possible reasons for the patterns you have observed.

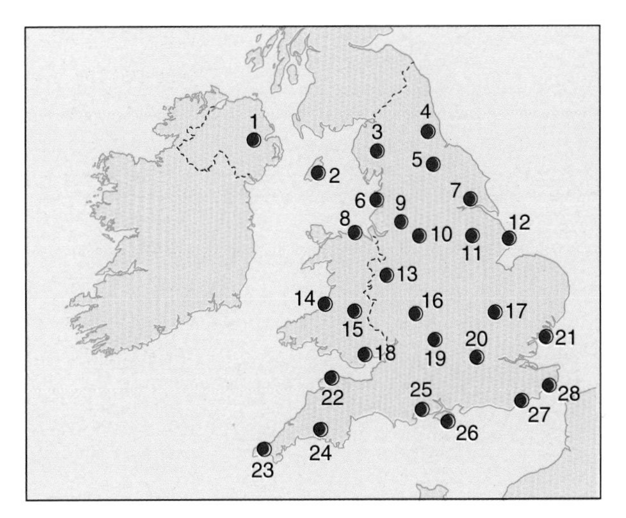

▶ **Figure 4.2** This map shows the location of places referred to in Figure 4.1.

Drawing an isoline map

This is a useful technique for revealing patterns when data are presented as a series of points at particular locations. It is very effective in showing spatial variations which are continuous over a large area – for example, sunshine or rainfall. The same technique is used by cartographers in producing contour maps.

Isolines are lines that join points of equal value. **Isotherms** are lines that join places of equal temperature, and **isohyets** are lines that join places with the same rainfall.

To draw an isoline map, follow these steps.

1 Study your data and the range of figures from highest to lowest.

2 Decide on a suitable regular interval between each isoline: for example, every 5 °C or every 50mm of rainfall.

3 Draw the isolines as shown in the example below.

Example

Assume that you start with all the locations and values shown in Figure 4.3(a). The lowest value, 3, is in the north-west and north-east corners of the map, and the highest value, 5.6, is towards the south-east. For this example isolines will be drawn at 0.5 intervals, from 3 to 5.5.

To complete the map:

1 Join all the points with a value of 3 as shown in Figure 4.3(b).

2 Draw the 3.5 line. In most cases you will have to guess where the line should go. Between two places, one value 3 and the other 4, the line will pass halfway between them. Between 3.3 and 4.2, the isoline will pass closer to 3.3 than to 4.2.

3 Figure 4.3(d) shows all the lines drawn in. Notice how the 5.5 line has been drawn to show isolated areas.

4 Shade between the lines to show patterns more clearly. Select shades of one colour for this, or a sequence of colours, such as yellow–orange–red–brown. Use light shading for low values and darker shading for the higher values. This technique of shading is used for most atlas maps, so look at some examples for suitable colour schemes.

▼ **Figure 4.3** How to draw an isoline map.

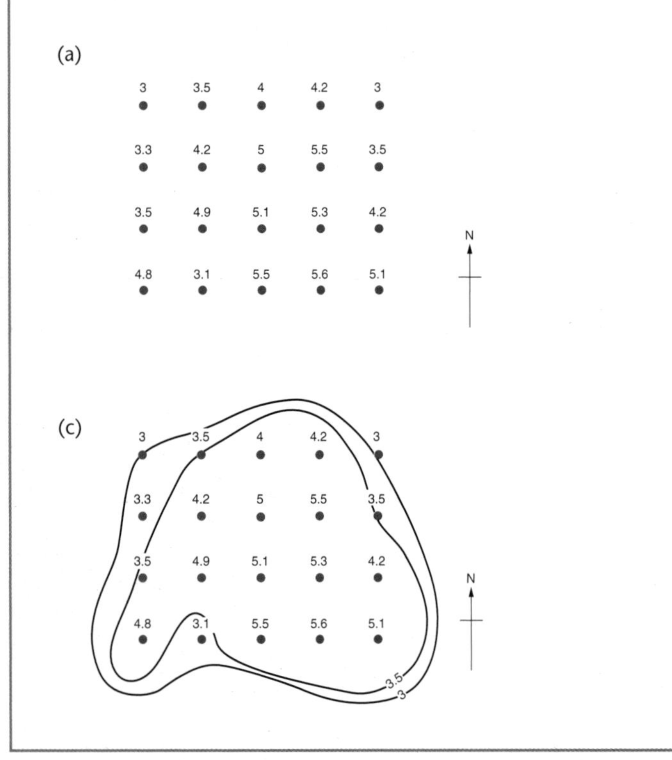

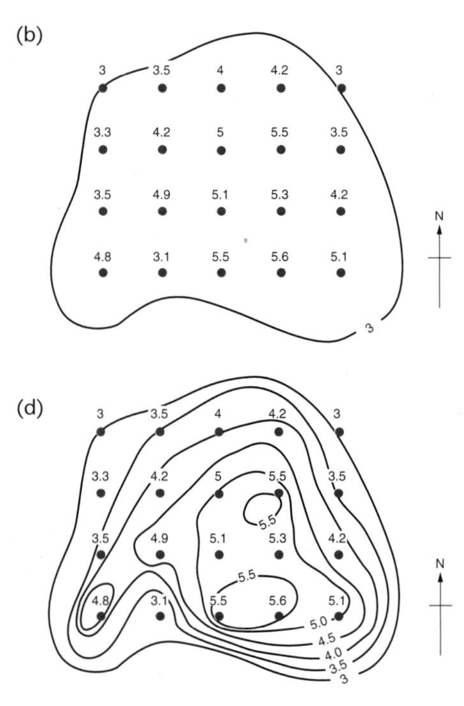

Explaining Britain's varied climate

The daily weather records of the British Isles from 1861 to 1971 were studied by the climatologist H. H. Lamb. He found that there are different air patterns affecting British weather over a typical year. He identified seven types of airflow, which correspond with different airmasses. These are shown in Figure 4.4. Five of these types of airflow (1 to 5 in Figure 4.4), are based on compass directions from which the airmasses have come. The final two refer to those times when weather is determined by depressions and anticyclones. Figures 4.5–4.7 are recent examples of weather maps showing the three most common types – westerly, cyclonic, and anticyclonic. Between them, the seven airflows account for almost all weather patterns across the British Isles and Ireland.

▼ **Figure 4.4** Lamb's airflow types and associated weather characteristics.

Airflow type	General weather characteristics	% frequency
1 Westerly	Unsettled weather with variable wind directions as depressions cross the country, giving most rain in northern and western districts, with brighter weather in the south and east. Mild in winter with frequent gales; cool and cloudy in summer (associated with mP, mT airmasses).	27.6
2 Northerly	In winter the weather is cold with snow and sleet showers, especially along the east coast; blizzards may accompany deep polar lows. In summer the weather is cool and showery, especially along the east coast (mA).	8.1
3 North-westerly	In winter, cool, showery, changeable conditions with strong winds. The weather in summer is cool with showers on windward coasts; southern Britain may be bright and dry (mP, mA).	6.1
4 Easterly	Cold in the winter period, sometimes with severe weather in the south and east with snow and sleet, but fine in the west and north-west. Warm in summer with dry weather especially in the west; occasionally thundery (cA, cP).	7.1
5 Southerly	Warm and thundery in summer. In winter it may be associated with a low in the Atlantic giving mild, damp weather especially in the south-west, or with a high over central Europe, in which case the weather is cold and dry (mT or cT in summer; mT or cP in winter).	6.6
6 Cyclonic	Rainy, unsettled conditions over most of the country, often accompanied by gales and thunderstorms. Wind direction and strength are variable. Conditions normally mild in autumn and early winter, cool or cold in spring and summer and cool in late winter (mP, mT).	12.8
7 Anticyclonic	Mainly dry with light winds; warm in summer with occasional thunderstorms; cold often with frosts and fog in winter, especially in the autumn (mT, cT in summer; cP in winter).	17.9

Key: cA = Arctic continental; mA = Arctic maritime; mP = Polar maritime; cP = Polar continental; mT = Tropical maritime; cT = Tropical continental.

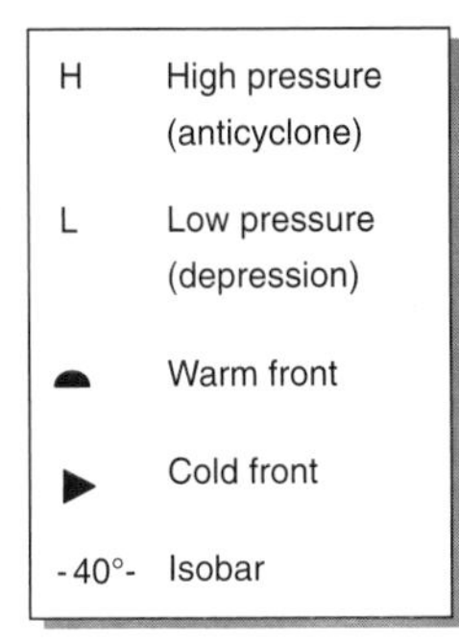

▼ **Figure 4.5** Westerly-type airflow.

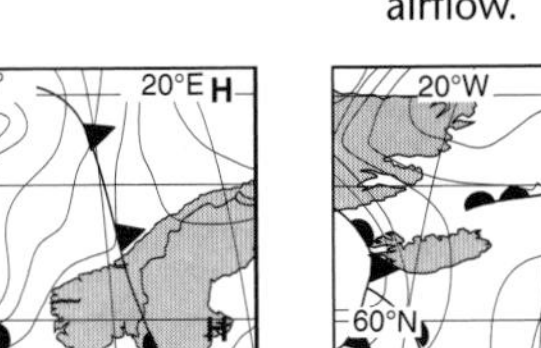

1200 GMT 18 september 1990

Low pressure to the north of the British Isles with high pressure to the south. Sequences of depressions and their associated fronts travelling eastward across the region denote this type of airflow.

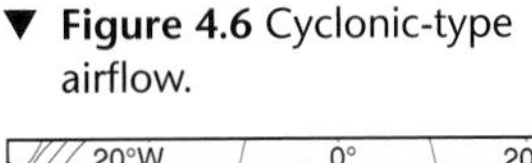

▼ **Figure 4.6** Cyclonic-type airflow.

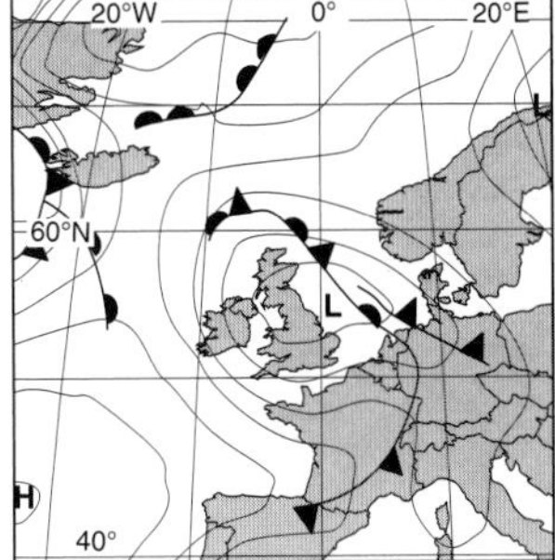

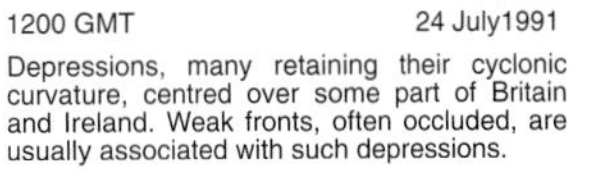

1200 GMT 24 July1991

Depressions, many retaining their cyclonic curvature, centred over some part of Britain and Ireland. Weak fronts, often occluded, are usually associated with such depressions.

▼ **Figure 4.7** Anticyclonic-type airflow.

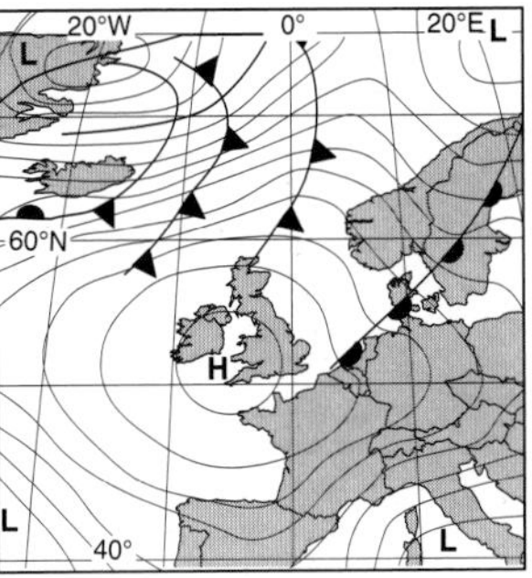

1200 GMT 7 December 1990

High-pressure cells centred on or extending over the British Isles. Lower pressure cols between two anticyclones are also included.

Using airflow types

Record the various elements of the weather in your local area over a week. You can use the following data sources.

- Primary data which you collect yourself using your own instruments, or those from your school or college.
- Secondary data from newspapers, Teletext, Oracle, or the UK Met Office daily weather service using the fax link METFAX.
- Your school or college may also have a weather satellite receiver or weather station with computer link.

1 Produce a short summary report to accompany your maps and statistics, outlining the weather changes during the week.

2 Do these patterns of weather correspond with the airflow types put forward by Lamb? Were all parts of the British Isles affected by the same airflow types at the same time?

3 How useful do you find this classification in helping to explain the variability in British weather?

4 **a)** In your atlas, select a location in the southern hemisphere which is close to or on the coast, such as Melbourne, Santiago, or Cape Town.
b) In your atlas, study climate maps which show airflow patterns for January and July. Produce a sketch map entitled 'Airflow types' for the place you have chosen.
c) What kinds of weather might the airflows that you have identified bring to the place you have chosen? How is the pattern different from that in the northern hemisphere?

Depressions and anticyclones

Depression, or **cyclone**, is a term that refers to a whole range of atmospheric disturbances which are characterized by low atmospheric pressure. 'High' and 'low' pressure are relative terms. They mean that the air is at a higher pressure or lower pressure than in the surrounding areas. **Atmospheric pressure** is a term used to describe the total weight of air molecules in a given column of air. Air pressure is measured in **millibars**, or **mb**. 1000mb is about average air pressure, 1050mb is exceptionally high, and 950mb is very low. The air in most pressure systems is usually between these two figures.

The level of air pressure in different places depends on differences in temperature. Air that is being warmed expands and rises, and weighs less, thus reducing the pressure it exerts on the Earth's surface. This air is described as being at 'low pressure', and forms depressions. Air that is rising is unstable. It takes water vapour with it into the upper atmosphere, which is much colder. On cooling, the vapour condenses and forms water droplets. At high levels, these droplets form clouds. Depressions are therefore associated with cloud and, often, precipitation.

Each time air becomes lighter and rises, or becomes heavier and falls, it allows more air to replace it. In general terms, depressions are linked to warmer rising air at lower pressure, and anticyclones are linked to sinking air at higher pressures. Winds are created when air moves from areas of higher pressure towards areas of lower pressure. The force of the wind depends upon the difference between the two pressure systems. If the pressure systems are far apart, and the difference in pressure is small, air movement (or wind) is slight. On a weather map, this situation is characterized by widely spaced isobars. On the other hand, if the pressure difference is great and the systems are close together, winds are strong. This situation is indicated on a weather map by isobars being grouped closely together.

Anticyclones

When air is cooled, its volume contracts and it becomes heavier. This forces it to sink under gravity. Often, large bodies of cooler air subside and collect as dense airmasses. These form large areas of stable air, known as areas of high pressure, or anticyclones. Once an anticyclone is established, it often remains for a long period. Because it consists of stable cold air, it contains very little moisture and is therefore dry. Although cold, therefore, the clear dry air allows the Sun's radiation through to the Earth's surface. In summer, anticyclones are therefore associated with settled, warm, sunny weather. Figure 4.8 is a clear example of this – a satellite photograph taken on a day when almost all of north-west Europe was visible from space. In winter, the clarity of the air allows bright sunny days, but the lack of cloud means that heat energy escapes at night. Sometimes, the repeated loss of heat during long winter nights enables temperatures to drop further and further. Some high-pressure areas remain as fixed features of the world climate pattern, such as the high pressure of the mid-northern Atlantic, known as the Azores High.

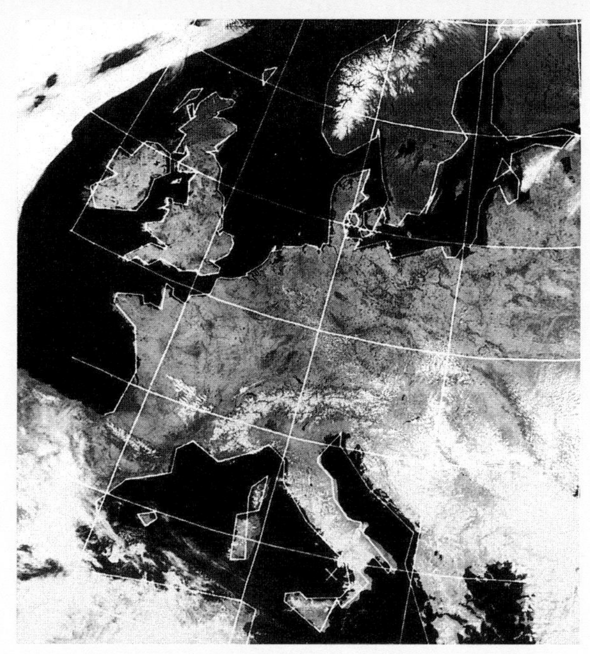

▲ **Figure 4.8** Satellite photograph of North-West Europe at 1245 GMT on 2 May 1990.

Depressions and British weather: a case study of the storms of 25 January 1990

The depression of 25 January 1990 was typical of events when storm-force winds pass over the whole of the British Isles. Similar events included the 'Great Storm' of 14–15 October 1987. Like most depressions that affect Britain, it began life as a shallow area of low pressure off the eastern coast of North America, on 23 January 1990. Most depressions begin in this way, as relatively small areas of low pressure off the shores of north-eastern USA. This is shown by a band of cloud stretching across the middle of the Atlantic south of Iceland, as you can see in the satellite photograph Figure 4.9, for example.

At this point, the pressure of the air is not very low. By midnight GMT on 24 January, the depression had started to deepen, that is, air pressure was falling. This, again, is typical. Pressure falls further as the air passes over the warm North Atlantic waters. When this happens, the pressure difference between high and low pressure increases. In all situations like this, wind speed increases – the bigger the pressure difference the greater the wind speed.

By midday on 24 January the central pressure had fallen to 992mb and wind speeds were high. Britain is at the end of the transatlantic route for such winds, and receives their full force. By the time the full storm had broken at 1800 hours on 25 January, pressure had fallen as low as 948mb. The sequence of events in the formation of a typical deep depression and its link with the upper atmosphere is shown in the theory box on page 97.

The satellite photographs which show the passage of the storm over the British Isles are shown in Figures 4.15–4.16. Notice the following features of the photographs.

- The whole depression is shown by a cloud mass which has a hook-shape. This is caused by the circling anticlockwise winds which take moisture and cloud in towards the centre of the depression.
- The cloud mass coincides with the position of the fronts shown on the weather map.
- The cloud mass moves from west to east.
- Once the depression has passed over, Britain is affected by shower clouds.

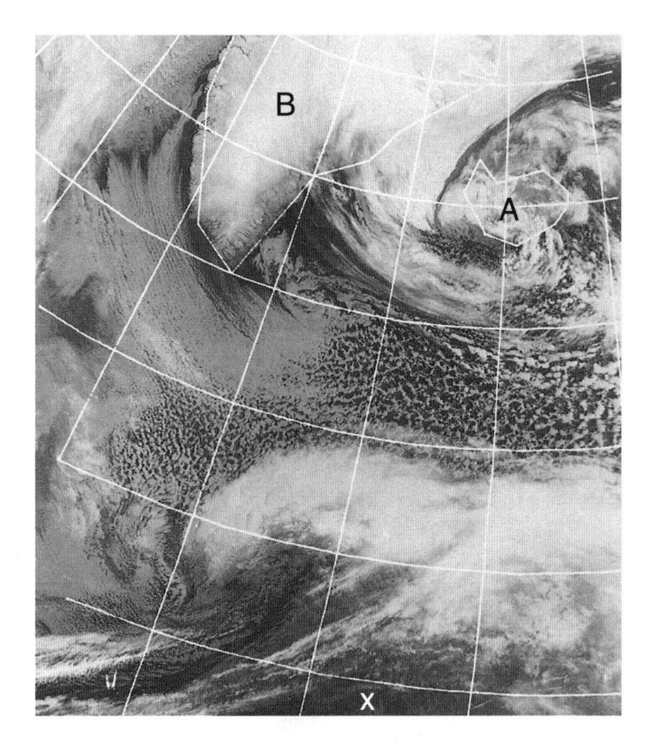

▶ **Figure 4.9** Satellite photograph of the North Atlantic at 1517 GMT on 24 January 1990. Iceland is shown at A, and Greenland at B, in white outline. To the south of the developing depression is a black/grey area at X, which is affected by the Azores high pressure. Clouds are thin, and reflect very little, hence their grey colour. The depression is shown in clear white bands of cloud. Further to the north-west, shower clouds are shown as speckled blobs of cloud. This photograph is too far west to feature the UK.

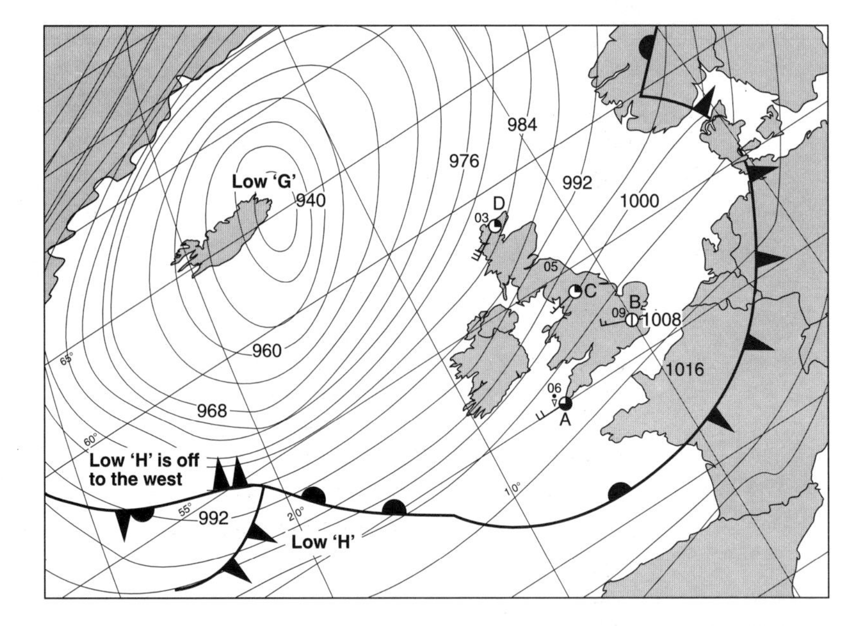

▶ **Figure 4.10** Synoptic chart for 1200 on 24 January 1990.

▼ **Figure 4.11** Synoptic chart weather symbols: these are the symbols used on all weather maps.

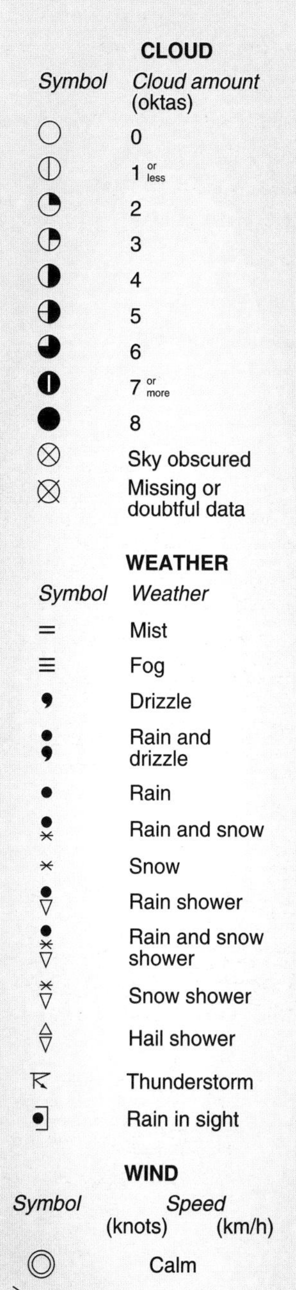

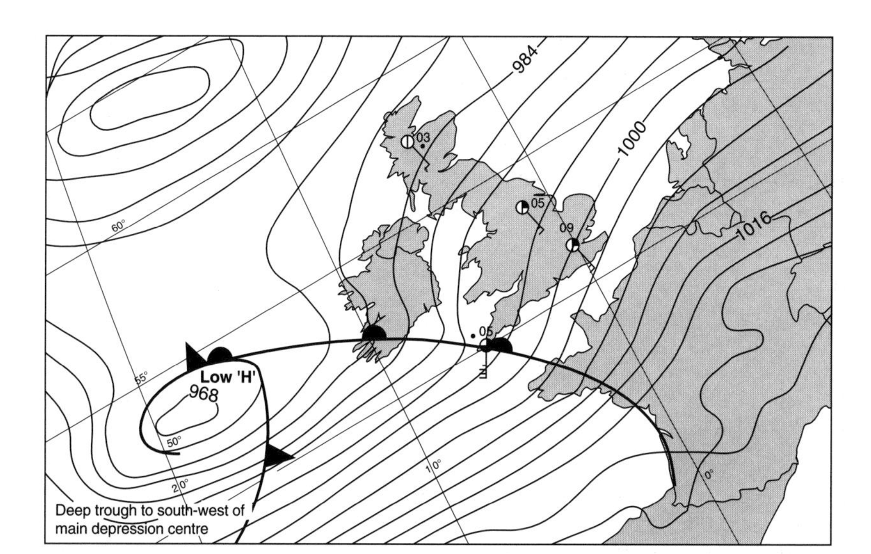

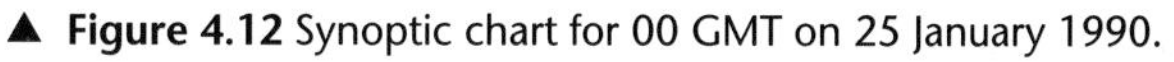
▲ **Figure 4.12** Synoptic chart for 00 GMT on 25 January 1990.

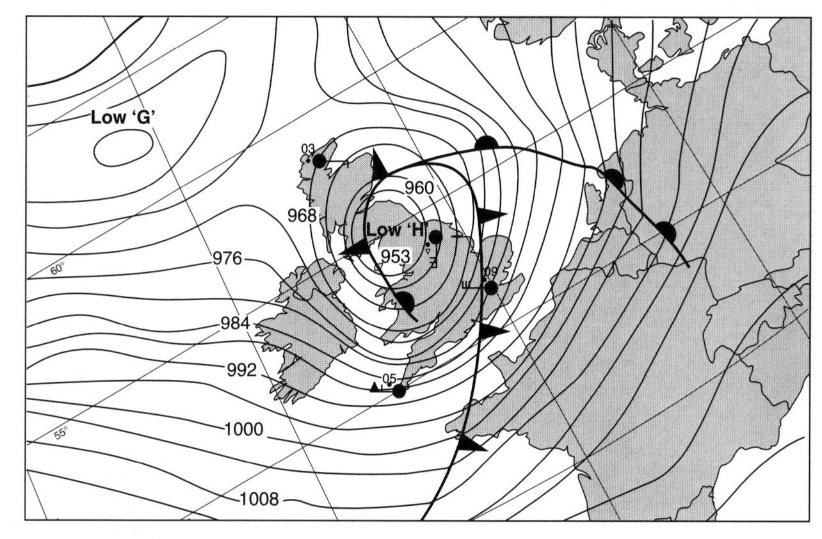

▲ **Figure 4.13** Synoptic chart for 1200 GMT on 25 January 1990.

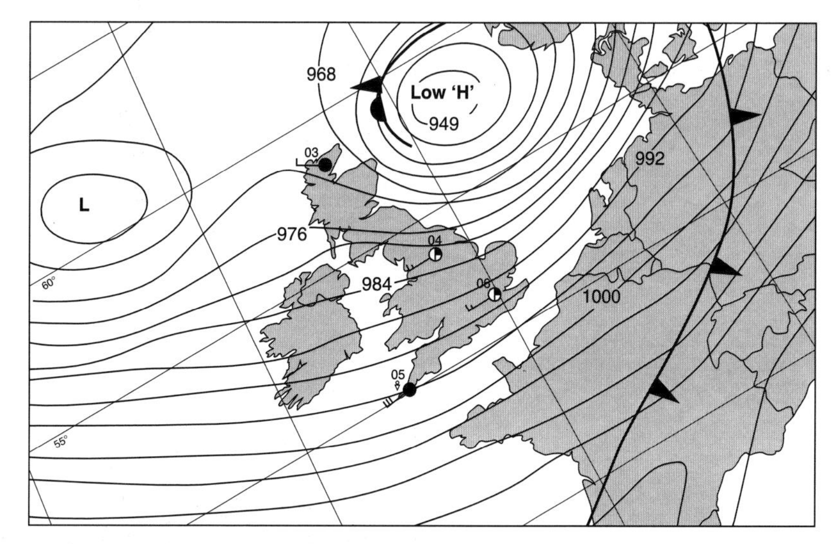

▲ **Figure 4.14** Synoptic chart for 2400 GMT on 25 January 1990.

The diary of a storm

This exercise will help you to observe and interpret some of the changes which take place when a depression passes over the British Isles. The pressure system to which it refers is pressure system 'H' on Figures 4.10 and 4.12 to 4.14, which caused the storm of 25 January 1990. Although it was much stronger than many systems which affect us, the main difference between this and other systems was that it occurred further south. The North Atlantic and northern Scotland frequently get storms of this kind.

1 Copy and complete the following table, and record the changes in air pressure at Land's End (location A), London (location B), Leeds (location C) and northern Scotland (location D) on Figures 4.10 and 4.13 to 4.14. The completed table will show the development of the pressure system over a 36-hour period between midday on 24 January and midnight on 25 January.

	1200 24/1	00 25/1	1200 25/1	2400 25/1
Land's End				
London				
Leeds				
Northern Scotland				

2 Now complete similar tables for each of the following, using Figure 4.11 to help you: wind speed, temperature, cloud cover, precipitation.

3 Radio 4's midnight news each day includes a 3–4 minute summary of the day's weather over the whole of the UK. It notes such features as changes throughout the UK in temperature, precipitation, and wind speed, and tries to link these to the movement of depressions or anticyclones. Write a report of the 36-hour period shown in Figures 4.10 and 4.12 to 4.14 which could have been used for the midnight news on 25 January 1990.

4 What was happening to the pressure of the low-pressure system by the time it reached northern Europe? What would eventually happen to it within the next few days?

▼ **Figure 4.15** Satellite photograph of a depression crossing the British Isles at 0330 GMT on 25 January 1990.

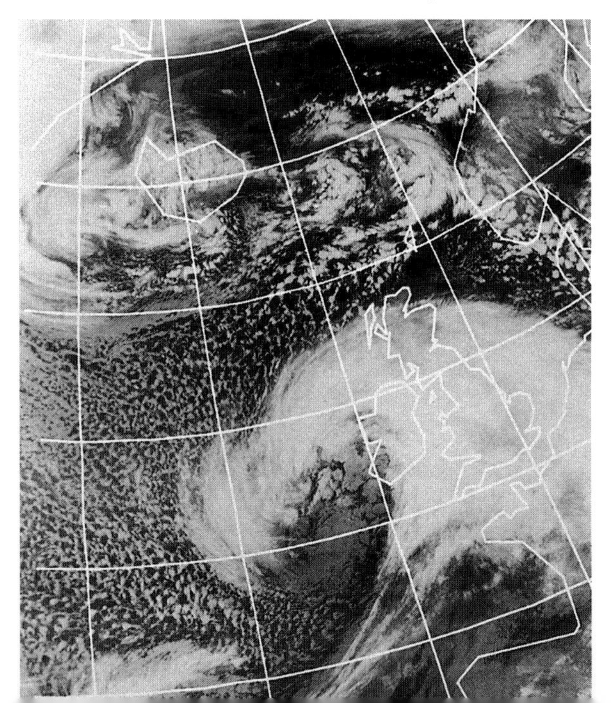

▼ **Figure 4.16** Satellite photograph of a depression having passed over the British Isles at 1324 on 25 January 1990.

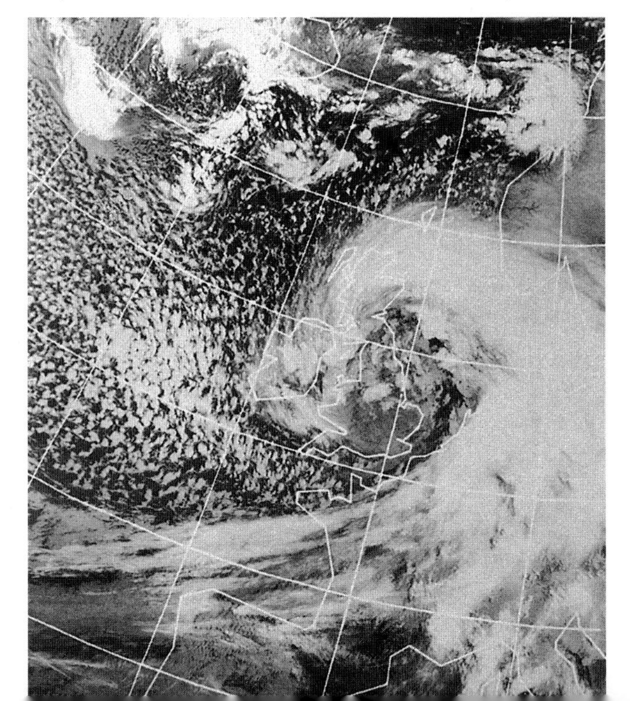

Interpreting satellite photographs

1 On a blank outline map of north-west Europe, sketch the outline of cloud for 0330 GMT on 25 January 1990, as shown in Figure 4.15. Using the synoptic charts for 1200 GMT on 24 January 1990 (Figure 4.10) and 1200 GMT on 25 January 1990 (Figure 4.13) to help you, mark the following features on your map:

a) the position of the fronts
b) the position of low-pressure area 'H'
c) the Azores high-pressure area
d) the area of showery cloud affecting the north Atlantic
e) an arrow showing the direction in which the low-pressure system seemed to be moving.

2 How and why do weather forecasters use satellite photographs such as these? How can these be used in predicting weather?

3 Identify the links between these photographs and the stages in the development of a depression, shown on page 96.

The effects of the storm

Storm damage was considerable. For many people in southern England, this was the second time in less than three years that storms had brought severe damage to people, property, and places, after the storm of 1987. In northern England, accidents on motorways such as the M62 trans-Pennine route included lorries falling over. Figures 4.17, 4.18, and 4.19 show the effects of the storms in Surrey. Figure 4.19 shows that the storm brought political conflict too, at a time when local councils were being hard pressed for money.

▲ **Figure 4.17** Storm damage in Dorking, Surrey.

The Great Gale 1990
25th January

The tremendous gale that caused so much destruction across Surrey on Thursday 25th January 1990 seemed inconceivable, coming so soon after the October 1987 'hurricane'.

No one believed that winds of such violence would strike in Surrey again for many a decade. But while foresters and National Trust gardeners were still clearing up after the 'storm of a lifetime', grim warnings were flashed on television as a ferocious area of low pressure looked set to unleash its pent-up fury over wide areas of England.

Savage winds increased in intensity throughout the day and by mid-afternoon, these reached speeds of more than 120 kph. Thousands of pounds' worth of damage was caused in virtually every Surrey street; countless trees were toppled, roofs blown off, many roads blocked and because it struck during the day, unlike the last great gale, a number of people were killed.

A schoolgirl, aged 11 from Banstead, was one of the victims. She was crushed by a tree outside St Philomena's Convent in Carshalton at about 3.45 pm. At Epsom Downs, a 75-year-old man was struggling to put up a rose trellis which had blown down, when he collapsed and died. It was the second tragedy in Epsom. A man died after falling while trying to secure loose roof tiles. In Haslemere, a man was crushed to death by a falling tree while trying to clear branches blown down.

Across Surrey, the damage in public places amounted to at least £5 million, the County Council reported. But while buildings could be reconstructed, the woodlands could not – at least for several decades.

◀ **Figure 4.18** A description of 'The Great Gale 1990', *Surrey Weather Book*.

An ill wind

After the second hurricane within three years swept through Greater London last week causing millions of pounds' worth of damage, the question again arises as to who pays what.

Fortunately most individuals whose properties suffered are covered by their household policies, but local authorities like Hounslow and Richmond face many bills that fall outside the ambit of insurance.

This chiefly involves the payment of overtime to council employees or fees to private contractors for clearing up after the damage – particularly to dislodged trees. Borough councils are urged by the Government to put aside the equivalent of a penny rate against national disasters. After this has been spent the Government provides three-quarters of the remaining cost from central funds. In reality it appears few have been able to do this.

Richmond argues that in view of all the other cutbacks it has faced it has been unable to create such a nest-egg.

▲ **Figure 4.19** Editorial from the *Richmond and Twickenham Times*, 2 February 1990.

▶ **Figure 4.20** A comparison of maximum gusts for storms of 25 January 1990 and 16 October 1987.

Station	Maximum gust (km/h) 25/1/90	Return period (years)*	Previous highest gust (km/h)	Year of occurrence	Maximum gust (km/h) 16/10/87
Aberporth (Dyfed)	172	120	152	1981	106
Bedford (Beds)	128	15	130	1976	98
Benson (Oxon)	135	48	122	1976	111
Boscombe Down (Wilts)	156	150	139	1947	124
Brawdy (Dyfed)	150	20	146	1965	83
Brize Norton (Oxon)	117	7	131	1976	91
Elmdon (W. Midlands)	107	5	135	1976	68
Farnborough R.A.E. (Hants)	143	500	113	1987	E89
Farnborough (1945–68) (Hants)	143	120	128	1947	–
Finningley (Yorks)	130	9	146	1976	72
Gatwick (Sussex)	133	30	159	1987	159
Gwennap Head (Cornwall)	167	7	191	1979	150
Heathrow (Gr. London)	141	210	122	1987	122
Hemsby (Norfolk)	128	4	154	1984	146
Herstmonceux (Sussex)	157	–	167	1987	167
Hurn (Dorset)	133	110	117	1977	113
Langdon Bay (Kent)	130	–	174	1987	174
Larkhill (Wilts)	139	30	135	1974	115
London Weather Centre (Gr. London)	139	22	152	1987	152
Lyneham (Wilts)	137	65	120	1984	106
Manston (Kent)	139	22	144	1987	144
Middle Wallop (Hants)	139	100	144	1987	144
Plymouth (Devon)	156	75	141	1987	83
Portland (Dorset)	141	11	144	1987	144
Rhoose (S. Glam)	156	60	200	1989	94
Ringway (Gr. Manchester)	113	4	144	1943	61
St Mawgan (Cornwall)	154	35	152	1976	131
Shoeburyness (Essex)	137	15	161	1987	161
Shoreham (Sussex)	167	–	185	1987	185
Stansted (Essex)	133	55	130	1976	120
Thorney Island (Hants)	130	13	167	1987	167
Watnall (Notts)	128	19	141	1976	80
Wattisham (Suffolk)	148	24	156	1976	133
Wyton (Cambs)	122	7	141	1976	109
Yeovilton (Somerset)	124	5	124	1984	113

* Return period is the average occurrence of a wind of specific strength based on historical records. The longer the return period, the stronger the gust.

1 Study Figures 4.17–4.19. In pairs, assess the impact of the storm and its effects in southern England. Classify these effects as either social, economic, or environmental.

2 Why would large insurance companies employ people whose job it is to monitor weather events and other physical phenomena?

3 Use Figures 4.20 and 4.21 to assess the force of the winds that struck Britain during this storm, and indicate those areas worst affected. How did the 1990 storm compare with that of October 1987? You may like to make use of an atlas and outline maps of the UK.

Force	Description	Specifications for use on land	Specifications for use at sea	Equivalent speed at 10m above ground Km per hour Mean	Limits	Description in forecasts	State of sea	Probable height of waves (metres)
0	Calm	Calm; smoke rises vertically	Sea like a mirror	0.0	0–1	Calm	Calm	0.0
1	Light air	Direction of wind shown by smoke drift, but not by wind vanes	Ripples with the appearance of scales are formed, but without foam crests	2.8	1–5.5	Light	Calm	0.1 (0.1)
2	Light breeze	Wind felt on face; leaves rustle; ordinary vane moved by wind	Small wavelets, still short but more pronounced. Crests have a glassy appearance and do not break	8.7	5.5–12	Light	Smooth	0.2 (0.3)
3	Gentle breeze	Leaves and small twigs in constant motion; wind extends a light flag	Large wavelets. Crests begin to break. Foam of glassy appearance. Perhaps scattered white horses	15.5	12–19	Light	Smooth	0.6 (1.0)
4	Moderate breeze	Raises dust and loose paper; small branches are moved	Small waves, becoming longer; fairly frequent white horses	24	19–29	Moderate	Slight	1.0 (1.5)
5	Fresh breeze	Small trees in leaf begin to sway; crested wavelets form on inland waters	Moderate waves, taking a more pronounced long form; many white horses are formed. Chance of some spray	33.5	29–38	Fresh	Moderate	2.0 (2.5)
6	Strong breeze	Large branches in motion; whistling heard in telegraph wires; umbrellas used with difficulty	Large waves begin to form; the white foam crests are more extensive everywhere. Probably some spray	44	38–50	Strong	Rough	3.0 (4.0)
7	Near gale	Whole trees in motion; inconvenience felt when walking against wind	Sea heaps up and white foam from breaking waves begins to be blown in streaks along the direction of the wind	56	50–62	Strong	Very rough	4.0 (5.5)
8	Gale	Breaks twigs of trees; generally impedes progress	Moderately high waves of greater length; edges of crests begin to break into spindrift. The foam is blown in well-marked streaks along the direction of the wind	68	62–74	Gale	High	5.5 (7.5)
9	Strong gale	Slight structural damage occurs (chimney pots and slates removed)	High waves. Dense streaks of foam along the direction of the wind. Crests of waves begin to topple, tumble, and roll over. Spray may affect visibility	81	74–88	Severe gale	Very high	7.0 (10.0)
10	Storm	Seldom experienced inland; trees uprooted; considerable structural damage occurs	Very high waves with long overhanging crests. The resulting foam, in great patches, is blown in dense white streaks along the direction of the wind. On the whole the surface of the sea takes a white appearance. The 'tumbling' of the sea becomes heavy and shock-like. Visibility affected	95	88–102	Storm	Very high	9.0 (12.5)
11	Violent storm	Very rarely experienced: accompanied by widespread damage	Exceptionally high waves (small and medium-sized ships might be for a time lost to view behind the waves). The sea is completely covered with long white patches of foam lying along the direction of the wind. Everywhere the edges of the wave crests are blown into froth. Visibility affected	110	102–118	Violent storm	Phenomenal	11.5 (16.0)
12	Hurricane	–	The air is filled with foam and spray. Sea completely white with driving spray; visibility very seriously affected	–	>118	Hurricane force	Phenomenal	14.0 (–)

▲ **Figure 4.21** The Beaufort scale of wind speed.

World pressure systems

Large-scale high- and low-pressure systems are caused by temperature differences. In the northern hemisphere winter, land cools more quickly than the surrounding sea. As it chills, so large areas of cold dense air collect, forming some of the largest high-pressure systems. Typically, Eurasia and North America have the largest high-pressure systems at that time of year because they are the largest landmasses.

The northern Atlantic and Pacific Oceans lose their heat less quickly. Circulation of heat through the oceans ensure they remain warmer than the surrounding landmasses for longer. Average temperatures in the Atlantic Ocean off Cornwall are 12–13 °C in January, while the UK landmass only is 6–8 °C or less. Distance from the sea means that the central parts of Russia, which are unaffected by the warming influence of the oceans, may have temperatures as low as –40 to 50 °C.

In July, the pattern is reversed. Land heats up more quickly than sea, because ocean waters circulate and take time to heat. As a result, the northern landmasses heat quickly during summer. Air patterns are disturbed and low-pressure areas form. The relative coolness of the oceans creates large areas of cooler high-pressure air, while in the continental interior, temperatures reach 30 °C or more.

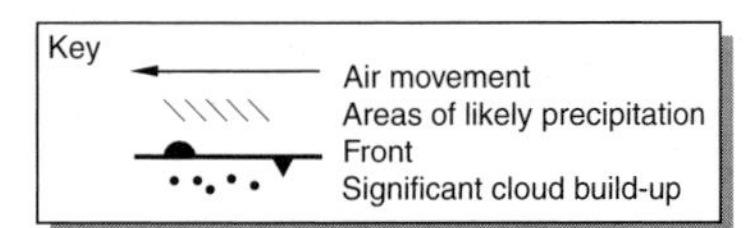

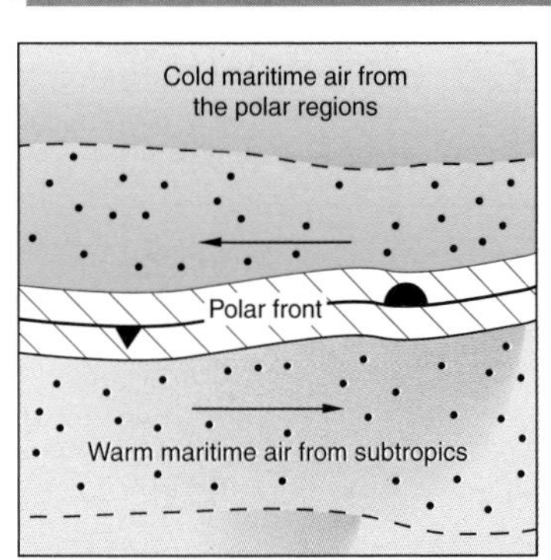

▲ **Figure 4.22** Cyclogenesis stage.

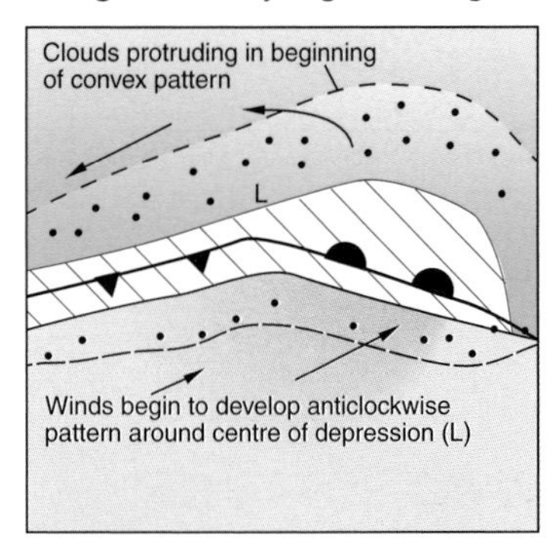

▲ **Figure 4.23** Wave depression stage.

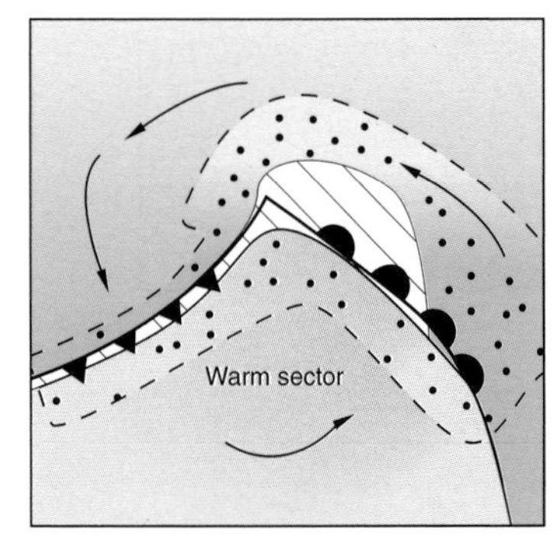

▲ **Figure 4.24** Warm sector stage.

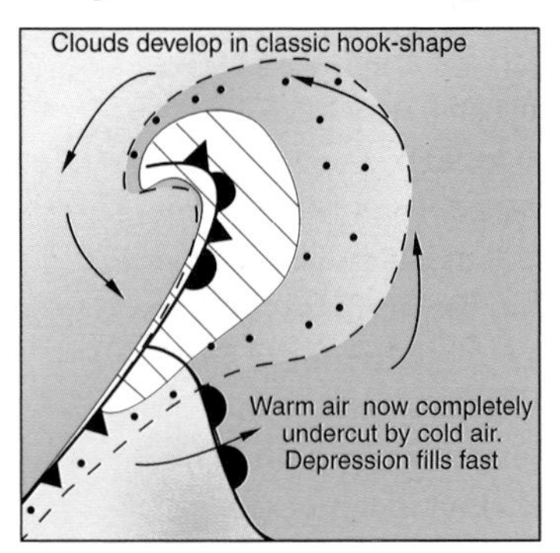

▲ **Figure 4.25** Occluding stage.

The development of depressions in the northern hemisphere

Most depressions which affect the British Isles exist for no more than three or four days. They begin their short life in the northern Atlantic, along a boundary between warm Tropical maritime air to the south, and cold Arctic maritime air to the north. This boundary is called the Polar front. The trigger for a depression is the development of upper-air movement ahead of a trough. The theory box on pages 97–98 explains how this happens.

In general the process can be divided into four stages (see Figures 4.22–4.25).

Stage 1: Cyclogenesis (Figure 4.22)

Surface pressures fall rapidly as Polar and subtropical air converges and spirals upwards into the upper air – this is known as the 'vacuum cleaner' effect. This brings greater mixing of warm and cold air at the polar front. Airstreams begin to change from their normal parallel pattern into a spiral as air is sucked into the centre of lowest pressure.

Stage 2: The development of a wave depression (Figure 4.23)

The rotation of the Earth produces a force called the Coriolis force. This causes air to blow in an anticlockwise fashion around the centre of the depression, as dense cold air undercuts the less dense, rising warmer air to the west. This is called the cold front. Warm air slides up and over the denser, colder air to the east of the depression centre. This is called the warm front. Pulled along by the upper-air jet stream, the cold front moves south-east, whilst the warm front moves north. Satellite photographs show a dense area of cloud to the north of the depression centre with a convex 'hook' shape.

Stage 3: Warm sector depression (Figure 4.24)

As pressures continue to fall, circulation strengthens and wind speeds increase. The cold front rapidly undercuts the warm air, which rises even faster. The wedge of warm air, known as the warm sector, progressively shrinks in size.

Stage 4: An occluding depression (Figure 4.25)

The cold front eventually catches up with the slower-moving warm front and the two fronts meet. The warm air which was originally at the surface is now lifted above it, first at the centre, then further away. This is called an occlusion. It occurs after two or three days and may persist for several more as it fills. The depression may achieve its lowest pressure in this phase, and satellite photographs show their characteristic hook-shape. This is the most destructive phase, as the 1987 and 1990 storms showed, when temperature differences between tropical and polar air were extreme. In 1987, the depression deepened by 50mb in one day and temperatures rose by 9 °C within an hour as the warm sector passed over southern Britain.

Depressions and upper-air movements

The formation of depressions and anticyclones is linked to air movements in upper levels of the atmosphere, at about 8000–12 000 metres above sea-level, the height at which most commercial passenger aircraft fly. Some idea of this can be gained from looking at transatlantic flights between London Heathrow and New York's JFK airport. Between London and New York, a Boeing 747 is scheduled for 8 hours' flying time; on its return journey, the same aircraft is scheduled for 6 hours and 45 minutes. Research carried out after 1945, using measurements by high-flying aircraft and balloons with weather sensors, discovered that rapid air currents move from west to east around the Earth in waves. These rapid air currents move between upper-air troughs of low pressure and ridges of high pressure. Aircraft travelling from North America to London are helped by these tailwinds, while those flying in the opposite direction may meet strong headwinds. During periods of fast-moving low-pressure systems such as those involved in the January 1990 storm, planes may arrive much earlier than scheduled into Heathrow airport, having had such a tailwind across the Atlantic.

Above the Earth's surface there are also higher-pressure and lower-pressure systems. High pressure exists where there is colder stable air, and low pressure where there is unstable warmer air. Movement of air takes place, causing high-level winds. Without the effect of the rotation of the Earth, these winds would blow in much more direct paths, and because they are less affected by the Earth's friction, they would blow at very high speeds. These winds are known as **jet streams**. They transfer seasonally. Figure 4.26 shows the average position of jet streams and their speeds over the northern hemisphere in July. The development of depressions and anticyclones is very closely linked to the speed and position of these jet streams. The continual west-to-east movement of upper-air winds causes the regular west-to-east movement of depressions and anticyclones that dominate British weather.

There are two distinct features of these winds. One is that the winds move in a series of westerly flows: they blow from the west towards the east. Second, these flows are not straight but move in a series of waves around the planet between latitudes 40° and 65°. These are called **Rossby waves**. Their formation is shown in Figure 4.27. They are caused by differences in pressure between Polar and Equatorial air. In Figure 4.27, the waves become greater in size with time. In diagram (a), the wave is gentle and situated a long way north of the Equator. Warm tropical air is therefore able to extend a long way north. As wave development increases, so Polar air is able to extend much further south, and Equatorial air much further to the north. Just as ox-bow lakes form in a river meander, so cells of low- and high-pressure air are 'cut off' from their original source – see diagram (d). Waves tend to have a limited life, and new ones begin as older ones die.

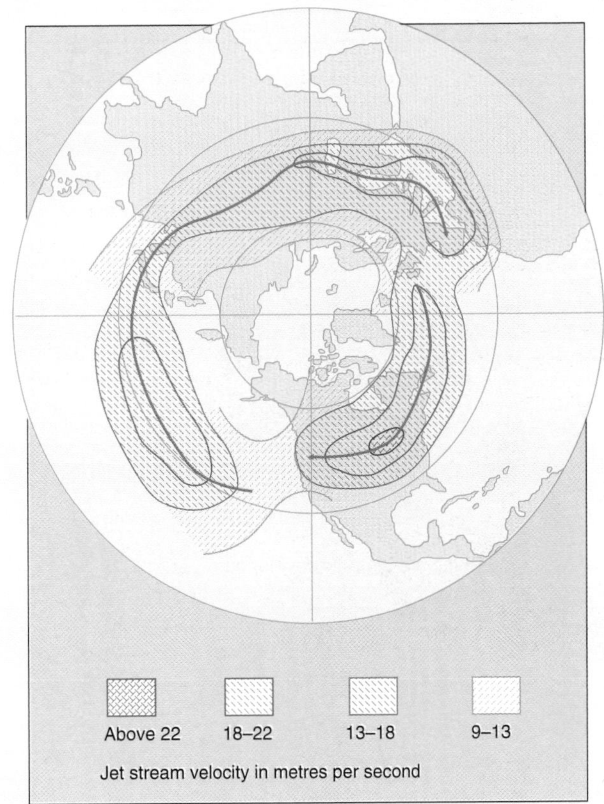

▲ **Figure 4.26** Jet streams and their speeds over the northern hemisphere in July. This map shows the average position of jet streams during July. Notice that the UK is right in the track of these high-velocity winds. They are closely associated with the development of fronts and depressions, and are part of the driving force which directs depressions from west to east across the Atlantic towards Europe.

Continued on page 98

Continued from page 97

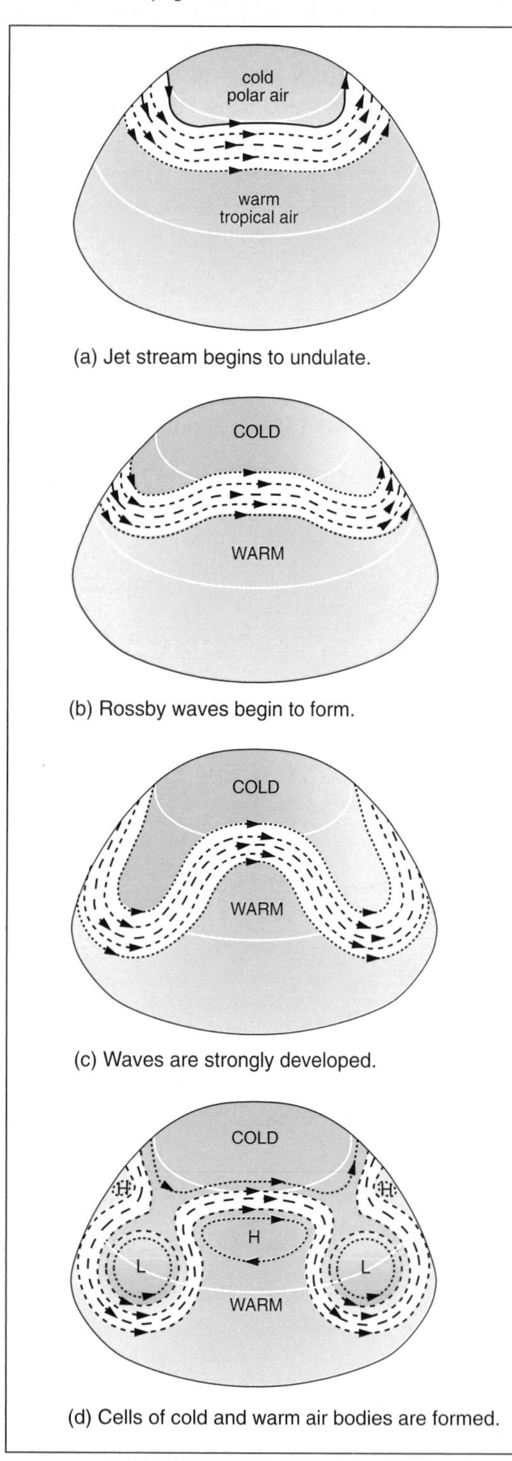

▲ **Figure 4.27** The formation of Rossby waves. These fast-moving currents of air operate in the upper troposphere and lower stratosphere where friction is at a minimum. Notice how the waves develop in size and depth over time, only to cut themselves off by stage (d). They enable Polar air to move further south than it might otherwise, and subtropical air to move further north.

Pressures in the atmosphere have to adjust to differences between warmer Equatorial air and colder Polar air. These are shown in Figure 4.28. If you look at the upper-air movements and compare them with those at the surface, you will see that surface low-pressure systems are found just to the east of an upper trough of low pressure. In the same way, surface high pressure is found just to the east of upper ridges of high pressure.

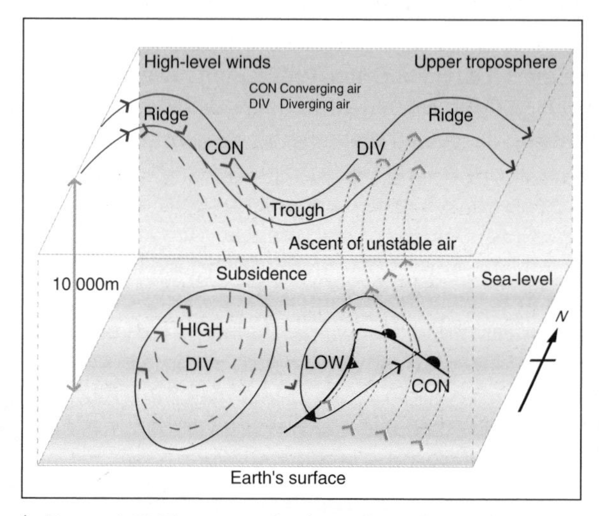

▲ **Figure 4.28** The connection between surface and upper-air movement.

Try to imagine Rossby waves as a series of meandering passages of air moving around the globe at 8–12km above the ground surface. Then add into these patterns rising air from depressions at the surface and colder air subsiding into anticyclones. These are shown in Figure 4.28. Rossby waves are therefore rather like a series of Mexican waves across the upper atmosphere. The *effects* of Rossby waves are perhaps more important to geographers than an understanding of their formation. When fully formed, Rossby waves create irregularities in air movements between the Equator and the Poles. As Figure 4.27 shows, fully developed waves allow Equatorial air to penetrate northwards, taking warm air further north than usual. Meanwhile, Polar air forces its way further south to other areas with reverse effects. One of the reasons why some winters may be much warmer than others in the UK is the northern penetration of warmer air.

The study of Rossby waves and jet streams is very complex. You may find it useful to read further in order to gain insight into how each system works. The 'Further reading' suggestions at the end of this chapter will help you.

Patterns of precipitation along fronts

Depressions may be deep, and Polar and Equatorial airstreams may mix very actively as a result. This causes air to rise, often vigorously, and creates what are known as anafronts. Where upper-air movement is less strong, and the temperature difference between airstreams is smaller, less active and slower-moving katafronts develop.

Clouds seen at the Earth's surface give signs of a developing depression – see Figures 4.29 and 4.30. Typically, twelve hours before the surface front arrives, high upper-air winds are shown by high wispy cirrus (Ci) cloud, followed by sheets of cirrostratus (Cs) and altostratus (As). Condensation of water vapour at these high altitudes produces ice-crystals, as a plane journey at around 8000 metres can reveal. Cloud thickens with the approach of the warm front, and the intensive uplift causes air to rise, cool, and condense. Nimbostratus (Ns) is typical, with the development of very high cloud columns and heavy rain.

In a katafront, descending warm air greatly restricts the development of high cloud, and flatter stratocumulus cloud is common, with steady light rain and drizzle. As the warm front passes, winds change direction from the south and south-west to north-westerly, pressure stabilizes or rises, clouds break up, and there are brighter intervals. As the cold front approaches, rapid undercutting of the warm air by the cold leads to the development of thick, steep, tall cumulonimbus clouds (Cn), which bring brief, intense bursts of rainfall. In summer, with additional rising warm air, there may also be thunderstorms.

Depressions and the future – a footnote

Chapter 6 looks at the possible causes and effects of global warming. Any future rise in average sea-surface temperatures may affect the intensity of depressions and the paths that they take. Current research using computer simulations is at an early stage, but three ideas are possible:

1 Storminess from depressions may increase over Britain and the western edge of Europe.

2 The centre of storm tracks may move northwards.

3 There may be relatively fewer but more intensive periods of storms.

(C. A. Senior, 1993)

The outcome will depend on whether the latent heat released when condensation takes place can compensate for the reduction in temperature differences at the Polar front, since global warming may be greater nearer the Poles.

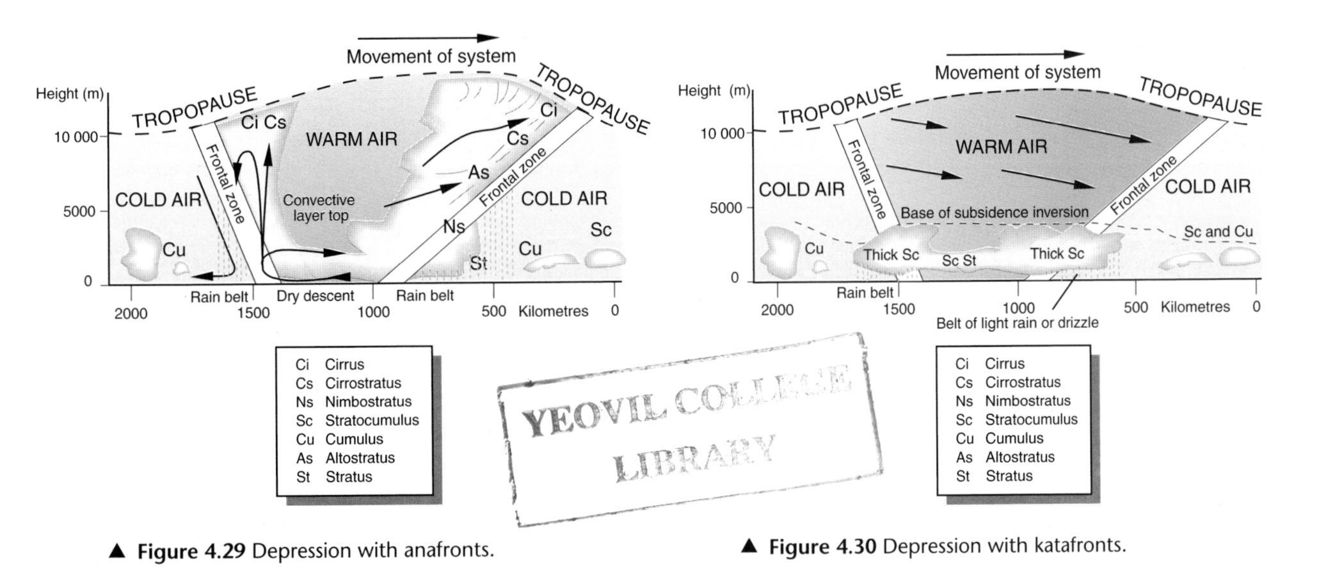

▲ **Figure 4.29** Depression with anafronts.

▲ **Figure 4.30** Depression with katafronts.

Study Figure 4.31 which gives details of the National Severe Weather Warning Service.

a) Describe briefly how this service works, and show the passage of information as a flow diagram in stages.
b) Refer to the 1990 storm as an example. Into which category would you place the storm?
c) Refer back to the synoptic charts in Figures 4.10 and 4.12–14. At what time would you have issued warnings from the NSWWS?

For better or worse? The costs and benefits of weather forecasting

Many people in the UK remember the outcry in 1987 about the failure of the Meteorological Office to predict the devastating storm in October that year. It led to the development of the National Severe Weather Warning System (NSWWS) which was first used in 1990. According to its public satisfaction questionnaire, some 93 per cent of those surveyed were quoted as being satisfied or very satisfied with the overall service.

There is now a developing forecasting service, which provides specialist weather information to the public, as well as to farmers, shipping companies, and the tourist industry, among others. Data are compiled from satellite observations, upper-air observations, surface weather stations, and weather radar. These are fed into computers at the Central Forecasting Office (CFO) at Bracknell, Berkshire. Twice a day, the computers perform calculations to produce high-speed forecasts for up to six days ahead. Other computers produce forecasts for local areas for up to eighteen hours ahead, which are then passed to local newspapers, radio, and TV companies. Future weather forecasts may provide predictions of up to 30 days, though the margin of error in forecasting is greater.

The ability of planners, organizations, and individuals to modify activities as a result of accurate forecasting can bring huge reductions in costs and social disruption. The following activity is designed to help you assess the costs and benefits of weather forecasting in Britain.

▼ **Figure 4.31** Organization of the National Severe Weather Warning Service.

There are two tiers of warnings within the NSWWS.

Tier 1 – warnings of severe or exceptionally severe weather:
1a: Early Warnings of major severe weather events likely to result in widespread disruption and/or to present a danger to life.
1b: Flash Messages of severe weather likely to result in considerable inconvenience to a large number of people and/or present a danger to life.
Tier 2 – warnings of hazardous conditions which might present the emergency authorities with potential operational problems.

Recipients of Tier 1a Early Warnings include the county emergency services, local authorities, some government departments (e.g. Home Office, Cabinet Office, DOE, MOD, MAFF, DTp), and other large organizations, such as BT, which may need to take action to prevent or deal with emergencies arising out of severe weather. They also receive Tier 1b Flash Messages which are sent to radio and television stations for broadcast to the public. Tier 2 warnings are mostly sent to police and fire services and to the BBC Travel Centre for inclusion in their motoring bulletins.

The warnings are issued by a cascade system to ensure effective and efficient distribution. They are sent first to focal points at the national and county levels who then cascade them down to other recipients. The focal points at the county level are mainly the Emergency Planning Units, but the county Fire Service may carry this responsibility instead, and so too, exceptionally, may the county Police. The Meteorological Office's responsibility is to these primary focal points, who subsequently distribute the warnings to other interested parties.

Early Warnings are issued by the Central Forecasting Office (CFO), Bracknell, when the forecasters have reasonably high confidence that severe conditions will occur. This may be for lead times of a few hours to several days. Once issued the warnings are updated each subsequent day until the event occurs or the warning is cancelled. Reference to the issue of an Early Warning will be made in the Synoptic Review issued as guidance to the regional Weather Centres, although on occasions the Chief Forecaster may decide that a Special Synoptic Review is necessary. A Press Release may be issued in conjunction with an Early Warning if the Chief Forecaster considers it appropriate.

Flash Messages are issued nearer the onset of the conditions, normally within six hours, by the regional Weather Centres, although CFO maintains a watching brief over the operation of the service, discussing the situation with Weather Centres whenever necessary. CFO also provides guidance on whether Flash criteria are likely to be exceeded in the short-period forecasts issued every six hours. The warnings will often be based upon actual reports of severe weather, providing greater detail than Early Warnings on location, duration, and severity. When the severe weather is widespread, CFO may issue a composite Flash Message for national dissemination to avoid proliferation of warnings. Tier 2 warnings are also normally issued within six hours by the regional Weather Centres.

Flood warnings that were not heard

No sea wall, however high and however well built, is inviolable. So for much of the British coastline, a system of flood warnings is essential – just in case.

But when the storm warnings went out one winter night in North Wales three years ago, everything went wrong, and hundreds of people were left in their homes as the water rose around them.

This is the sequence of events, as recorded in a report by the House of Commons Welsh Affairs Committee, that left the citizens of Towyn unprotected and unevacuated when the tide burst their sea wall on 26 February 1990.

25 February

1600 hours: Met Office phones National Rivers Authority, warning of heavy onshore winds and expected tidal surge at noon next day.

26 February

0340 hours: Met Office faxes detailed tidal predictions to NRA.

0600: NRA opens flood-warning room and reads fax.

0620: NRA telephones flood 'standby warning' to North Wales police HQ.

0706: NRA telexes tide predictions to North Wales police HQ.

0728: North Wales police HQ passes on telex to its divisional offices at Prestatyn, which in turn notifies Deeside police station, and at Llandudno, where a civilian operator 'did not appreciate its immediate significance'.

0750: Deeside police contact Delyn Borough Council, where the only person available was the council caretaker. The 'nominated officer' was contacted at his home at 0800, where he was given the flood warning.

0817: Llandudno divisional police contact Colwyn Bay police station.

0853: After several attempts, Colwyn Bay police pass on the message to Colwyn Borough Council. As the MPs' report remarks: 'Colwyn Borough Council was therefore informed five hours after the Meteorological Office had informed the NRA, that one of the highest water levels on record would be expected' – in just over three hours.

1000: First flooding recorded within Colwyn borough at Towyn, two hours before high tide.

Colwyn had decided, from experience of the sea wall being overtopped, that no evacuation of homes was necessary. But along with the overtopping, a hole appeared in the sea wall – 'a situation that could not have been foreseen', the council said later. The MPs said: 'From the above, it is not clear what, if any, predictions would have led Colwyn to evacuate people'.

At Delyn council, the confusion was even worse. Because a fax machine had broken down, the council never received details of the flood surge forecast. The council said it received a 'standby' warning at 0800, but no 'alert'. It was, a council officer told the MPs, 'when the alert comes that we mobilise people'. The NRA told the MPs: 'There is no such thing as an "alert".'

▲ **Figure 4.32** From the *New Scientist*, 2 January 1993.

1 Read through Figure 4.32. What went wrong in the operation of the National Severe Weather Warning Service on this occasion? What implications does the event have for planners in areas that are at risk of invasion by the sea?

2 Read through Figures 4.33–4.35. Summarize the costs and benefits of weather forecasts for the various 'consumers'. Does your analysis suggest any conclusions about the value of weather forecasting?

The retail trade saves up to £27 million a year courtesy of the weather forecast, a new report has claimed. The report highlighted anticipating increased demand for alcoholic and soft drinks during a heatwave as one of the main benefits of the Met Office's services.

One chain estimated that a three-day hot spell could force up drinks sales by 350,000 cases.

The Met Office said this pattern repeated nationally, without accurate forecasts to allow retailers to prepare, could cost retailers some £11.4 million. The managing director of the Weather Initiative said, 'Manufacturing, retailing and distribution companies spend more than £1 million on weather intelligence each year, to provide them with a competitive edge. For the first time, we have been able to quantify the savings more precisely.'

▲ **Figure 4.33** From the *Off-Licence News*, 29 September 1994.

Weather forecasts save the country millions of pounds by alerting farmers, oil rig operators, or building companies to conditions, according to a survey commissioned by the Met Office.

The Bramshill Consultancy in Hampshire claims forecasts are worth £1 billion a year to the British economy but the public weather forecasts cost £38 million. Examples in the survey include giving farmers information on when best to go in for hay making, silage production and harvesting. Forecasts of rainfall and humidity are helpful in predicting disease in crops and animals. Wind speeds need to be calculated when spraying to avoid drift to other areas. Advance warning of high wind chill means farmers can guard against lambing losses by giving them shelter. The total annual saving to the farming industry is calculated at £122 million.

◀ **Figure 4.34** Article by Paul Brown from *The Guardian*, 23 September 1994.

▼ **Figure 4.35** From the London *Evening Standard*, 12 August 1994.

A holiday theme park claims it lost £30,000 because a large black cloud was wrongly slapped over it on a national TV weather map. Park managers say the attendance was down by 3,000 on the day of the faulty forecast.

GMTV forecaster Sally Meen predicted rain and even gave a flood warning for the East Coast. And a large black cloud hovered on the map over Lowestoft, Suffolk, where the Pleasurewood Hills theme park is based. But the 4,000 visitors who did turn up to Pleasurewood Hills on Wednesday enjoyed a mild, warm day and basked in afternoon sunshine.

Ideas for further study

1 Conduct your own piece of research on weather forecasting. How accurate are forecasts? What social and economic benefits can improved forecasting bring? Use some of the sources below and keep a record of news articles on forecasting.

2 Identify differences between seasonal weather across different parts of the UK, or for different parts of another country for which you are able to obtain data. For example, how does summer weather vary between the south-west of England, one of the UK's most popular holiday destinations, and south-east England? Is it possible to determine the causes of any differences?

Summary

- British weather is highly changeable from day to day, from week to week, and between a single season in one year and the same season the following year.
- This changeability is closely linked to the influence of competing airmasses and to alternating high- and low-pressure systems.
- The impact of short-term events such as the 1990 storm on people's lives emphasizes the need for accurate national systems of weather forecasting and severe-weather warnings.
- The patterns of weather we experience on the Earth's surface are in many cases dictated by upper-level atmospheric conditions.
- Computer technology is helping to improve knowledge and understanding of the atmosphere. The quality of forecasting is commercially significant to a large number of companies.
- Weather forecasting still includes margins of error. Changing climate from global warming could complicate the process further.
- Even accurate forecasting cannot prevent ineffective human responses to weather hazards.

References and further reading

R. G. Barry and R. J. Chorley, *Atmosphere, Weather and Climate* (sixth edition), Routledge, 1992.

G. O'Hare and J. Sweeney, 'Lamb's circulation types and British weather: an evaluation', *Geography*, January 1993, pp.43–60.

C. A. Senior, 'The impact of increased carbon dioxide on the frequency of storms', unpublished note 1993, Hadley Centre for Climate Research, UK Meteorological Office, Bracknell.

The Meteorological Magazine, UK Meteorological Office, Bracknell.

Weather, the monthly journal of the Royal Meteorological Society, is available on subscription from 104 Oxford Road, Reading, Berks RG1 7LJ.

5 Weather and climate in tropical areas

Is there a 'tropical climate'?

Tropical areas are easy to define on a world map. They lie between the Tropic of Cancer in the northern hemisphere and the Tropic of Capricorn in the southern hemisphere, between 23° north and 23° south of the Equator. The climates of tropical areas are less easy to define, though there are very similar weather conditions between latitudes 0° (the Equator) and 30° north and 30° south. Five tropical climate zones are shown in Figure 5.1, out of a total of thirteen zones worldwide. The tropical zones are distinguished by three main criteria, as follows.

1 *Rainfall totals and distribution* For instance, rainforest regions, savanna, and hot desert are distinguished from each other by the amount of rain they each receive, and the seasons during which most rain falls.

2 *Temperature patterns through the year* At the Equator, temperatures remain almost constant all year. Away from the Equator, there is more evidence of warmer and cooler seasons.

3 *Maximum temperatures* With at least some cloud cover at almost all times of the year, equatorial regions rarely reach very high daytime maximum temperatures, above 35 °C. For the same reason, night-time temperatures rarely fall far below 25 °C. In hot deserts, the reverse is true. Daytime temperatures may climb to 45–50 °C, while night-time frosts are not uncommon.

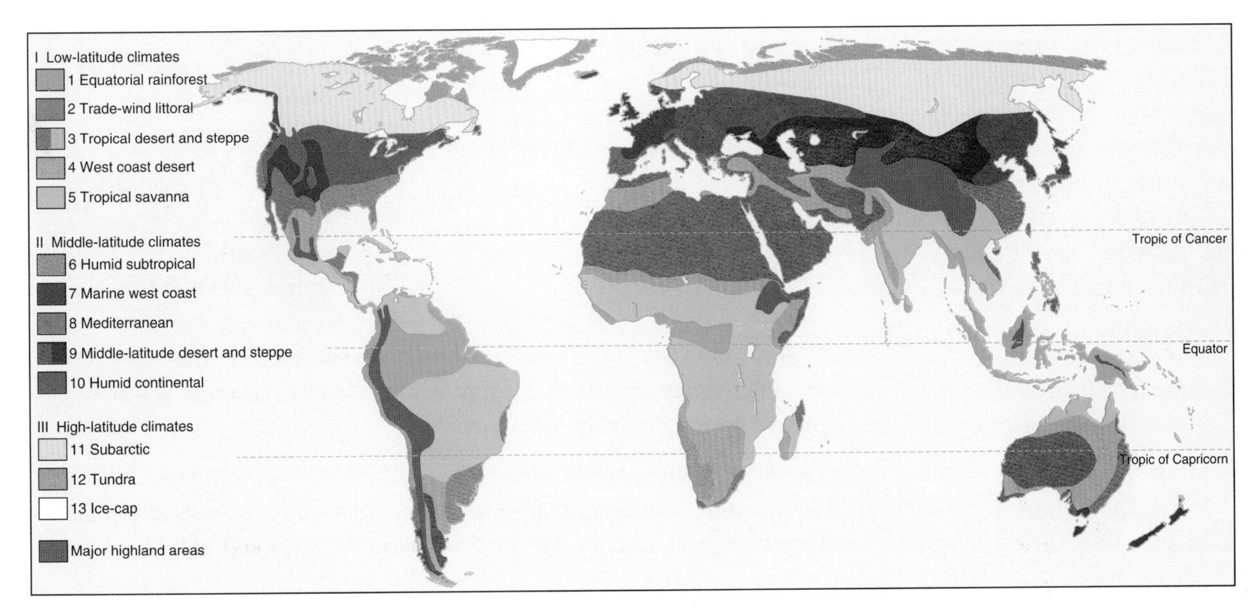

▲ **Figure 5.1** The world's major climate zones. This map shows the distribution of thirteen major climate zones in the world. This chapter focuses upon zones 1–5 (low-latitude climates).

Three-quarters of the world's population live within this broad climatic belt. The weather and climate of such areas are of vital interest to the geographer. Seasonal patterns of weather and climate, and the prospect of longer-term changes in climate, are of great importance to the populations of Asia, Africa, and South America. In many parts of the Tropics, most of the people work on the land, and many subsistence and commercial economies are based on farming and forestry. Climate is therefore a vital part of people's lives.

At first glance it is easy to think of tropical areas as regions where the climate is, by European standards, almost predictable. Three features seem to stand out.

1 Compared with temperate zones, temperatures are much higher throughout the year, and to someone from Europe it might seem as though even the cooler periods in tropical areas are better than many north European summers!

2 The length of daylight is less variable, 11–13 hours all year round, because of the high altitude of the Sun throughout the year. As a result, seasonal differences are less marked than those in temperate zones.

3 The distribution of rainfall through the year is linked to the movement of the overhead Sun between the Tropics. In places north of the Equator, there is a single maximum rainfall peak at the time of the summer solstice at 23° north on 21 June; south of the Equator, the peak occurs in a similar way on 21 December. The passage of the Sun twice over the Equator produces a double peak around the times of the equinox on 21 March and 21 September.

Now that data from surface weather stations and satellite images have become widely available, we know much more about tropical climates. We also know that winds transfer heat energy towards the Poles from the Tropics. This is explained in the theory box on pages 107–108. There seems to be a link between energy transfer and air movement which results in climatic variation between one year and another, and between periods of several years. The study of the Sahel later in this chapter explores this further.

This chapter will help you to identify patterns of tropical weather and climate, and explain the processes which produce them. It focuses on the following three areas.

1 Climatic variations in Africa.

2 Drought in the Sahel – this section looks at the effects of human and physical influences on the semi-arid zone known as the Sahel.

3 Study of an Atlantic hurricane. Tropical disturbances known as hurricanes occur frequently in the western Atlantic. The study focuses on Hurricane Andrew in August 1992.

1 Study Figure 5.3. Figure 5.2 helps to identify similarities and differences between climates of different places in Africa. Copy it and complete it using Figure 5.3.

2 How do these observations match up with the climate zones presented in Figure 5.1?

3 What factors might explain the difference in:
a) temperature at Libreville and Nairobi
b) rainfall totals and distribution at Freetown and Lagos?

4 What general rules seem to apply:
a) about rainfall totals, as you move from the coast inland
b) about temperature variations in different parts of Africa
c) about seasonality within the different climate zones in Figure 5.1?

5 How accurate would you think it is to use the terms 'winter' and 'summer' to describe the African climate? Are there better ways of identifying the seasons?

Climatic variations in Africa

Africa is a huge continent, and spans the whole of the Tropics. Its northern coast fringes the Mediterranean at about 37° north of the Equator, while its southern coast at Cape Agulhas is 35° south. For 7 hours of the 11 hours' flight to Cape Town from London, the plane is flying over the African continent. From Dakar on the west coast to the eastern tip of Somalia is almost as far. Three-quarters of the landmass lies within the Tropics. Yet Figure 5.1 shows that there are only five climate categories within this huge area (by comparison, Scandinavia has four). Does this mean that the climate of Africa is uniform? If not, how do climates vary in different parts of Africa?

	Rainfall total	No. of peaks in rainfall distribution	No. of months over 50mm of rain	No. of months 10mm of rain or less	Wettest 2–3 months	Temperature range through year	Average diurnal (daily) range of temperature	Average air pressure
Freetown								
Lagos								
Libreville								
Zungeru								
Khartoum								
Timbouctou								
Nairobi								
Harare								

▼ **Figure 5.3** Climate graphs for selected stations in Africa.

▲ **Figure 5.2** Climate matrix.

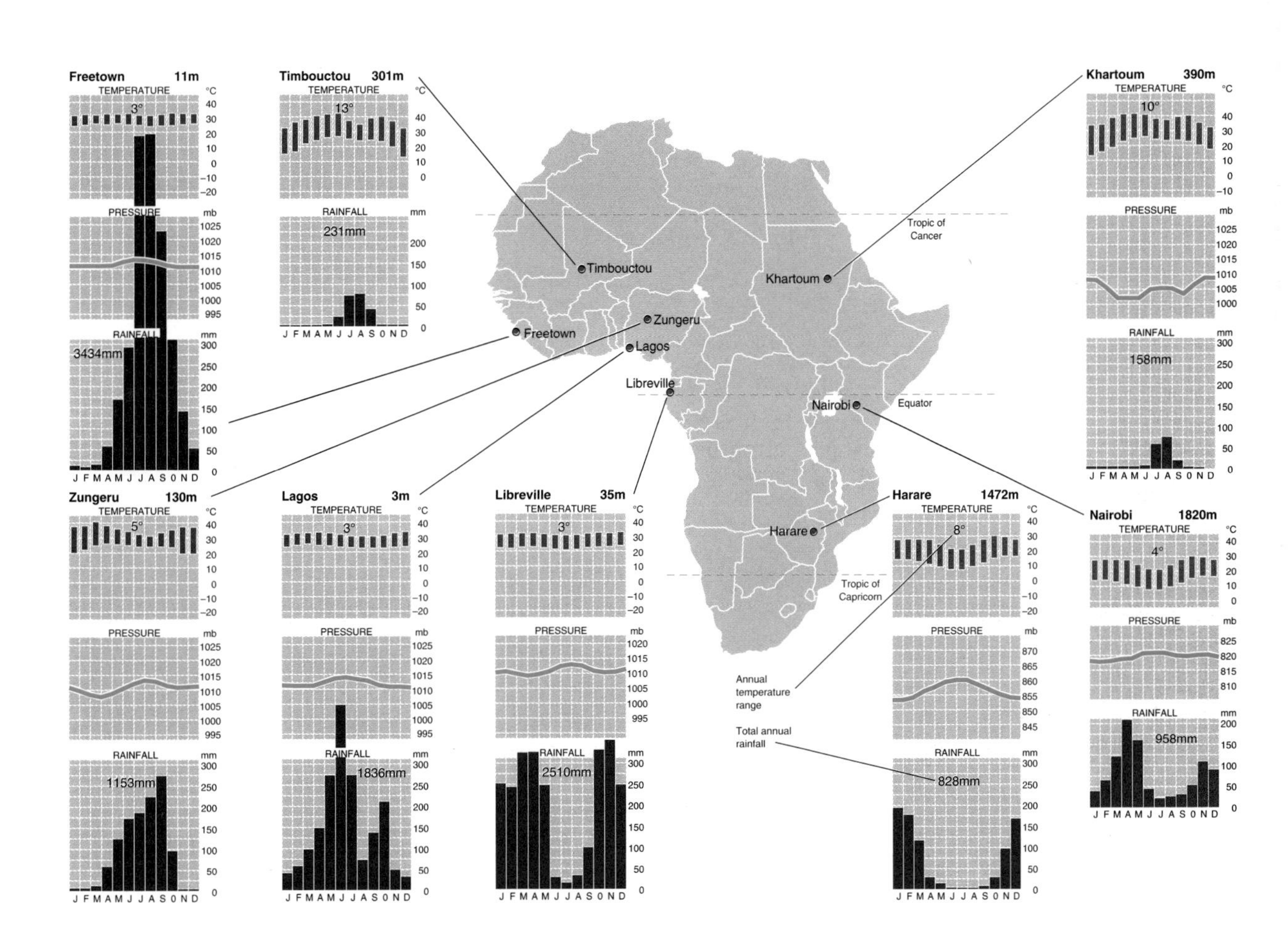

1 Draw two sketch maps of West Africa, one for December–February and the other for June–August. On each map show the wind directions and pressure systems. Label these. Give each of your maps a title.

2 Add labels to your maps to show the rainy seasons for Lagos, Zungeru, and Timbouctou. When do they occur, and how much rain is there at each place?

3 Explain the differences in rainfall total and duration of the rainy season at Lagos, Zungeru, and Timbouctou.

One common feature which you have probably identified in Figure 5.3 is that each of these places has at least one rainy season during the year. This is in marked contrast to the drier conditions that exist across much of Africa during the long dry season. Figure 5.4 shows mean rainfall totals across Africa. Other places have two seasons, according to the passage of the Sun. Such rainy seasons are referred to as 'monsoons'. Several areas of the world have well-defined monsoon seasons, including the Indian subcontinent, and African monsoons are equally marked. What causes these monsoons?

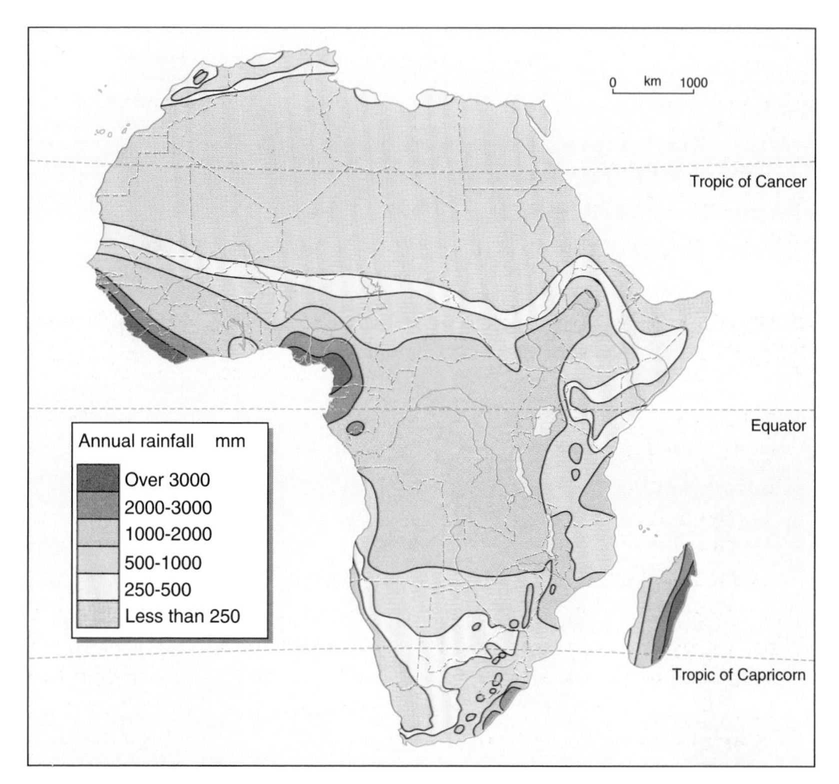

▲ **Figure 5.4** This map shows the 40-year average rainfall totals for Africa. It has been drawn using isohyets (see Chapter 4, page 86) which are lines joining those places that have equal precipitation. Remember that this is a map of annual *totals*, and does not show seasonal rainfall. This map should be compared with Figure 5.3 and with Figure 5.5.

▼ **Figure 5.5** Major pressure belts and wind circulation in Africa.

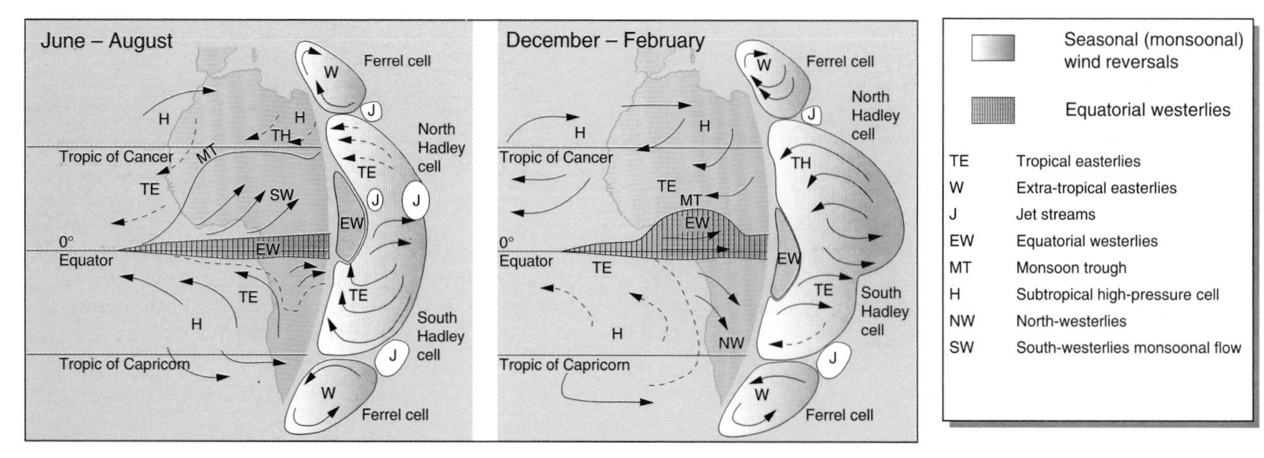

In Chapter 4, you learned about global pressure systems, and the ways in which these depend upon the seasons. Variations in pressure in Africa are linked to the passage of the Sun. In the northern hemisphere, the continental interior is relatively cooler than the surrounding tropical seas during the December–February period. This leads to the formation of areas of higher pressure over land and lower pressure over the seas. At the same period, the continent south of the Equator is experiencing its summer, with low pressure over the heated landmass and high pressure over the relatively cooler seas.

In West Africa, therefore, during the December–February period, high pressure lies to the north, over the Sahara, and there is low pressure over the Gulf of Guinea. Winds blow towards the Gulf of Guinea from the Sahara, so there are dry winds over the whole of West Africa at this time. The graphs of Timbouctou and Zungeru in Figure 5.3 show a similar pattern of drought at this time of the year.

In the June–August period, the pattern is reversed. Heating of the landmass north of the Equator produces low pressure, drawing winds inland from the Gulf of Guinea. These winds carry a considerable amount of moisture inland. Again, the graphs of Timbouctou and Zungeru in Figure 5.3 show that this is the period of heavy seasonal monsoon rains. This pattern is part of a much broader system of atmospheric circulation. It is complex and involves circulation of air in the upper atmosphere. This is explained in the theory box below.

The general circulation of the atmosphere

Taking the Earth as a whole, there is a surplus of heat energy at the Equator and a deficit at the Poles and in the upper atmosphere. The circulation of air in the atmosphere achieves more of a balance by transferring heat from areas of surplus to areas of deficit. If the Earth did not rotate, and consisted only of either land or water, air would rise at the Equator into the upper atmosphere, and create low pressure which would draw in colder, higher-pressure air from the Poles. This movement is called a **convection current**. Air rising at the Equator would eventually

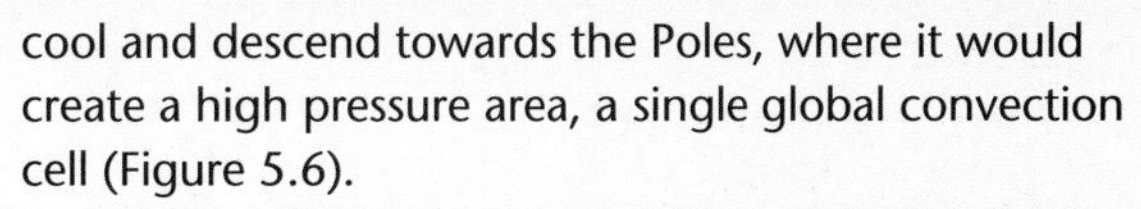

cool and descend towards the Poles, where it would create a high pressure area, a single global convection cell (Figure 5.6).

However, the Earth rotates, and there is an irregular coverage of land and water on its surface. This leads to different airflows and to different rates of heating and cooling between land and sea. The force exerted by the Earth as it rotates from west to east is called the **Coriolis force** (see page 96). It deflects winds to the right in the northern hemisphere and to the left in the southern hemisphere (see Figure 5.7).

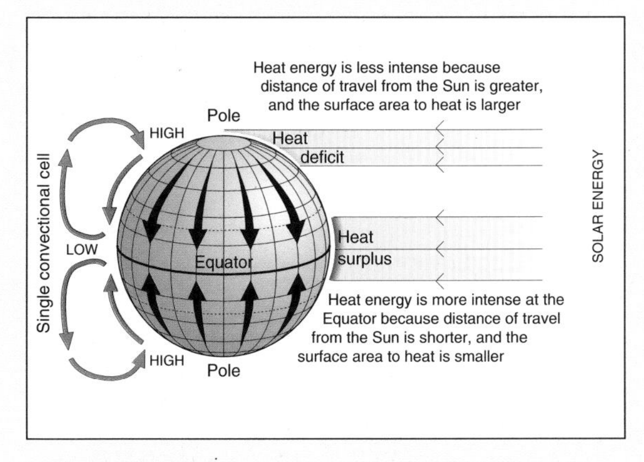

▲ **Figure 5.6** Air movement on a theoretical Earth which has no rotation, and a surface that is all land or all sea.

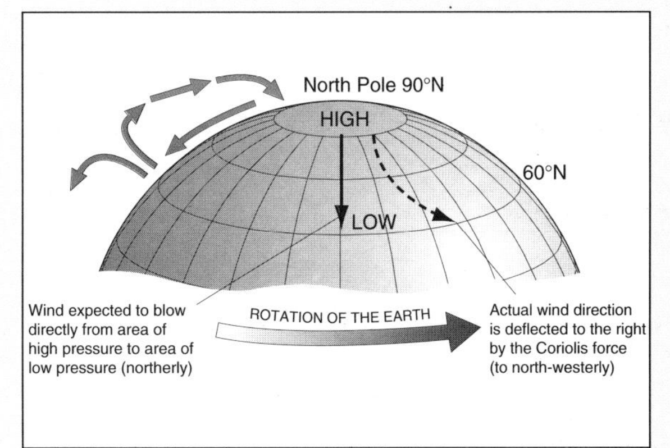

▲ **Figure 5.7** The effect of the Coriolis force in the northern hemisphere. Note how the rotation of the Earth causes any wind to blow towards the direction of rotation.

Continued on page 108

Continued from page 107

In the upper troposphere, where air movement is unaffected by friction with the Earth, there is a balance between the Coriolis force and the pressure gradient. This produces a **geostrophic wind**, which flows around the Earth within the troposphere at high altitudes, and which runs parallel to the isobars. Nearer the surface where friction is greater, the Coriolis force is reduced and winds move at a gentle angle across the isobars into a low-pressure area (Figure 5.8).

▼ **Figure 5.8** The formation of a geostrophic wind and the effect of friction. Geostrophic winds occur at high levels in the troposphere. In this diagram, there is an equal balance between the movement of wind caused by pressure and the opposite force created by the Coriolis force as the Earth rotates. Without any relief features to cause friction or disturb the flow of air, winds therefore blow at right-angles to the isobars.

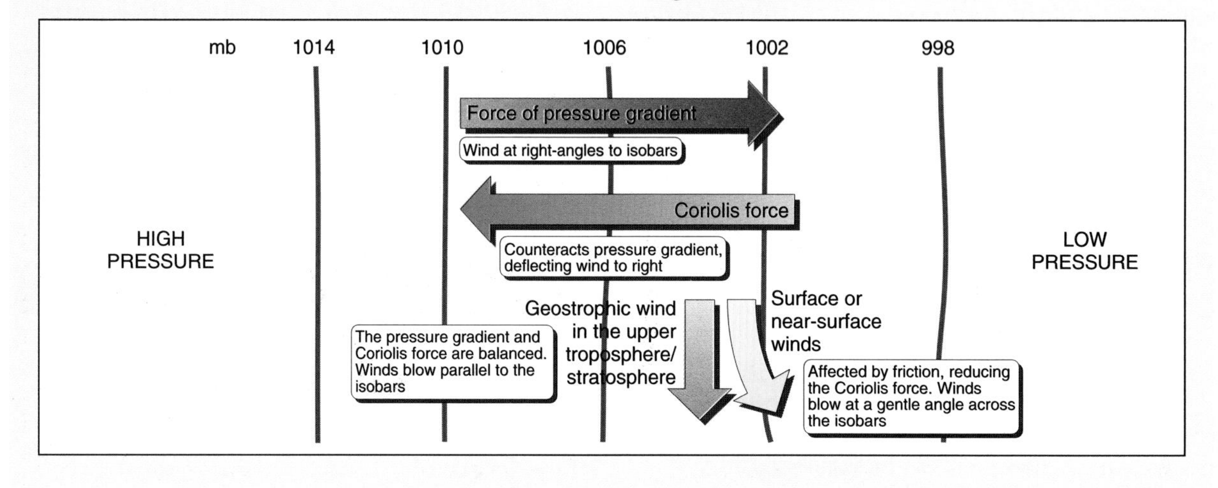

The three cells of global circulation

Two climatologists, Ferrel in 1846 and Rossby in 1941, constructed a model of global atmospheric circulation based on three convectional cells. The driving force for the whole model is the central or Hadley cell which develops from intense heating at the Equator. Winds which blow from the Equator are called trade winds. Trade winds blow into the low-pressure system called the inter-tropical convergence zone (ITCZ), which moves northwards and southwards with the overhead Sun during the year. The ITCZ is an area between the Tropics where two airstreams meet, or converge. Along the ITCZ, the Monsoon Front brings heavy rains.

▼ **Figure 5.9** The Hadley cell and the ITCZ over the African continent.

(a) June–August. Trade winds bring moisture into the West African landmass, with the ITCZ front positioned over the interior producing the rainy season.

(b) December–February. The ITCZ has migrated south with the passage of the Sun. As a result, offshore trade winds are drawn from the continent into the ITCZ, creating a dry season.

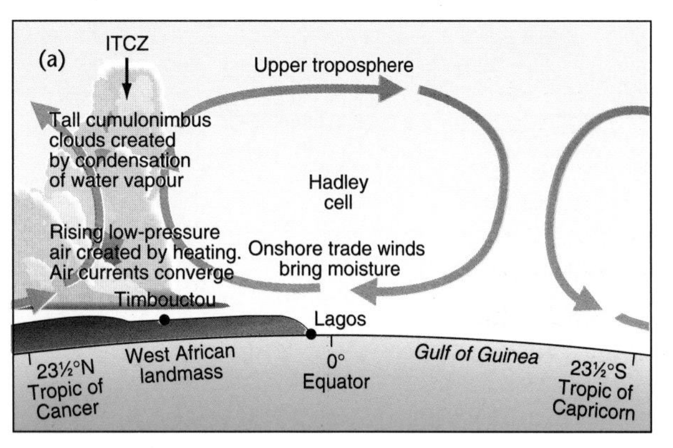

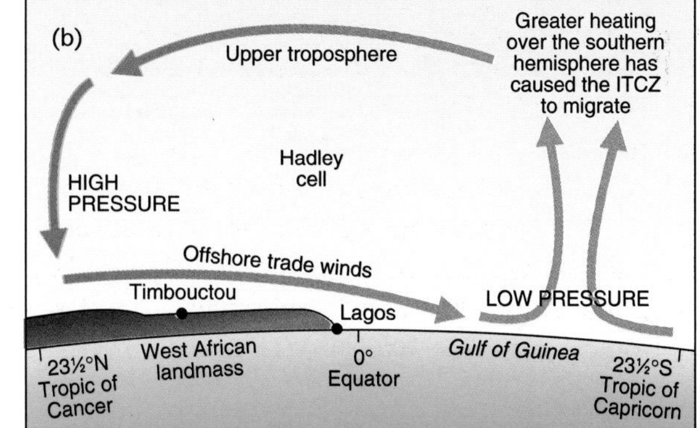

Trade winds pick up water vapour through the process of evaporation as they cross warm, tropical seas. Inland, they are forced to rise by convectional currents which are intensified when heat is released through condensation of water vapour into droplets. This process is often associated with towering cumulonimbus clouds which give rise to frequent afternoon thunderstorms in the equatorial zone. At ground level, winds associated with the ITCZ are light and variable. In the upper atmosphere, however, further cooling causes an increase in air density, and it begins to subside approximately 30° north or south from the ITCZ. This dense air forms the subtropical high-pressure belt. Here skies are clear, winds light, and conditions are generally dry and stable.

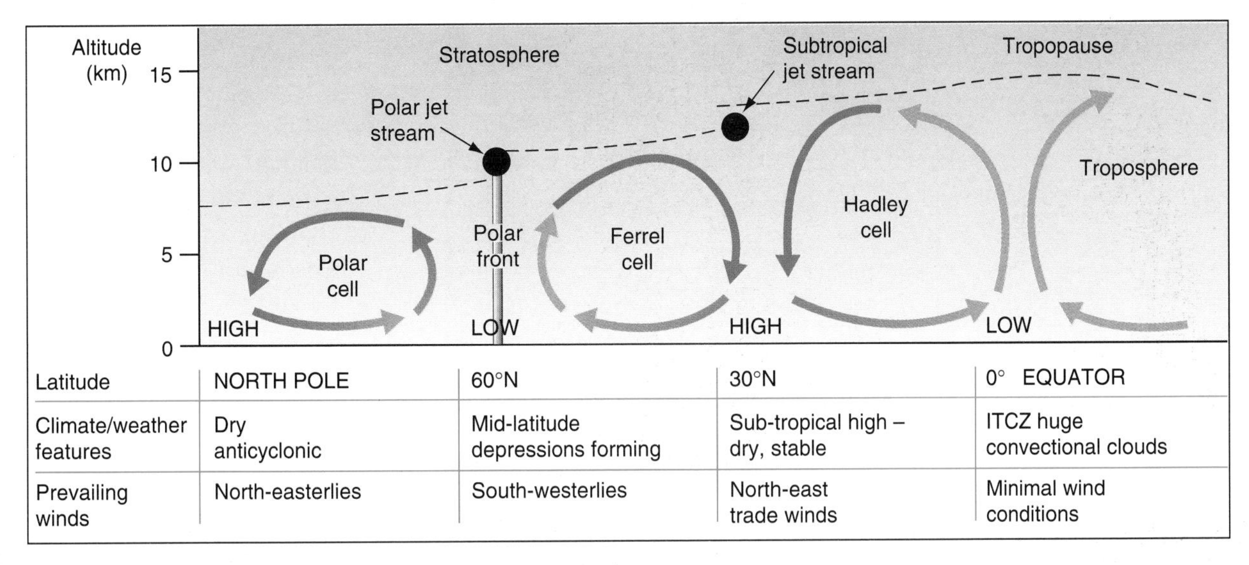

▲ **Figure 5.10** The three-cell model of global atmospheric circulation.

Figure 5.10 shows how the Hadley cell is just one of three cells that are formed in similar ways across the northern hemisphere. In terms of its size, the Hadley cell is the largest because of the immense amount of heating at the Tropics. This is also reflected in the thickness of the troposphere above the Tropics. In the northern latitudes, where the air is colder and more dense, the troposphere is less thick and the Polar cell is smaller.

In Figure 5.10, some air returns near the surface into the ITCZ, but some migrates as warm south-westerly winds towards the Pole. These warm south-westerly winds form the basis of the second major cell, the Ferrel cell. The winds collect moisture and meet southward-moving Polar air between 50° and 60° north. At this point, the release of latent heat produced by condensation forces them to rise, laying the foundation of a new mid-latitude depression. This sequence of depressions is described fully in Chapter 4. One of the rising limbs of this Ferrel cell cools and descends towards the subtropical high, the Hadley cell. Here it is warmed by compression, making it very stable.

A final, weaker Polar cell consists of the diverging, cooling air of the Polar front slowly descending over the Pole, giving dry, high-pressure conditions. A weak movement of cold dense Polar air moves south towards the Polar front low-pressure area. The Polar front is formed where this colder Polar air meets the northward-moving subtropical air from the Ferrel cell.

The climates of Africa – a general review

You should now be able to review the climate zones of Africa shown in Figure 5.1, and have some understanding of the formation of such zones. Write an essay entitled 'Africa's climate regions', describing and explaining the following:

- the locations of the three major African deserts – the Sahara, the Kalahari, and the Namib – their climate characteristics, and reasons for these
- the distribution of savanna regions
- the distribution of tropical rainforests.

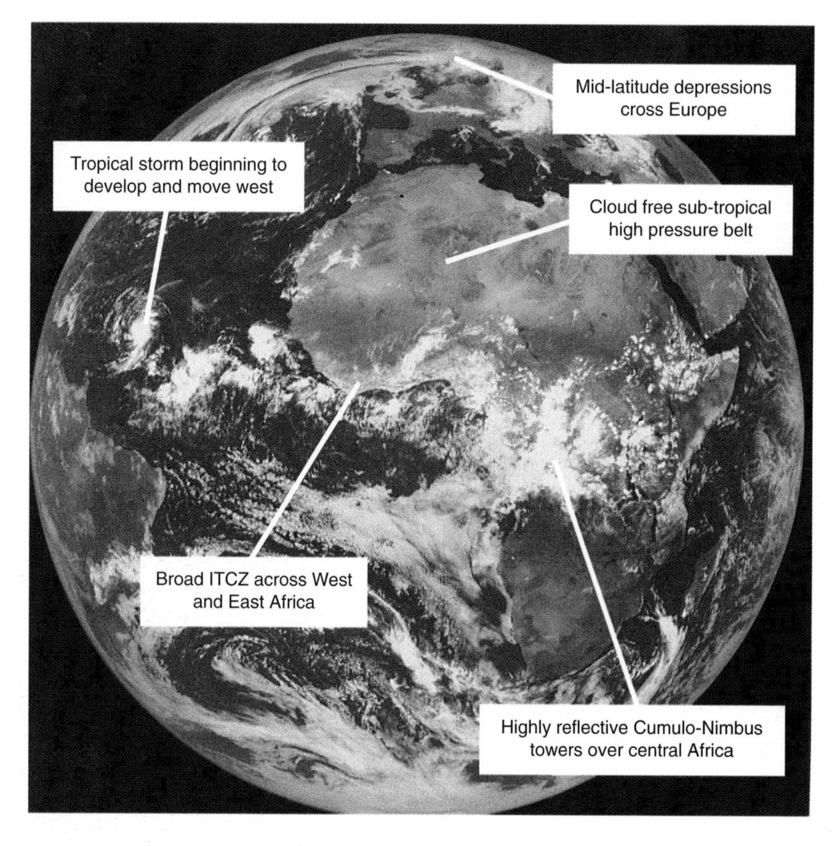

Figure 5.11 Visible image captured by Meteosat satellite on 25 September 1983 – this satellite is positioned 35 900km above the Equator.

Drought in the Sahel

Every so often, a drought shocks the world with its effects. In the mid-1980s, the droughts of Ethiopia and Sudan racked the consciences of people living in affluent countries, which in turn raised millions of dollars to fund aid programmes. These droughts were not isolated. They formed part of a general pattern which has become more and more apparent since the 1960s. Sudan is part of a broader zone of Africa known as the Sahel (Figure 5.12).

The Sahel is a region on the edge of the Sahara Desert in Africa, forming a transition between tropical hot desert to the north and dry savanna to the south. It is a belt of land some 3800km from east to west and 700km from north to south, stretching from Mauritania to central Sudan. It is vitally dependent on the monsoon for its rainfall, but because of the daily variation in the position of the wandering low-pressure system, the monsoon trough, rainfall is variable from day to day and between years (see Figure 5.13). Figures 5.14 and 5.15 show some aspects of this problem.

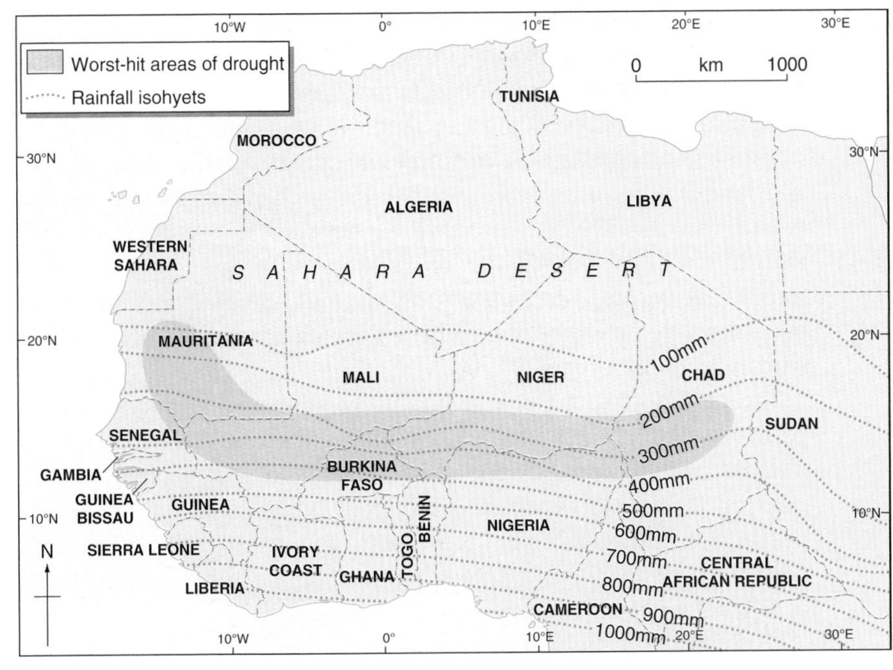

▲ **Figure 5.12** The Sahel region of West Africa: note the very tight rainfall gradient from less than 100mm north of 16° north, to over 700mm south of about 10° north.

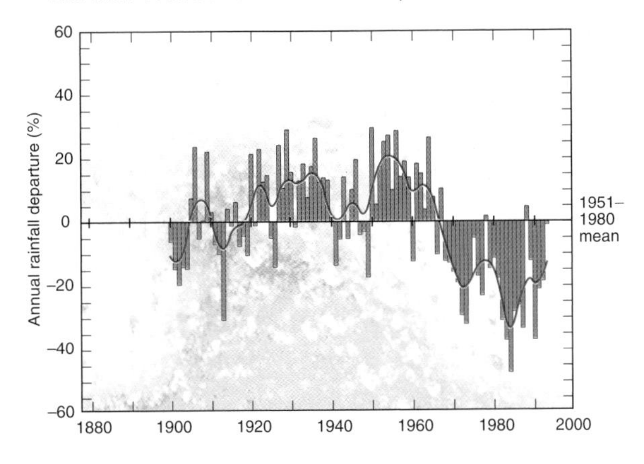

▲ **Figure 5.13** This graph shows the annual departure from the mean of the rainfall in the Sahel between 1900 and 1990. The red line shows the long-term trend based on running ten-year averages.

▲ **Figure 5.14** A convective storm developing over central Sudan. Air is forced to rise as land is heated intensely by the Sun, developing a low-pressure area. The water vapour in the rising columns of air condenses at great heights, forming anvil-shaped cumulonimbus clouds. Flash floods may result from intense storms, and soil erosion is widespread.

◀ **Figure 5.15** A small village in central Sudan. Note the dried wells in the foreground. The exodus of people to the major cities of Khartoum and Omdurman is prompted by the need to survive the 20-year drought, and the encroaching Sahara Desert.

The effect of this variation on human activity is considerable. Variations in rainfall already threaten plant survival. This is made worse by the overuse of resources such as fuelwood from rapidly disappearing trees, and grazing land which is not regenerated by summer rain. Populations continue to increase in this desert fringe and land use becomes more intense. For nomadic tribes such as the Fulani who head into the Sahel in summer to escape mosquito-ridden lowlands, the failure of rains puts further pressure on a traditional way of life. Already, their semi-nomadic way of life is threatened by neighbouring countries which are suspicious of people who cross their borders.

The period of drought and famine in the Sahel over the last 25 years is well documented. For geographers, the challenge has been to find the real reasons for the drought, and to analyse human and physical influences which have led to the chronic human misery of the region. The drought represents the most substantial and persistent change in rainfall of any region on Earth during the period of reliable weather readings.

Using samples from former lakes, descriptions of the landscape, and historical accounts, climatologists have learned of other droughts in the past which lasted for one or two decades, but the current drought may be as severe as anything in the last thousand years.

Causes of the drought

Three possible factors may be responsible for this drought:

- land degradation within Africa
- natural climatic variability
- climatic change linked to global warming.

The changing land surface – land degradation in Africa

The theory of land degradation resulted from research in the 1970s in the Negev Desert in southern Israel, and in the Sahel (see Figure 5.16). The theory is that deforestation or overgrazing leaves bare, light-coloured soils which have a high albedo and reflect more of the radiation received from the Sun than would otherwise be the case. As a result the ground warms up less intensely, convectional storms are rarer, and so rainfall is reduced. This process would cause a permanent change in the atmosphere, increasing the subsidence of the subtropical high-pressure area.

However, there are two main problems with this explanation.

1. Computer models have shown that small-scale changes in rainfall may occur in the long term but not over the whole Sahel region. Observations in the Sahel suggest that rainfall changes have been over-estimated.
2. Other computer models suggest that sea-surface temperatures may have a greater role than land cover change in causing drought in the Sahel.

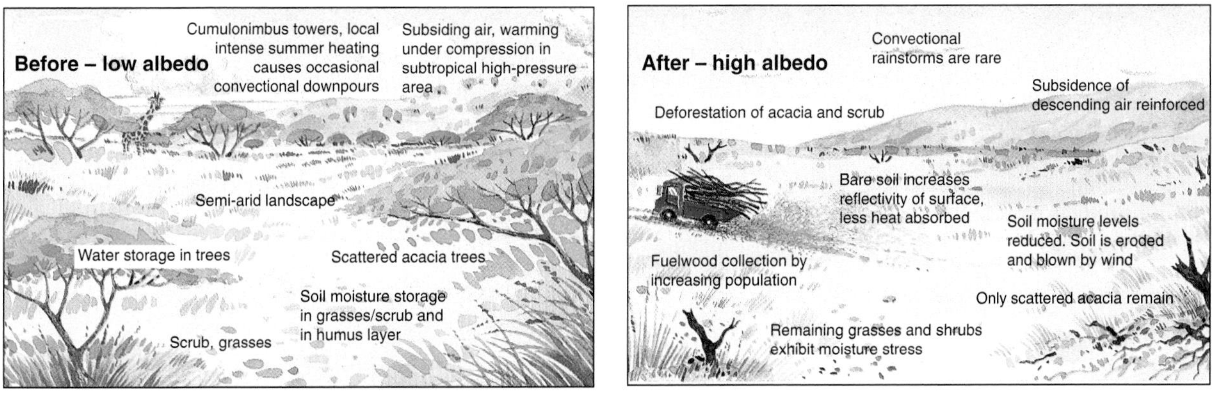

Before – low albedo

Cumulonimbus towers, local intense summer heating causes occasional convectional downpours

Subsiding air, warming under compression in subtropical high-pressure area

Semi-arid landscape

Water storage in trees

Scattered acacia trees

Soil moisture storage in grasses/scrub and in humus layer

Scrub, grasses

After – high albedo

Convectional rainstorms are rare

Subsidence of descending air reinforced

Deforestation of acacia and scrub

Bare soil increases reflectivity of surface, less heat absorbed

Soil moisture levels reduced. Soil is eroded and blown by wind

Fuelwood collection by increasing population

Only scattered acacia remain

Remaining grasses and shrubs exhibit moisture stress

▲ **Figure 5.16** The process of land degradation and its link with drought.

▲ **Figure 5.17** Evidence of deforestation of acacia trees in central Sudan. The changing surface characteristics are thought to have a permanent impact on climate.

1 Using information from CD-ROMs, atlases, and other recent data on the Sahel, create a folder of evidence or display material showing human impact on the climate. You might consider how such issues as population growth and changes in land use or land holdings might affect the climate, through increased pressure on resources. This exercise will complement your work on ecosystems in West Africa, and help you to evaluate the impact of people on the environment.

2 In pairs, assess to what extent the changes in the Sahel are
  a) the result of climate change, and
  b) a response to pressures brought about by war, and the search for food and fuelwood by the poorest members of affected countries.

Natural climatic variability – the link with the oceans and with global climate cycles

In the 1980s, geographers began to explore the strong statistical relationship that existed between Sahelian rainfall and sea-surface temperatures to the west of Africa, especially in the Atlantic. Higher temperatures south of the Equator and lower temperatures north of the Equator were associated with lower rainfall over much of tropical North Africa. It seemed that the ITCZ was moving south to the warmer southern hemisphere, leaving the subtropical high-pressure area of the Hadley cell (see pages 108–109) to dominate the Sahel for much of the year. The pattern seems to be linked to a widespread cycle across the Tropics. These are changes in air pressure known as the El Niño Southern Oscillation (ENSO), which is explained in the theory box on page 115.

The theory is that more frequent El Niño events in the last 20 years, and differences in sea-surface temperatures between the hemispheres, are linked to a global ocean conveyor-belt system (Figure 5.18). Warm tropical waters rise in the western Pacific and Indian Oceans, travelling north through the Atlantic, before returning as cold, deep water via the Antarctic or Southern Ocean. The whole circulation may take an ocean particle some 500 to 2000 years to complete.

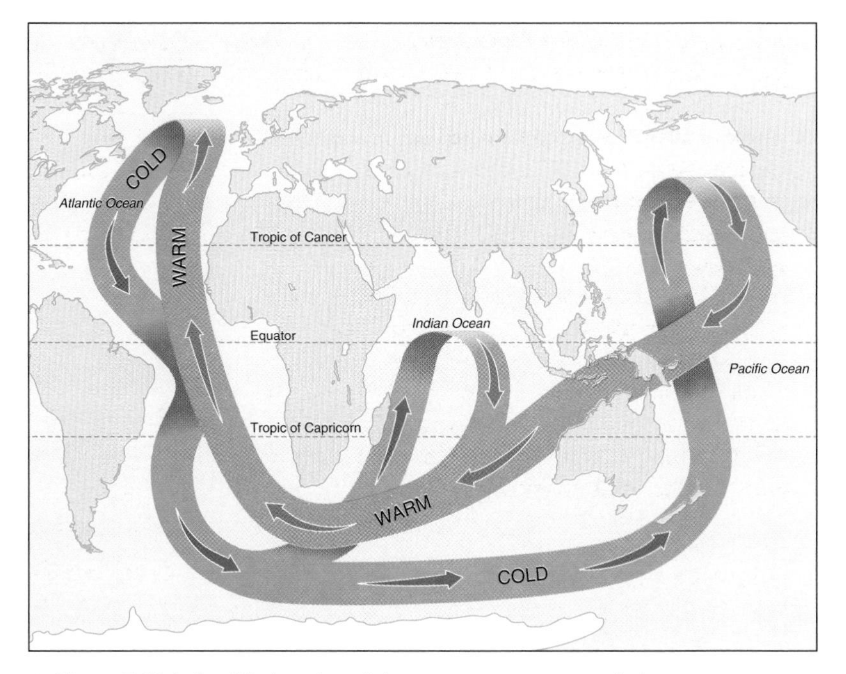

▲ **Figure 5.18** A simplified version of the great ocean conveyor belt.

(a) Non- El Niño years.

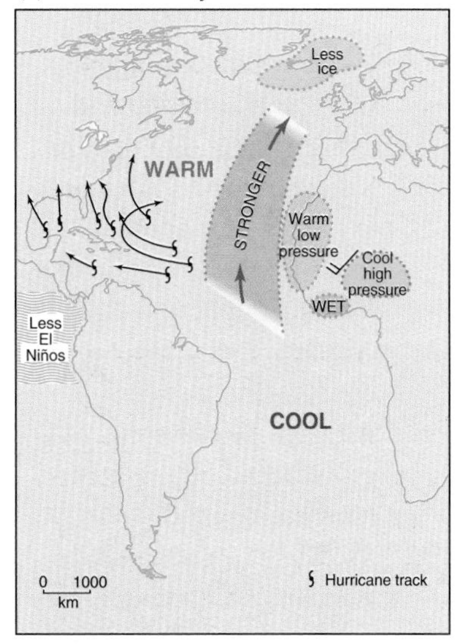

(b) El Niño years.

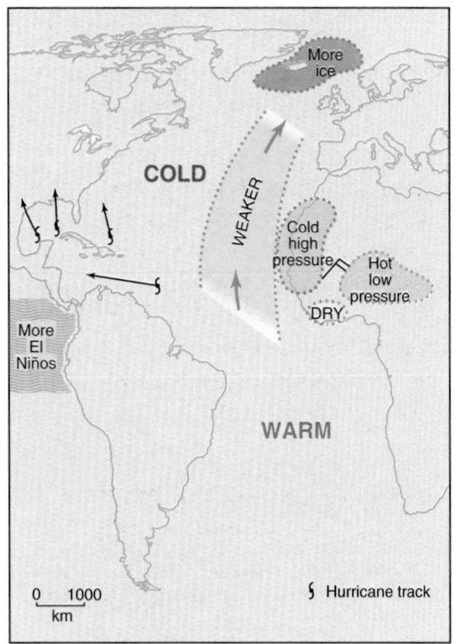

▲ **Figure 5.19** Changes in the ocean conveyor-belt system.

Recently, this circulation has slowed down. Figure 5.19 gives a possible explanation for this. Figure 5.19a shows how less ice near Greenland increases the salinity (salt content) of the extreme North Atlantic. This speeds up the conveyor-belt system because saline water is heavier, and causes these denser waters to sink. As a result, the North Atlantic becomes warmer and the South Atlantic is cooler in the Tropics. This causes a stronger African monsoon, more rainfall, and more Atlantic hurricanes. The stronger conveyor belt also tends to store less heat in the western Pacific Ocean, which leads to fewer El Niño events.

Figure 5.19b shows the situation where more ice means less salinity, a weaker conveyor belt, and less Sahel rainfall. There are also fewer Atlantic hurricanes and more El Niño events. The situation in Figure 5.19a was typical of the period 1940–69, while that in Figure 5.19b was typical of the 1970–94 period. If the theory is correct, the Sahel may see wetter conditions in future, enabling increased production from the land. Countries of the Caribbean, and the east coast of the United States, may experience more hurricanes like Hurricane Andrew (see pages 117–19).

Climatic variability and El Niño events

The term 'El Niño' is Spanish for 'boy-child'. It is used by Peruvian anchovy fishermen to describe the warm ocean current which appears in late December (around Christmas) off the South American coast every three to eight years. It replaces the usual cold Humboldt current, and prevents the upward draught of cold, nutrient-rich, deep ocean water on which plankton feed. Plankton are the food of the anchovy. During the 20th century, anchovy harvests have been ruined by four or five severe El Niño events – a considerable blow to the Peruvian fishing industry.

For climatologists, El Niño is linked to sea-temperature and air-pressure changes in the eastern Pacific Ocean and in the seas around Indonesia when warming is strong.

In normal years (known as La Niña), the circulation of air in the Pacific is based on the Walker cell which moves from east to west (Figure 5.20a). The cold Humboldt current brings cold water north from the Southern Ocean along the west coast of South America. The colder water then flows west along the Equator, where it is heated by the Sun.

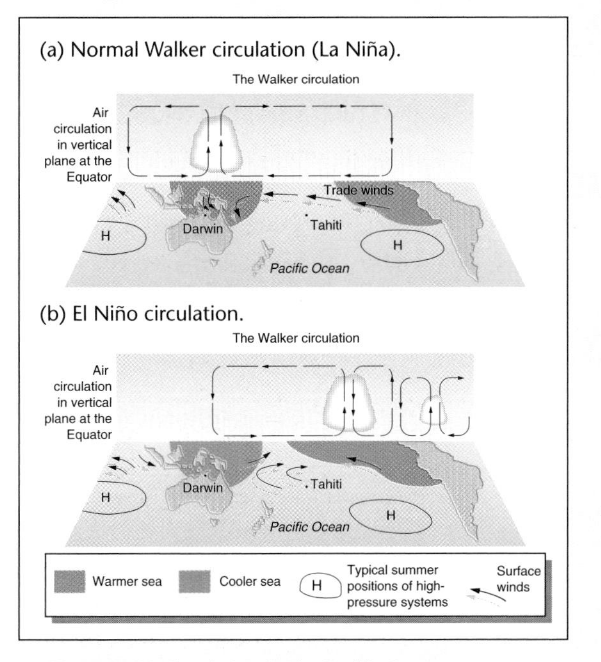

▲ **Figure 5.20** Circulation in the Pacific Ocean.

The western part of the Pacific Ocean is 3–8 °C warmer than the eastern part. A low-pressure area develops over Indonesia as warm, moist air rises to high levels, providing towering columns of cumulonimbus rain clouds. The low-pressure air travels east in the upper atmosphere, and sinks into the subsiding high-pressure area off the west coast of South America, producing dry, stable conditions in the eastern Pacific.

During an El Niño the pressure systems and weather patterns change (Figure 5.20b). Warmer waters develop in the eastern Pacific, with temperatures rising by 2–8 °C. Low pressure forms, drawing in westerly winds. Increased cloudiness results as warm, moist air rises. Around northern Australia and Indonesia, the seas cool and pressure rises, leading to lower rainfall. The change is known as the Southern Oscillation. Its strength and direction, and the speed at which change takes place, are together known as the Southern Oscillation Index (SOI), which is explored below.

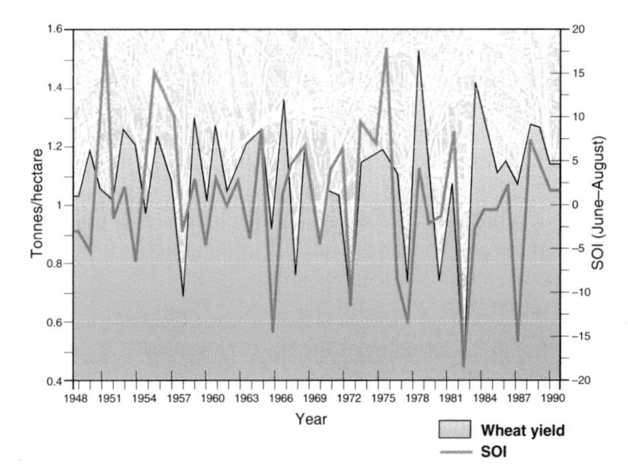

▲ **Figure 5.21** The Southern Oscillation Index (SO1) is shown by the blue line on this graph, and wheat yields are shown in orange. Wheat yields fall during drier periods. Often these coincide with periods of changing pressure, during El Niño years. Rising air pressure at Darwin, cooling sea-surface temperatures, and slackening easterly trade winds, are all signals of a possible El Niño event.

The regularity of such events has made it possible to predict the chances of an El Niño event, so that countries can prepare for reduced rainfall. In Australia, as Figure 5.21 shows, wheat yield has fluctuated closely in line with El Niño events. The key question in Australia is whether this is the responsibility of anyone other than the farmers or whether government agencies are responsible for the advice they give.

Climatic change linked to global warming

Some climatologists believe that the difference in sea-surface temperatures between the hemispheres cannot be put down to theories such as those outlined above. They believe that the greenhouse effect, described in Chapter 6, has led to faster global warming in the southern hemisphere. They believe that cooling by sulphate aerosols, a product of pollution, will balance global warming in the northern hemisphere, because pollution levels are much higher there. There is much less pollution in the southern hemisphere and so warming may be greater.

Computers have simulated the possible effects of a 1 °C rise in mean global temperature. They predict increases in rainfall for most areas of Africa, except for a large part of the Sahel, northern Africa, and the Mediterranean. If warming is greater, then there may be dramatic reductions in rainfall before the year 2100 (Figure 5.22).

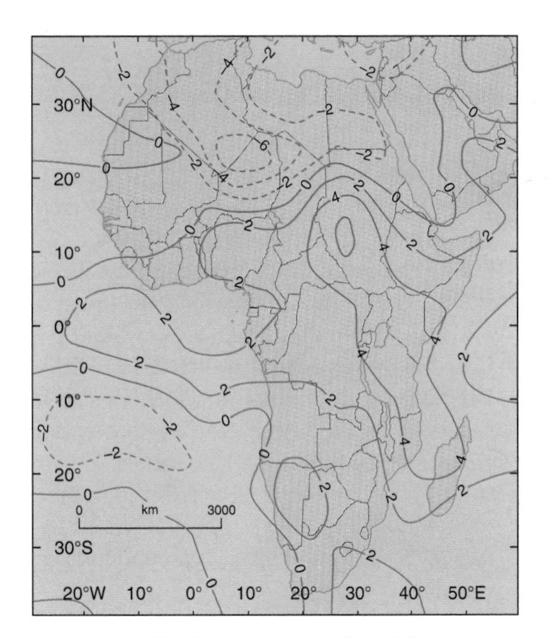

▲ **Figure 5.22** Changes in annual rainfall over Africa. Predictions are based on a 1°C rise in global temperatures. The isolines are of equal change in rainfall: 2′=2% increase in rainfall, –2′= 2% decrease in rainfall. Most of Africa is likely to have increases in rainfall except for the northern parts of the continent, including the Sahel.

The Great Drought pricked consciences. A hue arose, repeated almost every year since, in every part of Africa: We must switch from relief to development. And the outside world set out to save the Sahel.

Sunburned white men crisscrossed the desert in Cessnas and Land-Rovers, each clutching a water bottle and clipboard. Everyone had an idea and money to spend. Where engineers found grass but no water, they sunk a borehole. Soon the Sahel was scarred by devastated circles around abandoned boreholes, just like the Kalahari. Tractors were flown in to open new land; instead they tore up the fragile surface that farmers had protected with their short-bladed hoes. Winds carried off topsoil, leaving behind more desert. Irrigation schemes brought motor pumps to remote villages. But no one thought of spare parts, maintenance, fuel, or marketing. Farmers, encouraged to stay on land that could not feed them, cut down too many trees to cook what food they had.

Experts decided the Sahel needed more trees. The World Bank and other organizations spent up to $6,000 an acre to grow village wood-lots of exotic species that sucked down water levels, if they grew at all. Mostly, goats ate the seedlings. White men and sleek government officials were telling villagers to pour good water into the ground, on scarce land, for some vague benefit ten years in the future. Foreigners who stayed to listen learned the equivalent in Bambara, Tamachek, Fulani, and Hausa, of: "Gimme a break."

Foreign experts told Africans not to cut trees. Hearing that, I could imagine a Fulani in flowing robes at the pumps along the Santa Monica Freeway haranguing motorists: "Don't pour that liquid into your metal box on wheels. Its supply is limited." I have yet to meet an African person who does not know dwindling wood is a problem. But what can he do—besides eat raw grain? Rural families, in need of some cash for cooking oil, sugar, taxes, have few sources of income beyond selling firewood and charcoal.

▲ **Figure 5.23** Solving the Sahel drought. Extract from *Squandering Eden*, Rosenblaum and Williamson, 1990

1 Form small groups of two or three people. Prepare a ten-minute lesson on the following:
   - El Niño and its effects on people's lives.
   - La Niña and its effects on people's lives.
2 Produce a table to summarize the key points in each of the three theories explaining the Sahelian drought. Note evidence for and against each explanation.
3 What does Figure 5.23 tell us about the Sahel drought? What must be done if drought is to be successfully managed?

Study of an Atlantic hurricane

The Tropics are the site of some of the fiercest short-term weather systems on Earth, known as hurricanes. Hurricanes produce winds of terrific force and destructive power – much stronger than the midlatitude depressions and storms described in Chapter 4. Hurricanes occur from July to October in the Atlantic, the eastern Pacific, and the western Pacific north of the Equator. South of the Equator, off the eastern coast of Australia and in the Indian Ocean, they occur from November to March.

The term 'hurricane' is used only for intense tropical disturbances in the Atlantic; in the Pacific they are known as 'typhoons', in the Indian Ocean 'cyclones', and in Australia as 'willy-willies'. Each disturbance is given a name beginning with 'A', 'B', etc. in order of occurrence, and alternately male and female. The focus of the following case study is Hurricane Andrew which struck Florida, The Bahamas, and Louisiana in August 1992.

Hurricane Andrew

Hurricane Andrew reached land in south-east Florida in the early hours of 24 August 1992, when gusts of over 278km per hour were recorded, and a storm surge of 4.5m flattened coastal developments. It took less than four hours for Andrew to become the most expensive natural hazard in the history of the United States. In Dade County, Florida, damage alone exceeded US $25 billion. As many as fifteen people lost their lives in Florida, eight in Louisiana, and three in The Bahamas, and more than 250 000 people were made homeless. It is a weather hazard from which it will take years to recover. Many lessons can be drawn from the study of its short life about the impact of such systems on people's daily lives and activities.

1 Study Figure 5.24, which shows the annual cycle of intense hurricane activity. What does the graph tell us about the seasonality of hurricane activity?

2 Summarize the direction of the main tracks of hurricanes as revealed in Figure 5.25. Why do so few hurricanes develop south of latitude 5° north?

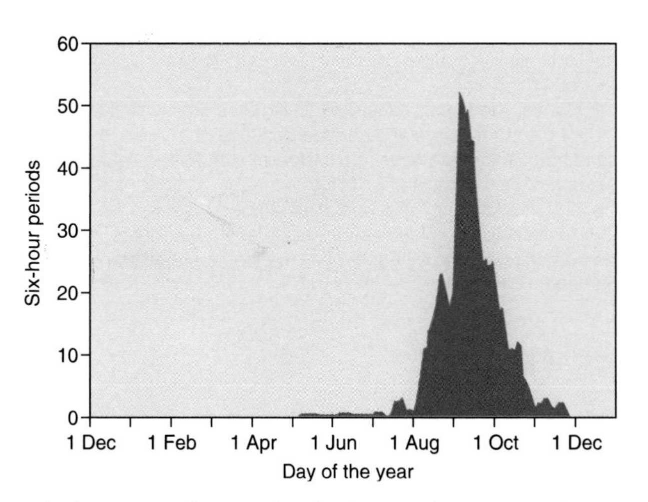

▲ **Figure 5.24** The annual cycle of intense hurricanes in the Atlantic basin, 1886–1989. Units are based on six-hourly reports of hurricane activity.

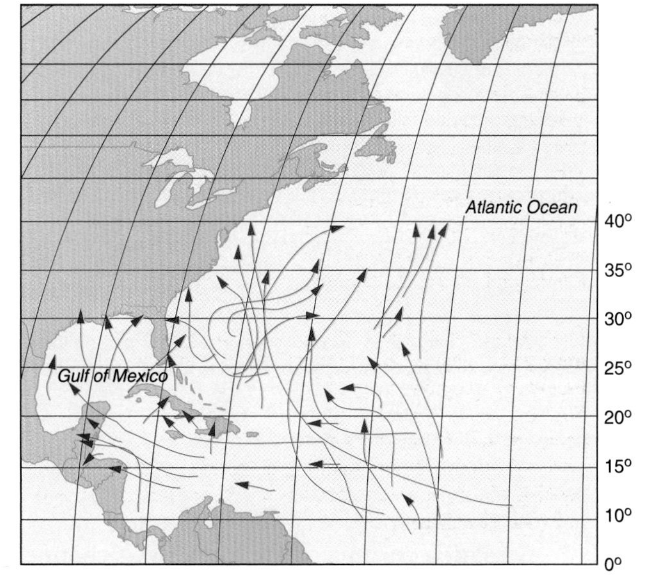

▲ **Figure 5.25** Intense hurricane tracks over the Atlantic, 1950–90. These tracks are based on more than eight days of continuous intense hurricane activity.

Hurricane Andrew 17–28 August 1992

Date	Latitude	Longitude	Status
17.8.92	11°N	37°W	Tropical depression
18.8.92	14°N	46°W	Tropical storm
19.8.92	17°N	53°W	Tropical storm
20.8.92	20°N	59°W	Tropical storm
21.8.92	23°N	64°W	Tropical storm
22.8.92	25°N	66°W	Tropical storm
23.8.92	25°N	71°W	Hurricane
24.8.92	25°N	78°W	Hurricane
25.8.92	26°N	85°W	Hurricane
26.8.92	29°N	90°W	Hurricane
27.8.92	31°N	92°W	Tropical storm
28.8.92	34°N	87°W	Tropical depression

All dates indicate position of hurricane centre at 0000 GMT.

Tropical depression – maximum wind speed 20–34 knots (37–63km per hour)

Tropical storm – maximum wind speed 34–64 knots (63–118km per hour)

Hurricane – maximum wind speed greater than 64 knots (118km per hour)

◀ **Figure 5.26** Data for plotting Hurricane Andrew, 17–28 August 1992.

1 a) Use an acetate sheet and an atlas map of the North Atlantic. Plot the position of Hurricane Andrew on different dates, using Figure 5.26.

b) Annotate the acetate to show wind and pressure changes, using Figure 5.27.

▼ **Figure 5.27** Minimum central pressures and maximum sustained wind speed in Hurricane Andrew.

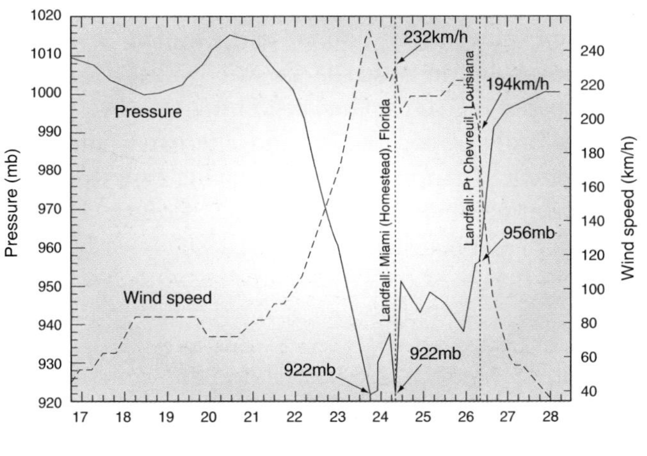

Rapid intensification

Andrew reached hurricane strength on 22 August. An eye formed that morning and the rate of strengthening became rapid. From 0000 GMT on the 21st (when Andrew had a barely perceptible low-level centre) to 1800 GMT on the 23rd, the central pressure fell 92mbar, down to 922mbar. In fact, in a period of 36 hours, Andrew intensified from a tropical storm to the threshold of a Category 5 hurricane on the Saffir/Simpson Hurricane Scale (SSHS).

About noon on 23 August, a reconnaissance aircraft at an altitude of about 2.5km encountered 170 knot (kn) winds in Andrew's eyewall. Andrew then had estimated sustained surface winds near 135kn and was centred about 100km east of the north-west Bahamas.

Andrew came ashore in the north-west Bahamas and then south-east Florida on the night of 23/24 August accompanied by a 50km wide zone of destructive winds and, near the coastline, decimating storm surges characteristic of a Category 4 hurricane on the SSHS. In the Bahamas, the wind speed indicator on the Harbour Island anemometer reached the top end of the scale, 120kn, and stuck there, malfunctioning. A combination of waves and storm surge reached 7m.

In Florida, wind gusts exceeded 150kn. The 5m storm tide (sum of the storm surge and astronomical tide) that headed inland from Biscayne Bay is a record maximum for the south-east part of the peninsula.

Andrew moved nearly due west over land and crossed the extreme south of the Florida peninsula in about four hours. The hurricane weakened about one category during its transit over land.

▲ **Figure 5.28** Extract from Rappaport, *Weather*, February 1994: Hurricane Andrew intensifies.

2 Produce a report of Hurricane Andrew to be presented on a major American news channel such as NBC. Provide a clear account of its path, its development into a hurricane, and its impact on human activity and the natural environment in Florida. Make use of the full range of materials presented here in Figures 5.24–5.32.

The Saffir/Simpson Hurricane Scale

Category 1

Winds 118–152km/hour (64–82 knots). Damage primarily to shrubbery, trees, poorly constructed signs, and unanchored mobile homes. No significant damage to other structures.

Storm surge 1–1.5m above normal tide. Low-lying coastal roads inundated, minor pier damage, some small craft in exposed anchorages torn from moorings.

Category 2

Winds 154–176km/hour (83–95 knots). Considerable damage to shrubbery and tree foliage; some trees blown down. Extensive damage to poorly constructed signs. Major damage to exposed mobile homes. Some damage to roofing materials of buildings; some window and door damage. No major damage to buildings.

Storm surge 2–2.5m above normal tide. Coastal roads and low-lying escape routes made impassable by rising water 2–4 hours before arrival of hurricane centre. Considerable damage to piers. Marinas flooded. Small craft in unprotected anchorages torn from moorings. Evacuation of some shoreline residences and low-lying island areas required.

Category 3

Winds 178–209km/hour (96–113 knots). Foliage torn from trees; large trees blown down. Practically all poorly constructed signs blown down. Some damage to roofing materials of buildings; some window and door damage. Some structural damage to small buildings. Mobile homes destroyed.

Storm surge 2.5–3.5m above normal tide. Serious flooding at coast and many small structures near coast destroyed; large structures near coast damaged by battering waves and floating debris. Low-lying escape routes made impassable by rising water 3–5 hours before hurricane centre arrives. Flat terrain 1.5m or less above sea-level flooded inland 13km or more. Evacuation of low-lying residences within several blocks of shoreline possibly required.

Category 4

Winds 211–250km/hour (114–135 knots). Shrubs and trees blown down; all signs blown down. Extensive damage to roofing materials, windows and doors. Complete failure of roofs on many smaller residences. Complete destruction of mobile homes.

Storm surge 4–5.5m above normal tide. Flat terrain 3m or less above sea-level flooded inland as far as 11km. Major damage to lower floors of structures near shore due to flooding and battering by waves and floating debris. Low-lying escape routes made impassable by rising waters within 3–5 hours before hurricane centre arrives. Major erosion of beaches. Massive evacuation of all residences within 500m of shore possibly required, and of single-storey residences on low ground within 3km of shore.

Category 5

Winds greater than 250km/hour (135 knots). Shrubs and trees blown down; considerable damage to roofs of buildings; all signs down. Very severe and extensive damage to windows and doors with extensive shattering of glass components. Complete failure of roofs on many residences and industrial buildings. Some complete building failures. Small buildings overturned or blown away. Complete destruction of mobile homes.

Storm surge greater than 5.5m above normal tide. Major damage to lower floors of all structures less than 4.5m above sea-level within 500m of shore. Low-lying escape routes made impassable by rising water 3–5 hours before hurricane centre arrives. Massive evacuation of residential areas on low ground within 8–16km of shore possibly required.

▲ **Figure 5.29** The Saffir/Simpson hurricane scale.

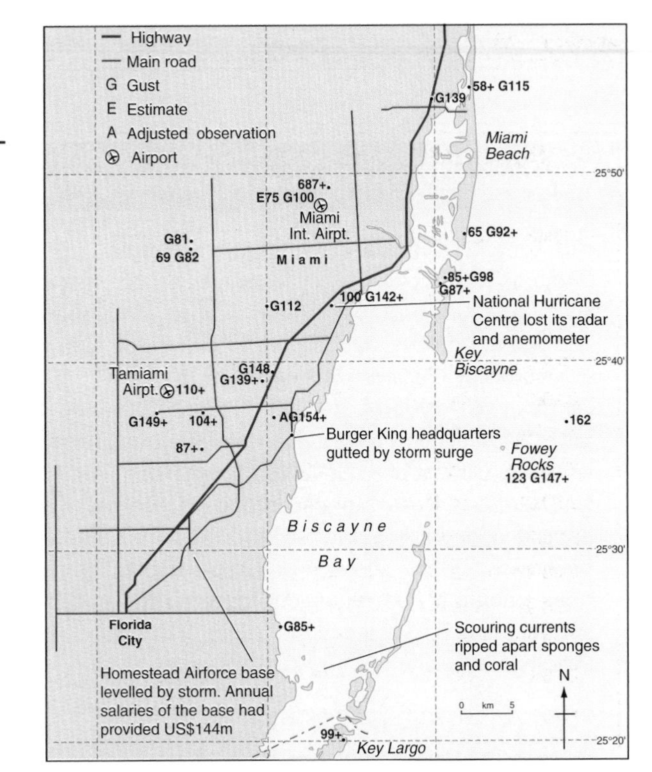

▲ **Figure 5.30** Hurricane Andrew brought sustained wind speeds and gusts over south-east Florida. This culminated in a 2.7m storm surge, and winds over 185km per hour flattened hundreds of hectares of Australian pine. Wind speeds are given in knots (1 knot = 1.85 km/hr).

▼ **Figure 5.31** Hurricane Andrew, as seen from a satellite.

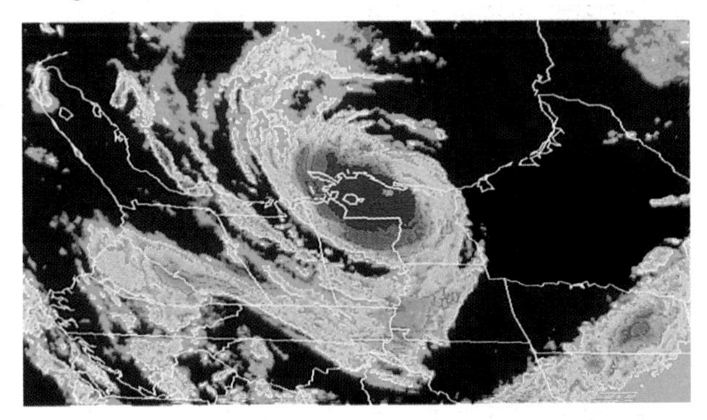

1. Form small groups. Half of each group should prepare a case *for* the statement: 'It is the private individual's responsibility to prepare their property for events such as Hurricane Andrew. The state of Florida cannot be held responsible for damage incurred and should not have to fund disruption to people's lives.' The other half should prepare a case *against* this statement. The group should then hold a debate.
2. What was resolved during the debate? Write 500 words to explore what happened.

Hurricane formation and development

Hurricanes, cyclones, and typhoons are a major hazard in the Tropics. Every year, some 80 occurrences claim, on average, 20 000 lives and cause immense damage to property, natural vegetation, and shipping.

A hurricane is a large rotating storm around a centre of very low pressure with values of 950mb or lower. Most systems have a diameter of about 650km, less than half that of a mid-latitude depression. Wind speeds often exceed 33m per second (119km per hour) and, in the most intense hurricanes like Andrew, may reach 50m per second (180km per hour). The huge amount of heat required to create and maintain the hurricane is reflected in the height of the clouds near its centre, often up to 12km above the Earth.

How do hurricanes form?

A number of trigger mechanisms, which rely on intensive rising convection currents, are required to transform a tropical disturbance into a more destructive hurricane. The development of a hurricane is dependent on combinations of particular mechanisms.

1. An extensive ocean area with surface temperatures greater than 27 °C for a significant period of time.
2. Sufficient spin from the Earth's rotation to trigger the vicious spiral in the centre of the hurricane, usually between latitudes 5° and 30° north and south of the Equator.
3. A lack of strong horizontal air movement or wind shear, near a jet stream for example, which causes the break-up of the spiralling winds.
4. Where upper-air winds in the troposphere cause a rapid ascent of air to be sucked in at the sea surface to replace that lost above. This causes a rapid upward draught of moisture-laden air, which in turn produces huge volumes of condensation to form massive clouds.

As trade winds rush into the centre of the storm, they spiral inwards and upwards, releasing heat and moisture. Earth's rotation twists the rising columns of air into a whirling cylinder around an eye of still, cloud-free descending air. The hurricane is fed by energy stored in water vapour brought upwards by air from evaporation at sea-level. As this air rises, 90 per cent of the stored energy is released as heat as condensation occurs. This heat generates further uplift and instability, so long as the rising air is warmer than the surrounding air. The energy released within hurricanes is enormous. See Figure 5.31.

Hurricane decay

As soon as the hurricane reaches land it loses its supply of heat and moisture and therefore its energy source. It can quickly become a mere tropical storm, especially when cold air is drawn in or when the upper atmospheric disturbance which caused the hurricane moves away.

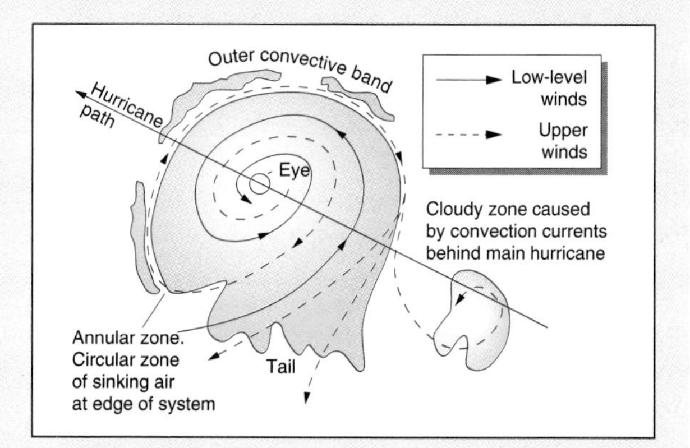

▲ **Figure 5.32** Typical hurricane structure. Notice the
▼ intensive convection and uplift around the centre of the storm and the calm 'eye' of the storm where warm air descends and produces clear skies.

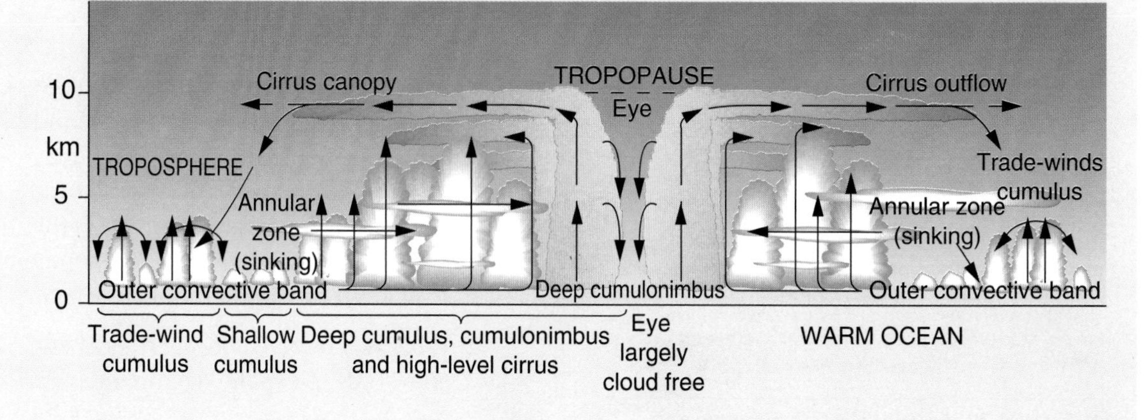

Ideas for further study

Write an essay of about 1500 words on one of the following titles.

- 'Long-term climatic hazards such as drought are rarely caused by human activity.' Discuss with relevance to a range of case studies.
- 'Tropical climates are characterized more by their daily variety in weather conditions than by their seasonal variations.' To what extent do you agree?

Summary

- Until recently, tropical weather and climate were thought to be much less complex than they are in the mid-latitudes. This is not the case.
- Tropical weather and climate are influenced by the role of the inter-tropical convergence zone (ITCZ). The ITCZ is a key part of the three-cell model of the Earth's atmosphere.
- Both short-term hazards such as hurricanes and long-term hazards such as drought have a profound impact on the lives and activities of people in the tropical zone.
- Careful management of weather and climate hazards requires in-depth knowledge of processes. This is gained from the operation of national meteorological services and a sharing of relevant data.

References and further reading

R. G. Barry and R. J. Chorley, *Atmosphere, Weather and Climate* (sixth edition), Routledge, 1992, especially Chapter 6.

R. Gore, 'Andrew aftermath', *National Geographic*, April 1993, Vol.183 No.4.

M. Hulme, 'Causes of climatic change and variability in the African Sahel', *SAGT Journal*, 1994, No.23, pp.12–21.

IPCC (International Panel on Climatic Change), J. Haughton *et al.*, *Climate Change, the IPCC Scientific Assessment*, Cambridge University Press, 1990.

C. W. Landsea *et al.*, 'Seasonal forecasting of Atlantic hurricane activity', *Weather*, August 1994, Vol.49 No.2, pp.273–84.

E. N. Rappaport, 'Hurricane Andrew', *Weather*, February 1994, Vol.49 No.2, pp.51–61.

M. Rosenblaum and D. Williamson, *Squandering Eden*, Paladin, 1990.

6 How is the atmosphere changing?

Concern about climate change has focused on the prediction of an increase in global mean temperature, or global warming. This may be caused by an increase in carbon dioxide and other trace gases, notably:

- chlorofluorocarbons (CFCs)
- methane
- nitrous oxides.

The predicted change has been based on General Circulation Models (GCMs). Large banks of data taken from past observations and measurements are fed into a computer and processed. Assumptions are made from the patterns that are produced. Based on these assumptions, the GCMs provide predictions or 'scenarios' of the future.

Much of what was decided at the Earth Summit in Rio in 1992 was based on these scenarios. The result was The Framework on Climate Change, a document signed by approximately 150 countries. The aim was to reduce the emissions of each greenhouse gas to 1990 levels by the year 2000. How did this international co-operation come about?

The greenhouse effect – how it works

The driving force for weather and climate is the Sun. The Earth intercepts solar radiation in different ways according to surface – oceans, ice, land, living matter, atmosphere. About 30 per cent of solar energy is reflected back into space by clouds, and by reflective land surfaces such as ice and snow. The ability of a surface to reflect back solar radiation is known as albedo. In the long term, energy absorbed from solar radiation is balanced by outgoing radiation from the Earth's surface. Long-wave infra-red radiation is crucial to this balance. It is partly absorbed and then re-emitted by a number of abundant and trace gases in the cooler atmosphere above. These gases are known as the greenhouse gases. The main natural greenhouse gases are:

- water vapour (the biggest contributor)
- carbon dioxide
- methane
- nitrous oxides
- ozone, which occurs in the troposphere up to 10–15km above the Earth's surface.

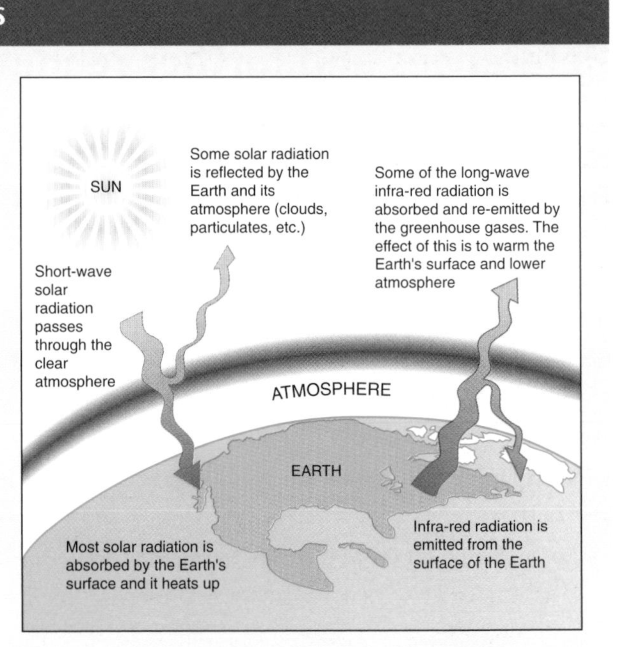

▲ **Figure 6.1** The greenhouse effect – how it works.

Without absorption by these gases the mean temperature at the Earth's surface would be 33 °C lower, and impossible for human habitation. Concentrations of these gases have been found in Antarctic ice cores which are 160 000 years old. However, levels of concentration vary, and suggest that changes in carbon dioxide and methane levels, in particular, are closely linked to large global temperature swings such as those that occur during Ice Ages.

An enhanced greenhouse effect?

Many scientists believe that human activities have increased concentrations of greenhouse gases to a point where there will be a marked increase in mean annual temperatures. Scientists describe this as 'enhanced', or 'increased'. The 'greenhouse effect' occurs naturally, but recent land use changes, industrialization, and urbanization have released natural stores of carbon and nitrogen. These increase the net gain of solar radiation from the natural greenhouse effect, when they oxidize in the atmosphere. It is argued that the main greenhouse gas, water vapour, will also increase in quantity because warmer air can hold more moisture than cold air. This will lead to greater condensation and cloud cover, which in turn will create a further heat trap. Other scientists dispute the water vapour theory and question the timing, extent, and impact of an enhanced greenhouse effect.

Global warming

The prospect of a 'greenhouse effect' has been widely reported in the media. This may give the impression that the link between increasing greenhouse gases and global warming is a cast-iron certainty. We have accepted a future scenario of melting ice-caps, changing climatic zones, increasing weather unpredictability, and a rising sea-level. In fact, there is much debate about this. The views in Figure 6.2 may cause you to re-think some of these ideas. There is much uncertainty about the cause of any warming, and little agreement among scientists about future climate changes. The process of climate change is complex, and the use of GCMs in predicting the future is still at an early stage. To complicate matters, every scientist, environmentalist, and politician looks at the evidence from a particular viewpoint, so that their predictions about the future are affected by their own beliefs and attitudes. The activity will help you to analyse where some of the main participants in the global warming debate stand, and to understand why they believe what they do.

Shades of green

1 Form small groups of two or three people. Read carefully the quotations from various speakers and writers on global warming (Figure 6.2). Discuss what each person is saying and what you believe their bias or standpoint is.

2 Study the values spectrum diagram (Figure 6.3) and place the different views on it.

3 Do particular people seem to form groups or clusters on the spectrum? If so, you have probably identified those who hold similar beliefs or values.

4 Discuss with the rest of the class how helpful you found the spectrum in clarifying and classifying the range of opinions. Do you think that this will change the way you look at the evidence on the causes of the greenhouse effect?

5 Do you think it is likely that different political parties (for instance in the UK or in the USA) would be divided on issues like this? Is the spectrum useful in helping you to understand why politicians differ in their opinions?

1

"To capture the public's imagination . . . we have to offer up some scary scenarios, make simplified, dramatic statements, and little mention of any doubts one might have . . . Each of us has to decide the right balance between being effective and being honest. I hope that means being both."

(Dr Stephen Schneider, adviser to US Vice-President Albert Gore Jr, 1989)

2

"Environmental pressure groups have for some years been promoting the apocalyptic vision; so successful have these agents of doom been that by the late 1980s the UN had been persuaded to commission a report on the scientific evidence for climate change – the Intergovernmental Panel on Climate Change (IPCC) report."

(Roger Bate and Julian Morris from the Institute of Economic Affairs Environment Unit, 1994)

3

"The data don't matter . . . Besides we [the UN] are not basing our recommendations [for immediate reductions in carbon dioxide emissions] upon the data; we're basing them upon the climate models."

(Dr Chris Folland, UK Meteorological Office, member of the IPCC)

4

"I promise you that global warming is nonsense, the latest example in a long line of doomsday predictions stretching all the way back to Noah and the Ark."

(Teresa Gorman, Conservative MP, 1992)

5

"If governments do not act to reduce greenhouse gas emissions (a scenario usually described as 'business as usual') global temperatures could rise to levels greater than any experienced over the last 10 000 years."

(Friends of the Earth, Climate Change, *1990)*

6

"Uncertainty should make us more rather than less wary of imposing limits on greenhouse gases . . . All energy subsidies and taxes should be eliminated – the market and its supporting institutions will then be able to adapt more readily and rapidly to a changing environment."

(Roger Bate and Julian Morris, IEA Environment Unit, 1994)

7

"There is no longer any significant disagreement in the scientific community that the greenhouse effect is real."

(Senator Albert Gore Jr, later US Vice-President, 1986)

8

"Developed countries should fundamentally change their urban structure, transportation system, and lifestyle from the current industrialized culture aiming only at economic growth for its own sake based on mass production, mass consumption and mass disposal, to one more environmentally sound."

(Japanese Global Environment Minister, 1992)

9

"Treading gently on the earth becomes the natural way to be. The global must [give way] to the local, since the local exists within nature, while the global exists only in the offices of the World Bank . . . . Any activity with potential impact on the local environment should have the consent of local people."

(Vandana Shiva, Indian environmental activist, 1992)

10

"Several scientists showed . . . that implementing selected policies today versus implementing these policies in 10 years' time makes little difference to future temperatures. They speak of a window of opportunity to get the science right and then make critical policy decisions."

(Dr R. C. Balling Jr, Associate Professor and Director, Office of Climatology, Arizona State University, USA)

▲ **Figure 6.2** A selection of views on global warming.

▼ **Figure 6.3** A values spectrum for analysing environmental issues (*after O'Riordan, 1992*).

GREEN LABEL

ECOCENTRIC PERSPECTIVE (nature centred)

Holistic[1] world view. Minimal disturbance of natural processes. Integration of spiritual, social, and environmental dimensions. Sustainability[2] for the whole Earth. Devolved, self-reliant communities[3] within a framework of global citizenship. Self-imposed restraint on resource use.

ANTHROPOCENTRIC PERSPECTIVE (people centred)

People as environmental managers of sustainable global systems. Belief in the 'no regrets' principle. Population control given equal weight to resource use. Strong regulation by independent authorities required.

TECHNOCENTRIC PERSPECTIVE (technology centred)

Technology can keep pace with and provide solutions to environmental problems. Resource replacement solves resource depletion. Need to understand natural processes in order to control them. Strong emphasis on scientific analysis and prediction prior to policy-making. Importance of market, and economic growth.

*[1]**Holistic** – belief that nature forms 'wholes' of living organisms that are greater than the sum of each part. Change in one part affects every other. Every organism is of equal worth.*

*[2]**Sustainability** – use of global resources at a rate that allows natural regeneration and reduces damage to the environment to minimal proportions.*

*[3]**Devolved community** - one that decides how it should progress and has power over decisions which affect it.*

Evidence for an enhanced greenhouse effect

The Intergovernmental Panel on Climate Change (IPCC) reports of 1990 and 1992 were sure that the concentrations of certain gases in the atmosphere had steadily increased over the past century. They calculated with confidence that 'some gases are potentially more effective than others at changing climate, and their relative effectiveness can be estimated. Carbon dioxide has been responsible for over half the enhanced greenhouse effect in the past, and is likely to remain so in the future' (IPCC, 1990). They were also confident that greenhouse gases are often long-lived, and respond slowly to reductions in emission. Therefore, if present rates of pollution continue, the reduction will have to be greater in the future to stabilize concentrations. For long-lived gases, immediate reductions of over 60 per cent are needed to stabilize concentrations at today's levels. Methane would require less drastic reductions of 15–20 per cent.

However, the IPCC was more concerned about future change in the atmosphere and how possible future scenarios seemed to depend upon the increase in greenhouse gases. The graphs in Figure 6.5 show these increases up to the year 2000.

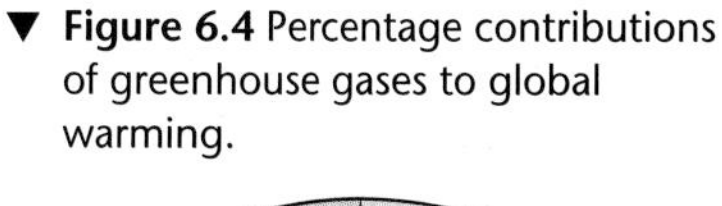

▼ **Figure 6.4** Percentage contributions of greenhouse gases to global warming.

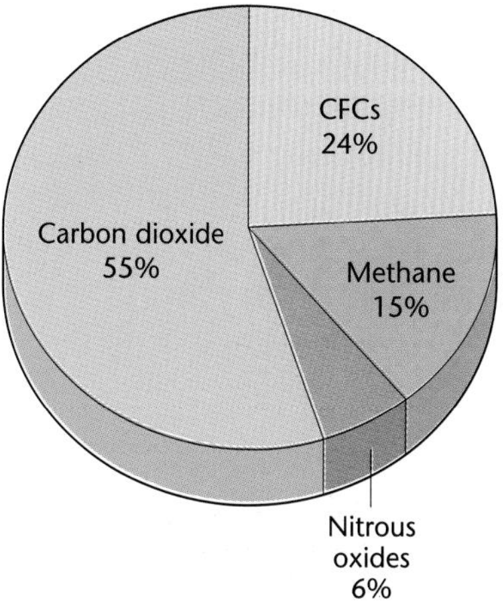

▼ **Figure 6.5** Predicted changes in greenhouse gases, 1750–2000.

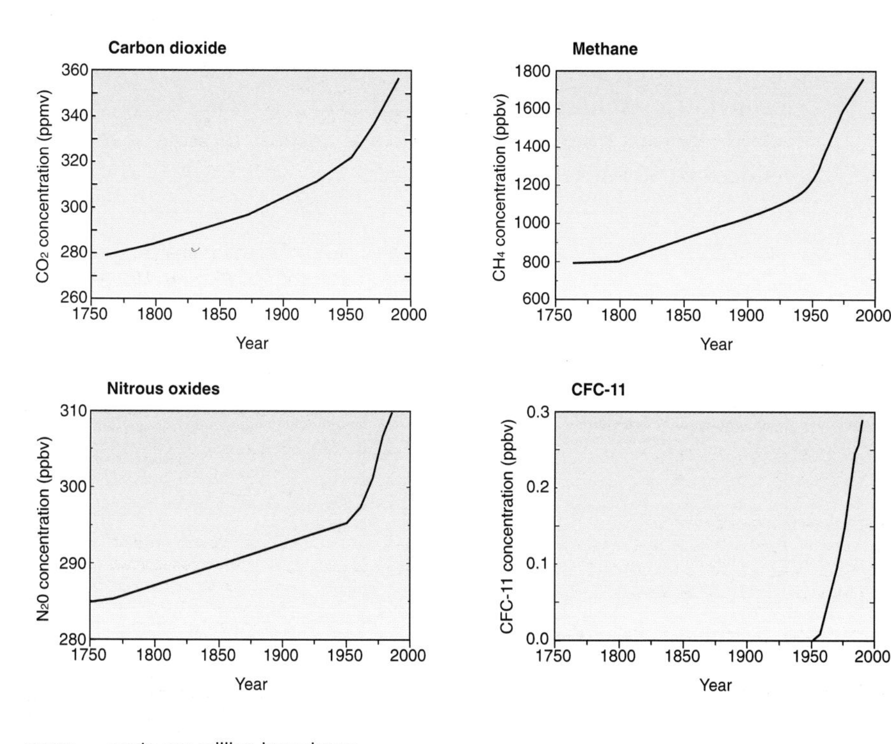

ppmv = parts per million by volume
ppbv = parts per billion (thousand million) by volume

1 What is meant by the term 'scenario'?

2 Work out the approximate rates of change in each of the graphs, in units per decade. Describe any trends you notice.

3 What assumptions did the IPCC make in producing their scenarios? Which scenario seems more likely, given your knowledge of present patterns of energy use?

4 Which groups of countries will need to change their energy use and lifestyles most, in order to achieve the reductions needed to reduce the threat of global warming? What changes might this cause for ordinary people and decision-makers?

Figure 6.6 shows four possible predictions of carbon dioxide and methane emissions between 1980 and 2100, and Figure 6.7 suggests the global temperature rises that could result from each. The four predictions have been made by the IPCC as follows.

1 ***Scenario A*** shows 'business as usual', with intensive use of coal, and only modest use of energy-efficiency measures in homes, offices, factories, and power stations. It assumes that controls on carbon monoxide are limited, deforestation continues until tropical forests are virtually destroyed, and emissions of methane and nitrous oxides from agriculture remain uncontrolled. The reduction in CFC emissions by 50 per cent by the year 2000 (as agreed in The Montreal Protocol in 1987) is partly successful. In this scenario, the world continues to operate as it is now.

2 ***Scenario B*** shows an energy supply shifting away from high-carbon fossil fuels to lower-carbon alternatives, especially natural gas. Large energy-efficiency increases are achieved, deforestation is reversed, carbon monoxide is strictly controlled, and The Montreal Protocol on CFCs is completely successful.

3 ***Scenario C*** shows a major shift towards renewable energy and nuclear power in the second half of the 21st century. CFCs are phased out and agricultural emissions of methane and nitrous oxides are controlled.

4 ***Scenario D*** assumes that a move towards renewable energy and nuclear power takes place in the first half of the 21st century, leading to a stabilization of carbon dioxide levels in the developed world. By the middle of the century carbon dioxide levels are reduced to 50 per cent of 1985 levels.

▼ **Figure 6.6** Simulated increase in carbon dioxide and methane based on the four IPCC scenarios, 1980–2100.

GtC = measure of carbon level, gigatonnes
Mt = measure of methane level, million tonnes

▼ **Figure 6.7** The simulated rise in global temperature based on the four IPCC scenarios.

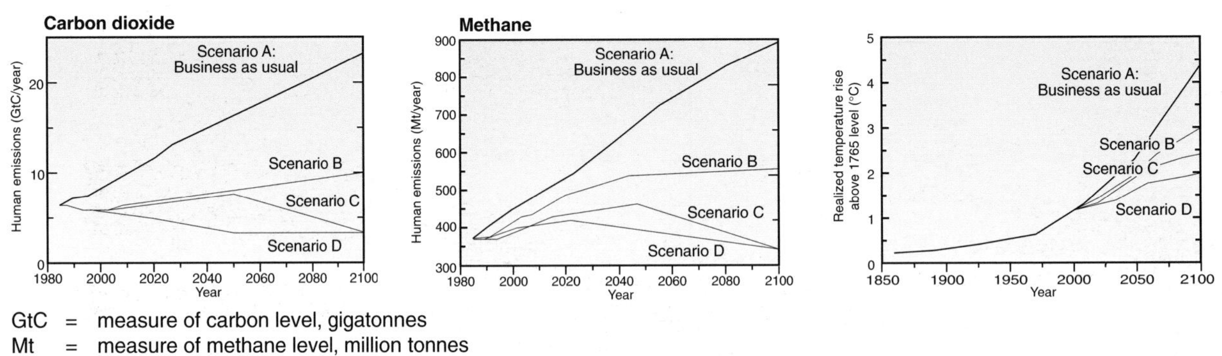

The main greenhouse gases and their link with human activity

Carbon dioxide

Carbon dioxide is one of the major greenhouse gases. It is controllable because its emission is caused by human activities, releasing carbon stored in a variety of 'sinks' in the Earth, as shown in Figure 6.8. Since 1750, when the British Industrial Revolution began, the concentration of carbon dioxide in the atmosphere has increased by 26 per cent as a result of the burning of fossil fuels in the developed world. In the United Kingdom in 1990, 50 per cent of emissions came from private cars alone, with fossil-fuel power stations, industry, domestic and commercial heating making up the remainder. In the developing world, deforestation is contributing most to global warming: cutting and burning each year releases an estimated 4 billion tonnes of carbon into the atmosphere. The carbon contained in the Amazon basin alone is equivalent to 20 per cent of the entire atmospheric concentration of carbon dioxide. It is likely that 80 per cent of this would be released into the atmosphere if all the Amazon basin forest were cut down and burnt, although half of this would be absorbed by the oceans. Rapid industrialization in the developing world may offset any reduction in carbon emissions in the developed world. Between 1950 and 1980, carbon dioxide emissions rose:

- by 586 per cent in the developing world (but from a low base)
- by 337 per cent in the former Soviet Union and Eastern Europe
- by 91 per cent in North America
- by 125 per cent in Western Europe (from an existing high level).

Chlorofluorocarbons

Chlorofluorocarbons (CFCs) were first released in quantity from industries and other uses in the developed world in the 1960s. Their use as propellants in spray cans, use in foam plastics (e.g. fast-food packaging), refrigerant fluids, and as solvents in the electronics industry became widespread. In the 1970s they were thought to be mainly responsible for cooling effects on the atmosphere as a result of their ability to reflect solar radiation, but more recently it has been found that they absorb solar radiation and thus contribute to global warming. The chemical stability of CFCs means that they have a long lifetime in the atmosphere, and their global warming potential outweighs their relatively low concentrations. In the mid-1980s, scientists discovered an extensive thinning in the ozone layer in the stratosphere 25km above Antarctica. This was attributed to high concentrations of CFCs which had travelled through the upper atmosphere from the northern hemisphere to the southern hemisphere since the 1960s.

▼ **Figure 6.8** Carbon stores, flows, and human activity.

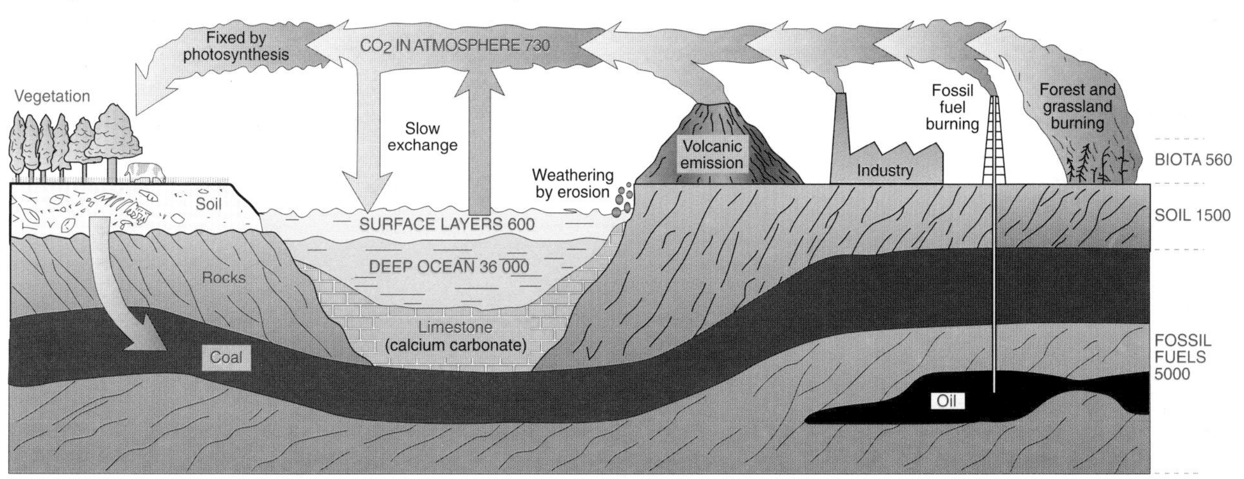

1 Explain the differences between a natural and an enhanced greenhouse effect.

2 Read and summarize the evidence to support the arguments for an enhanced greenhouse effect. You can extend your research to include the 1995 IPCC report, and details of weather events in the 1980s and 1990s from sources such as *Keesing's Archives* or from newspapers. You should also look at evidence about the retreat of glaciers and ice-sheets, and rises in sea level.

This discovery startled politicians, who had believed that pollution would not significantly damage the atmosphere over such a short time-span. The result was the first major international agreement on limiting damage to the atmosphere, The Montreal Protocol of 1987. Leading industrial nations agreed to reduce CFCs, and to conduct research to find substitutes. In 1989 the Protocol was strengthened to reduce CFCs by 50 per cent by the year 2000. In 1992, further amendments sought a phasing-out of CFCs and of other chemicals known to destroy ozone by the year 2000, and the participation of some developing nations in eliminating CFCs altogether. Reductions of CFCs of 40 per cent were achieved, and this encouraged campaigners against global warming to influence world leaders to take similar action at the Earth Summit in Rio in 1992. However, ozone depletion means that there is now less absorption of incoming short-wave solar radiation, and the atmosphere is therefore cooling in a way that offsets any warming effect from CFCs.

Methane

Since 1950 the concentrations of methane have risen by 1 per cent per year – four times the rate of increase for carbon dioxide – which could lead to methane becoming the main greenhouse gas within 50 years. The principal reasons for this are increased rice production worldwide, the burning of vegetation, coal mining, leakage of natural gas from distribution systems, and flatulence from termites and livestock (as a result of increasingly intensive cattle rearing). Pearce (1989) has estimated that the annual emission of methane from the world's cattle is close to 100 million tonnes. The trend towards intensive production reflects the demand for cheap mass-produced meat from fast-food outlets and supermarket chains. However, emissions from some sources remain uncertain and IPCC figures exclude other sources of methane, such as those at mid-ocean ridges where magma emerges from the mantle, and the breakdown of organic material buried in oceans at the edges of the continents.

Nitrous oxides

Human activities which release nitrous oxides into the atmosphere include the use of agricultural fertilizers, fossil-fuel combustion (especially from cars), and various industrial processes which involve the production of synthetic chemical substances such as nylon. Nitrous oxides help to trap infra-red radiation in the lower atmosphere, and change to nitric oxide in the stratosphere, which destroys ozone. They therefore contribute to global warming.

Ozone in the troposphere

At low levels, ozone is toxic to plants, to humans and other organisms, and also acts like a greenhouse gas. The most recent research shows that the warming effect of ozone is at its most significant 12km above ground level – the height at which most aircraft operate and where levels of pollution from nitrous oxides are at their peak. The total warming effect from ozone is difficult to calculate partly because it is so sensitive to the presence of other gases with which it reacts chemically.

Many scientists now believe that a doubling of carbon dioxide would create a rise in mean annual global temperatures of between 1.5 and 4.5 °C, with an average of 2.5 °C. This is supported by most of the signatories of the World Convention on Climate in 1992. It is also believed that this rise would offset any cooling trend caused by such events as the Pinatubo eruption in 1991, and pollution by sulphates in the lower atmosphere.

To support their theory of warming, scientists point to increasing weather extremes, such as increased storminess in mid-latitudes and the Tropics, drought in continental interiors, and increasing unpredictability of rainfall in arid areas. The two decades from 1972 to 1992 showed a strong warming trend; 1990 was the warmest year on record, globally, and eight of the ten warmest years on record have occurred since 1980.

Read through Figure 6.9 and the box below.

1 Summarize the assumptions on which the models are based.

2 What conclusions are stated in the article about future climatic change? How do these compare with the conclusions of supporters of global warming?

3 What problems are there in predicting or simulating climatic change using computer models?

Using computer models to simulate global warming

The IPCC uses data generated by computer simulations of future climate change. These computers use historical data of rainfall, temperature, seasonal changes, and wind patterns. They also use responses of land, atmosphere, and sea to these phenomena and to each other, to perform mathematical calculations that try to predict overall changes to the Earth's climate. The computers calculate changes at regular intervals (e.g. every 30 minutes), for a set time-period such as a century.

The program may run for several months before a simulation is complete. If the model is reliable it will produce results that resemble closely the behaviour of the real world and of the real atmosphere. To predict climatic change the program will be run again over the same time, but with increasing concentrations of greenhouse gases, in order to estimate what changes these might cause. The time needed for such calculations and the volume of data are leading to the development of supercomputers (Figure 6.9).

▼ **Figure 6.9** Extract from *The Independent*, 25 July 1994 by Nicholas Schoon.

A report soon to be published by the UN's Intergovernmental Panel on Climate Change (IPCC) will confirm that mankind is gradually altering the composition of the atmosphere, and thereby its heat balance.

The document's theories hold that volcanoes and variations in the sun's output are a bigger influence on climate than man-made pollution. But there remain huge uncertainties about when and where the warming will occur, how it will change rainfall and weather, how much harm it can do us. The best estimate is for an average temperature rise of between 1°C and 3.5°C midway through the next century.

One complication: pollution from the burning of fossil fuels is *cooling* large areas of the planet's surface as well as warming it. Emissions of carbon dioxide (CO_2) trap heat in the atmosphere that would otherwise escape into space, but sulphate aerosol – from burning coal and gas – acts as an atmospheric heat shield.

Scientists at the Hadley Centre for Climate Prediction have taken delivery of their new Cray C90 supercomputer, also used by the Met Office for weather forecasting.

Starting from around 1860, the Cray will make its own calculations of how the atmosphere and oceans have responded to gradually rising emissions of greenhouse gases. The supercomputer will march on into a future in which concentrations of 'greenhouse gases' continue to rise as the world burns more coal, oil and gas. Experiments using the centre's previous supercomputer showed that after 70 years of steadily rising greenhouse gas concentrations (1 per cent extra a year) average temperatures went up by 1.5° C. But this simulated global warming was highly uneven. In some places there had actually been a slight cooling, while parts of the Arctic had warmed by 6°C.

The Hadley scientists hope to get credible predictions by inputting this gradual, accelerating build-up of greenhouse gases which reflects recent history and the likeliest future. The Hadley scientists are just beginning to model the cooling effect of aerosols. Sulphate aerosol cools directly, by reflecting incoming sunlight into space, and indirectly by 'seeding' clouds and boosting the number of water droplets they contain. The clouds are then whiter and brighter and reflect more incoming sunshine back to space.

Computer modelling of these two cooling influences is in its infancy. But they may well mask some of the global warming which has already occurred and explain why supercomputer simulations of global warming predict a greater warming to date than appears to have been the case.

Clouds, which could have a critical influence on the progress of any man-made global warming, are just as confusing. A warmer atmosphere would probably contain more water vapour – itself a greenhouse gas – causing further warming. But more vapour could mean more clouds, which have both heating and cooling effects. They trap some of the Earth's outgoing heat radiation, but also reflect incoming sunlight, preventing solar radiation from heating the lower atmosphere and ground. Ice clouds above 35,000ft have an overall warming effect. Dense low cloud is a coolant during daytime. We need to know the fate of clouds in a warming world, but supercomputers can, as yet, give little help.

Evidence against an enhanced greenhouse effect

A vocal and well-informed minority of scientific opinion has produced a series of arguments to support the case that global warming is unlikely; or that, if it does happen, then it will occur with the speed or with the effects that these scientists think likely. The main arguments against global warming centre around:

- the length of global climatic records and the effects of natural climatic variability
- the slow response of global temperatures so far to a century of greenhouse gas increases and recent carbon dioxide emissions
- the cooling effect of aerosols from volcanic eruptions and from human activity
- the role of increasing water vapour levels and cloud formation in reducing solar radiation, thus offsetting any warming
- variability in incoming solar radiation
- the role of the oceans in controlling the rate of global warming.

These are discussed in more detail below.

Global climatic records

Records of global temperature only go back to the mid-19th century. Many scientists believe that a rise of 0.5 °C over this period may well be within any natural variation of the Earth's climate. Also, changes in observation methods, improvement in weather recording technology, gaps in the database for several locations, and local urban warming effects, produce uncertainty in the records by about the same amount, of 0.45 °C. Natural changes in climate, such as the El Niño event (page 115), and the Little Ice Age in the 17th century, are common over several decades and may give a mistaken impression of longer-term warming or cooling. It is possible that they simply represent temporary patterns of climate change which re-occur at regular intervals. Others see the trend to global warming as no more than a worldwide urban 'heat-island' effect, where measurements taken in urban areas reflect the effects of ozone, carbon dioxide, and nitrous oxides in polluted cities. Recent research by Balling and Idso (1992) for 961 rural measuring stations in the USA has shown a cooling of 0.15 °C between 1920 and 1990.

Responses to carbon dioxide and recent trends in emissions

Some scientists believe that the case for the warming effect of carbon dioxide is exaggerated, and global temperatures so far do not support IPCC predictions. Richard Lindzen (1993) of the Massachusetts Institute of Technology argues that, if computer models which predict a 4 °C warming for a doubling of carbon dioxide are correct, then we might have expected to see a 2 °C increase already, rather than the 0.5 °C actually recorded. He insists that feedbacks in the atmosphere–ocean system reduce any warming effects from increasing emissions of greenhouse gases. He believes that the small warming effect is nothing more than natural variability of global climate. In relation to the Tropics, he states: 'There is ample evidence that the average Equatorial sea surface has remained within 1 °C of its present temperature for billions of years, yet current models predict average warming of 2–4 °C even at the Equator. It should be noted that for much of the Earth's history, the atmosphere had much more carbon dioxide than is currently anticipated for centuries to come.' Lindzen, like other 'greenhouse sceptics', feels that the global warming supporters ignore the fact that most of the recent warming of the Earth occurred before 1940, before the largest emission of greenhouse gases from human activity took place. They are supported by the reduction in the rate of increase in atmospheric carbon dioxide, which is partly related to the reduction in rates of consumption of fossil fuels (see Figures 6.10 and 6.11). It is possible that the atmosphere's ability to absorb more carbon dioxide is slowing down as it reaches a saturation point. They believe that plant biomass will increase with more carbon dioxide, and that at higher levels of concentration plants need less moisture to survive. Increased water vapour levels will therefore produce more heat-reflecting cloud which in turn will produce a net cooling.

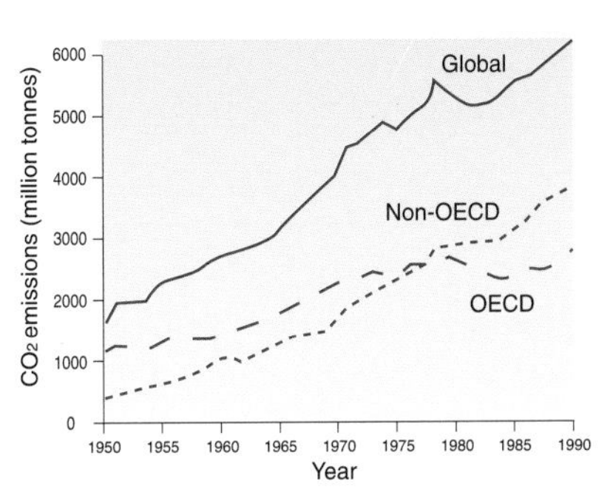

▲ **Figure 6.10** Trends in carbon dioxide emissions between 1950 and 1990. Most of the OECD (Organization for Economic Co-operation and Development) countries are in the economically developed world.

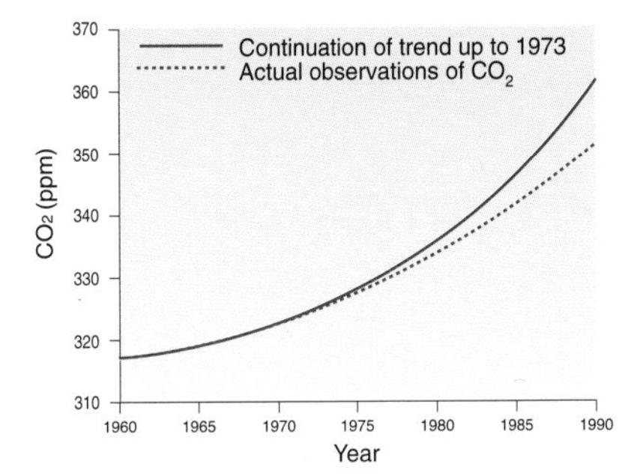

▲ **Figure 6.11** Increases in carbon dioxide concentrations at Mauna Loa Observatory, Hawaii. Note that after 1973 the trend is away from exponential growth which gets faster and faster, to a more linear pattern where increases occur by the same amount in each time-period.

Cooling effects of aerosols

The role of aerosols given out by large volcanic eruptions was highlighted by the eruption of Mount Pinatubo in the Philippines in June 1991. It injected some 20 million tonnes of sulphur dioxide into the atmosphere to heights of up to 25km, causing the largest global climatic disturbance of the century (see Figure 6.12). In the stratosphere the sulphur dioxide was dispersed by strong winds and transformed in sunlight to form sulphuric acid droplets. This helped to scatter incoming solar radiation back to space, whilst also absorbing re-radiated infra-red energy from the Earth, thus releasing it into the stratosphere and causing a warming of the upper atmosphere and a cooling of the lower atmosphere. The overall cooling of global temperatures by 0.4 °C in 1992 is mainly attributed to Pinatubo, although its effect was balanced a little by the El Niño event (see page 115). Some continental interiors (North America, Asia) experienced a 'year without a summer' in 1992, very similar to that of 1816, the year after the historical eruption of Tambora in Indonesia. Both events signal the fact that global climate is very sensitive to aerosols, but their effect is short-lived (two to three years) and may not delay any longer-term warming trends.

The role of sulphate aerosols produced by human activity presents a more continuous impact on global temperatures than volcanic eruptions. Levels of sulphate from human activities have tripled over the last century, with greatest concentrations in the northern hemisphere. Sulphate aerosols can prolong the life of clouds by acting as nuclei for condensation, and the increased albedo produces a cooling effect.

Increasing water vapour and cooling effects of clouds

'Greenhouse sceptics' disagree with the supporters of a 'warming Earth' especially over the role of water vapour, the most common greenhouse gas. The sceptics feel that too little emphasis is given to the argument that a warmer atmosphere can hold more water vapour than a cooler one, and thus the potential for more condensation and clouds is increased, particularly when combined with the effect of aerosols. Some sources say that increased cloud cover leads to lower daytime temperatures as solar radiation is reflected back to space, while at night higher temperatures are likely as cloud traps heat which would otherwise escape to space. There is some truth in this argument, but it is difficult to assess the overall impact. The greatest global warming seems to occur in the most northerly latitudes in winter when snowfall is highest, which ought to result in an increasing reflectivity, and therefore cooling!

But the most sobering evidence concerning the influence of water vapour comes from the planet Venus. Early in that planet's history, all water vapour rose into the upper atmosphere, releasing the hydrogen content which was lost forever to space. Since then, Venus, without its protective water vapour and cloud layer, has experienced a runaway greenhouse effect, and surface temperatures of 470 °C now exist there (Pickering and Owen, 1994).

Variations in solar radiation

Some scientists have argued that cycles of solar activity may account for variations in global temperature. Records of sunspot activity since 1700 show cycles of roughly 11 and 100 years; while carbon dating of wood fragments has identified a 9000-year cycle. The Little Ice Age of the 17th century corresponded with a so-called 'quiet sun' or

▶ **Figure 6.12** The eruption of Mount Pinatubo, 1991. Volcanic eruptions cause short-term but very large emissions of sulphate aerosols which have a temporary cooling effect on global temperatures. They also discharge volumes of chlorine and fluorine compounds into the stratosphere, which combine with CFCs to deplete the ozone layer further.

'sunspot minima' of a 100-year cycle, as did the low winter temperatures of the 19th century. Warming in the 20th century seems to coincide with 'sunspot maxima', and some believe that the 21st century may return to 'sunspot minima' and another Little Ice Age, unless global warming intervenes.

However, the link between global warming and sunspot activity is difficult to prove. Records of solar activity are short and the smoothed temperature record since 1860 shows a generally upward trend (Figure 6.13).

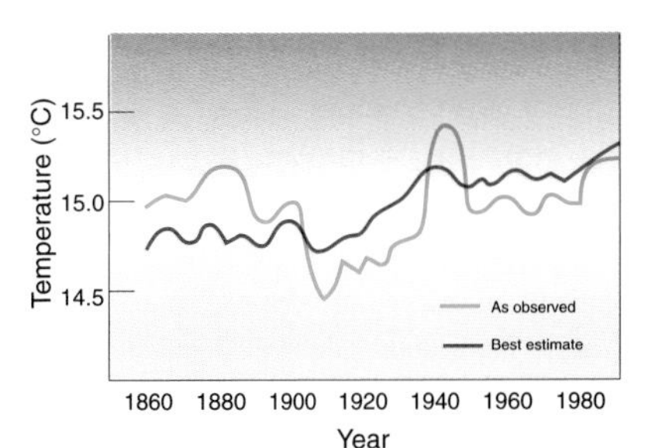

▲ **Figure 6.13** Global temperature change since 1860. The blue line shows observations from the historical records, and the red line shows the output from computer runs which adjust for events such as El Niño and the eruption of Pinatubo. The historical (blue) line is more complex than the steady rise shown by computer runs because it is based on real events.

The role of the oceans in controlling the pace of global warming

Oceans transport as much heat around the globe as the atmosphere. They can store as much heat in their top few metres as the entire atmosphere, and, when the whole depth of ocean is taken into account, they have a heat capacity that is 1000 times greater. The deepest oceans respond very slowly to atmospheric temperature change compared with land areas, and may absorb much of the warming that is currently thought to be taking place. Measurements in the North Atlantic over a 20 year period show some warming at depths between 500 and 2500 metres but cooling nearer the surface (Figure 6.14). In the Pacific, however, hardly any warming at depth has occurred. These observations are confirmed by a computer simulation run by the Hadley Centre – part of the UK Meteorological Office. Figure 6.15 shows that the oceans below a depth of about 2500 metres are not yet responding to increases in mean global temperature, although there are signs of warming at and near the surface.

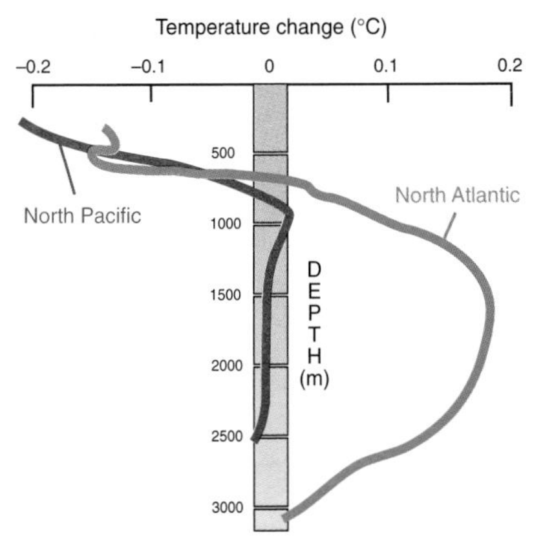

▲ **Figure 6.14** Temperature changes in the North Pacific and North Atlantic.

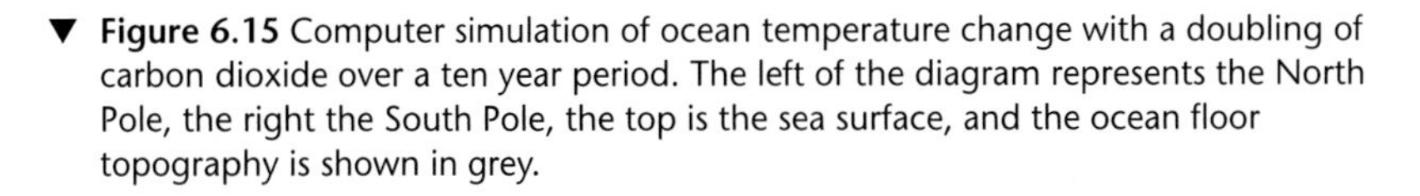

▼ **Figure 6.15** Computer simulation of ocean temperature change with a doubling of carbon dioxide over a ten year period. The left of the diagram represents the North Pole, the right the South Pole, the top is the sea surface, and the ocean floor topography is shown in grey.

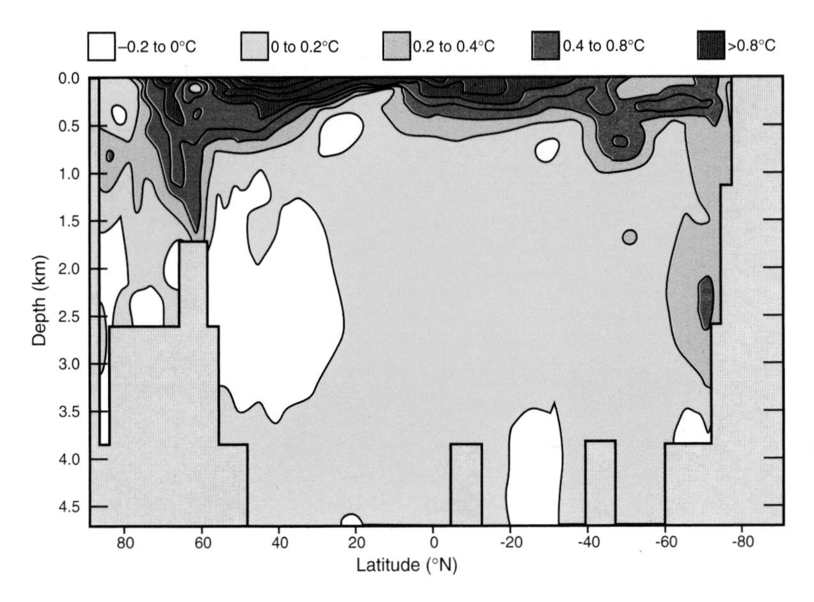

1 Read through the evidence against the enhanced greenhouse effect and try to evaluate the strengths and weaknesses of the case. On which side does the evidence seem stronger?

2 Discuss the two cases together in class. Convene a global warming court. You will need 'lawyers' for the prosecution and defence, a 'judge', witnesses for each side, and the rest of the class as the 'jury'. The charge is that 'Global warming is changing the world's climate'. The jury should be asked to reach a majority verdict on the basis of the evidence heard after a 'summing-up' by the judge of the key arguments presented. Afterwards, in a debrief, you may like to discuss 'out of role' how a decision was made. Was additional evidence needed by the jury? By the defence or by the prosecution?

a

International action on global warming

There are several possible options for world leaders in deciding if and how to respond to the scenarios presented in a global warming future. Each option reflects a particular value-judgement about the likely response of the atmospheric system to the whole range of controlling factors that have been described in this section. Three main categories of response are shown in Figure 6.16.

▼ **Figure 6.16** The three main policy options for world leaders.

1 A 'no regrets' option

Supported by most scientists and policy-makers who accept that doubling of carbon dioxide over the next century will lead to a temperature rise of 2–4 °C. They are influential in deciding global policy. They believe that the delay in response of the atmosphere to increases in greenhouse gas emissions is due to the absorption of heat by the oceans, and that this will hide warming for another 30 or 40 years. They urge research but also feel that there is no time to be lost in case the worst-case scenarios come true. They therefore urge major changes in lifestyle and economic policy. It is called a 'no regrets' policy because even if the worst effects of global warming are not felt for some decades, there will be great benefits. These include reduced levels of pollution (and therefore improved air quality in cities, and less acid rain), lower dependence on fuel imports, and preservation of energy sources. A range of solutions is put forward including:

a) Limiting the use of fossil fuels and using more sources of renewable energy. This began following the agreement at the Earth Summit in 1992 to reduce carbon dioxide to 1990 levels by the year 2000. To date, only 14 of the 160 signatories have confirmed that they will do this. Some including the UK are waiting for action from others before committing themselves. Some developing countries are struggling with the costs involved (see option 3). Others feel there is an urgent need to ensure that developing countries control their future use of fossil fuels. To stabilize emissions at 1986 levels and allow for rises in fossil fuel use by India and China, developed countries would have to reduce greenhouse gas emissions by 30 per cent by the year 2025 (Parikh *et al*, 1991).

b) Introducing a carbon tax, This proposal is gathering momentum in Europe. An agreement in the European Union in May 1992 to introduce a carbon tax depended on the USA and Japan doing the same, but this has not yet happened. However, several countries including Germany, the Netherlands, Sweden, Norway, and Finland, have introduced their own measures. In the UK the introduction of VAT on heating was seen by some as a response to this, but opposition groups see it as a tax on the least well-off and a cynical attempt to raise money.

c) Promoting research into climatic change and its impact. All scientific opinion accepts that computer modelling requires improvement. We need to develop the ability to simulate medium-scale changes in weather systems, such as the behaviour of depressions and anticyclones.

d) Setting up very expensive 'geo-engineering' projects such as:

- placing giant mirrors in space to reflect solar radiation
- seeding oceans with iron to fertilize algae and phytoplankton to increase their intake of carbon dioxide as growth is stimulated
- delivering huge amounts of heat-scattering dust to the stratosphere from high-flying aircraft in order to 'seed' clouds.

e) Introducing energy-efficiency projects to reduce waste in homes, factories, offices, and power stations.

2 A 'wait and see' policy

The supporters of this are the 'greenhouse sceptics'. They see little point in investing huge amounts of public money in projects that will radically change lifestyles, and will commit them to changes which may be difficult to reverse if the scenarios turn out to be inaccurate. Some scientists who support this view – largely those on the right-wing – believe that government policy has been 'hijacked' by environmentalists who are seen as left-wingers who predict a frightening future when the arguments for global warming are unclear. 'Greenhouse sceptics' believe that controlling emissions by means of carbon taxes will reduce economic growth, and they see this as a high-risk strategy. They believe that governments should do nothing until there is clear evidence that global warming is a fact. They also believe that the 'market' will help to avoid damage and adapt to environmental change much more quickly than government policies.

3 'Polluter pays' – a view of developing countries

Despite the concerns about increasing use of resources, and the demand for higher standards of living, the developing world is still behind the developed world in its emissions of greenhouse gases. For example, India emitted some 115 million tonnes of carbon in 1985 compared with the USA's output of 1.3 billion tonnes; in 2025 India's total emissions are likely still to be only 0.6 billion tonnes, a per capita figure of 0.36 tonnes compared with a world average of 1.2 tonnes. The developing world argues that it is the developed world that has built up carbon dioxide. On a principle of 'the polluter pays', those countries that have above-average emissions (the 'debtors') should transfer resources to those who are below-average contributors to global warming (the 'creditors'). This would pay for basic environmental improvements such as access to safe drinking water, proper sanitation, and adequate nutrition by securing support for small farmers. The detailed proposal is set out in Figure 6.17.

Every year the world pumps about 31,100 million tonnes of carbon and 255 million tonnes of methane into the atmosphere. The Earth's ecological system – its vegetation and its oceans – absorbs just over half the carbon and four fifths of the methane produced. Global warming is caused by overloading the cleaning ability of this system. Some countries are doing a lot more damage than others – and it is the whole world that will suffer. But environmentalists from the South have come up with a scheme – called 'tradable emissions' – which could inject some justice into this scenario. This is how it would work:

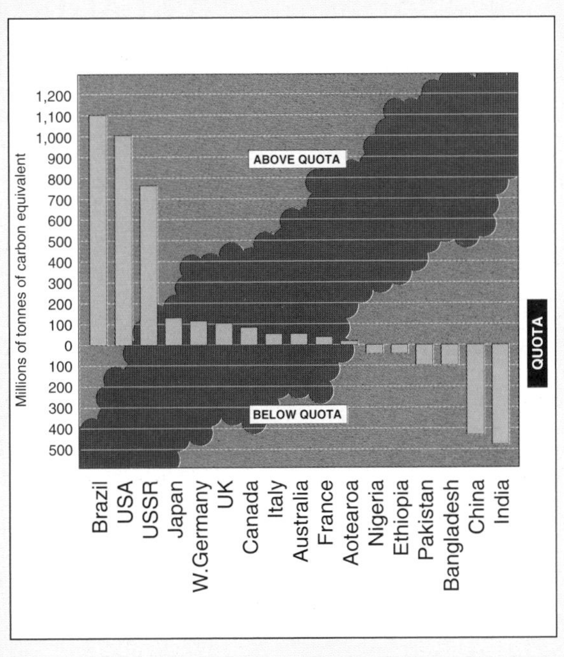

- Each country prepares a budget of greenhouse gas emissions to set against the size of its population and its carbon 'sinks', like veg-etation and soil, and the earth's total cleaning capacity.
- Each country's fair share of the oceanic and atmospheric sinks – a common heritage of human-kind – is calculated.
- Each country's permissible emissions are worked out.
- Compensation is paid by those who are over the permitted level to those who are under it.

In 1990 India, for example, had 16.2 per cent of the world's population but contributed just 6.0 per cent of the carbon dioxide and 14.4 per cent of the methane absorbed by the Earth's ecological system. By contrast, the USA with only 4.73 per cent of the world's population emitted as much as 26 per cent of the carbon dioxide and 20 per cent of the methane.

If the 'tradable emissions' scheme were in place with over-producers paying $15 per 1,000 tonnes of excess, the US would have had to pay $6.3 billion and India would have received $8.3 billion. India could have funded almost its entire education programme for one year with this money. Twenty developing countries would receive about $30 billion.

Industrialized countries together exceed their limit by 2,839 million tonnes of carbon equivalent. These countries also emit large quantities of chloroflu-orocarbons (CFCs), which do not get absorbed at all because there is no natural sink for them – the USA alone produces 25.8 per cent of the world's CFCs. All such chemicals should be added to the net emissions of individual countries. Developing countries provide space for about 1,459 million tonnes of carbon to be released by developed countries: India, China and Pakistan alone provide two thirds of this space.

▲ **Figure 6.17** Extract from *The New Internationalist,* April 1992. The developing world proposal for tradable emissions.

1 Using the data in this section and other supporting research, produce a seven-minute summary of the policy choices available to world leaders, for the *Newsnight* programme on BBC2. This will then lead to a studio discussion of the options by representatives for the different arguments, which different members of the class could role-play. You may wish to elect a chairperson to referee your discussion.

2 Using a recent CD-ROM of climatic change data such as *World Climate Disc* from Chadwyck-Healey Ltd (your school may already have this), produce a report on the global and regional trends in mean temperatures for a typical computer run of 100 years. Are variations local or global? How do your results compare with the data presented in this chapter? The software will enable you to produce maps of change with different projections to highlight particular areas, as well as tables of data, and a variety of graphs.

Changes in urban atmosphere

For a large and growing proportion of the world's population, the climatic modification of the urban atmosphere is their most immediate concern. By the end of the 20th century, the proportion of people living in urban areas worldwide will be 50 per cent, and some 25 cities will have more than 10 million inhabitants. The effects of cities on their immediate surroundings have been studied intensively by geographers since the late 1960s, and the findings are summarized below. By far the most disturbing aspect of climatic modification in cities is the build-up of gaseous emissions in the layer of the atmosphere closest to the Earth (up to 300m). For those urban planners, local authorities, and government departments faced with increasing public concern over deteriorating air quality, several key questions are high on the political agenda.

- What is the level of pollutants in urban atmospheres? Does this vary across a city?
- Are rural areas healthier places to live in than cities?
- What are the effects of urban pollution on people's health? How can these be reduced?
- How will global warming affect the meteorological conditions which in turn control pollution levels?
- What policy changes are needed in order to improve urban air quality?

How urban microclimates function

The construction of buildings and roads at high densities, combined with intensive human use, especially during a working day, produces great changes in the urban microclimate. Two aspects are most important:

- modification of temperatures
- modification of atmospheric gases.

Modification of temperatures

City temperatures indicate a gradient from higher values in central areas to lower values in surrounding rural areas. This 'heat-island' effect may result in minimum temperatures that are 5–6 °C higher in urban areas than in the surrounding countryside. It is particularly noticeable in mid-latitude cities during calm, clear, anticyclonic weather, especially in the early evening. This is when the heat stored by urban surfaces during the day is released, supplemented by space-heating in buildings. It is also caused by the relative heating and cooling of urban and rural areas. Rural areas, which generate less heat from human activities, cool down far more quickly.

Research shows that it is the density of urban buildings, and not the size of the city, that increases the heat-island effect. Many medium-sized cities, of 250 000 people, say, may therefore have the same warming effect as large cities of over 1 million people. To this is added the friction effect on air movement over densely packed buildings in cities, which causes increased air turbulence over cities. This produces instability in the atmosphere and results in more thunderstorms over cities – see the example of St Louis in Figure 6.19.

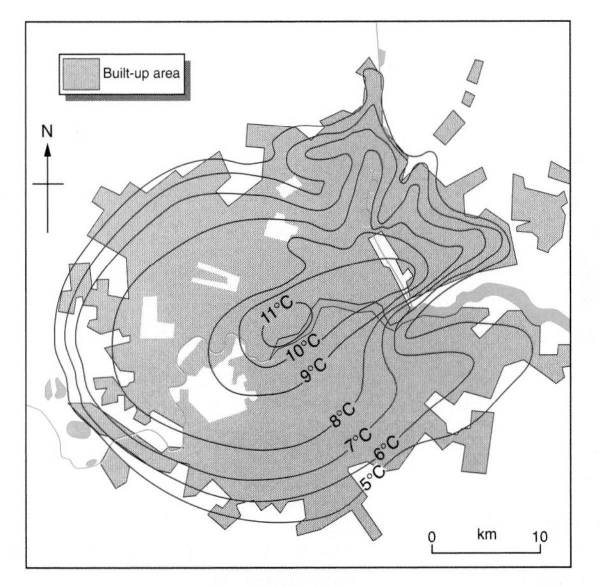

▲ **Figure 6.18** London's urban heat island, 14 May 1959.

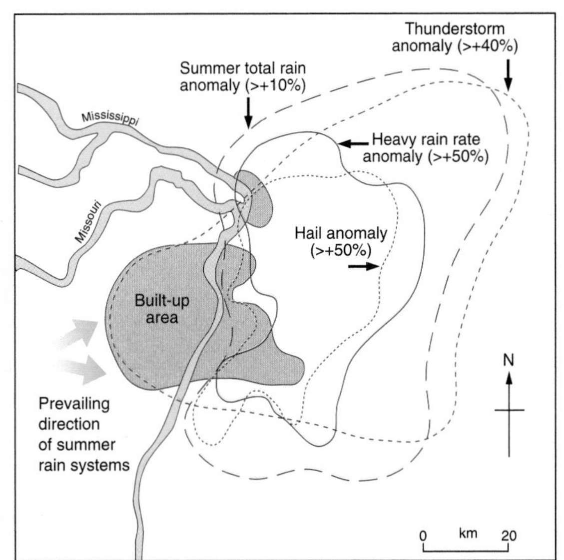

▲ **Figure 6.19** Increased rates (anomalies) of heavy rainfall, hail, and thunderstorms downwind of St Louis, USA.

Continued on page 138

Continued from page 137

Modification of atmospheric gases

The atmosphere of cities is modified by the addition of aerosols, which produce smoke pollution and fogs as a result of the greater number of particles on which condensation can take place. This is less characteristic of cities in Europe, the USA, and Japan where there are some laws concerning air pollution. In some cities of the developing world, China, and the former Communist bloc, the use of coal with a high sulphur content, particulates from heavy industry in city centres, and lack of regulation on emissions, cause health-threatening smogs. Vehicle emissions are the chief problem in Europe, the USA, and Japan. These react in sunlight to produce a 'photochemical smog' that contains high levels of ozone and nitrogen dioxide. Respiratory disorders (asthma and bronchitis especially), heart conditions, and cancers are all thought to be linked to the build-up of particulate and photochemical smogs.

The build-up of smog is greatest in late afternoon and evening and is made worse by calm, anticyclonic weather conditions. The formation of a pollution dome (Figure 6.20) is the result of rising warm air being replaced by denser, subsiding cold air.

A boundary develops between the stable air near the urban surface and the warmer air above. It is the opposite of the normal temperature pattern where cooler air replaces warmer air above, and it is therefore called an **inversion**. The trapping of pollution below the inversion is called a **pollution dome**, which can lead to a build-up of pollution. With morning sunshine after a clear night, convection currents develop, causing pollutants to rise and then fall with down-draughts as they reach the inversion boundary. Figure 6.21 shows how pollutants travel long distances in a horizontal plume in stable conditions, called **fanning**, and with light winds above the inversion layer, called **lofting**.

In cities such as Los Angeles and Mexico City, other factors maintain pollution plumes and smog for much longer periods of time. In Los Angeles (Figure 6.22), pollutants build up below an inversion layer up to 1000m thick, reinforced by subsiding air cooled by cold offshore ocean currents. Sea breezes are drawn into the urban heat island during the day and move pollution eastwards towards the barrier of the Rocky Mountains.

▼ **Figure 6.20** An urban pollution dome.

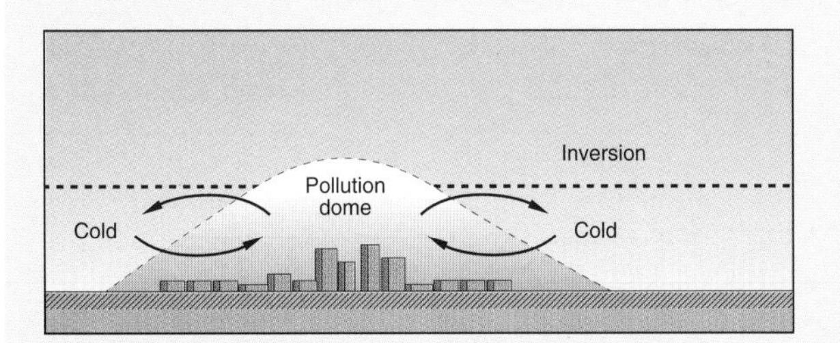

▼ **Figure 6.21** Movement of pollutants over a city in anticyclonic conditions.

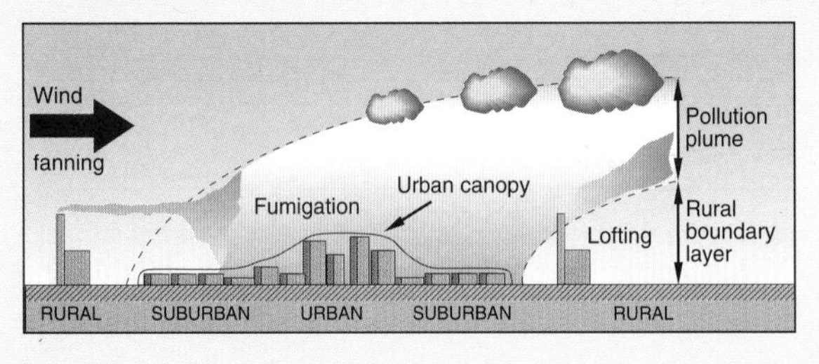

▼ **Figure 6.22** Pollution traps in Los Angeles.

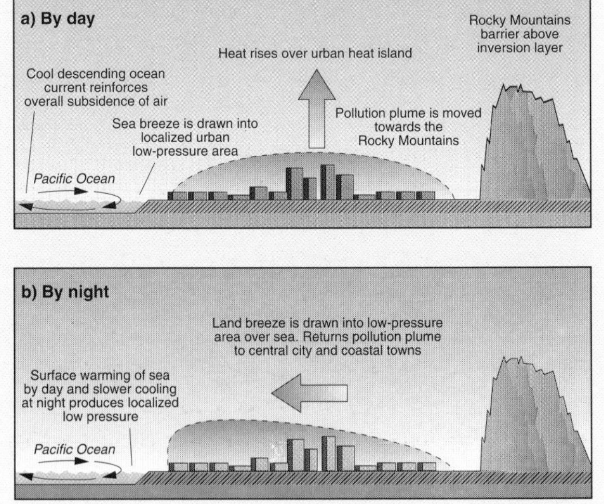

At night, when the sea cools down more quickly than the city, air is drawn back across the city towards the coast as a land breeze. In Mexico City, which is build on a flat lake bed and surrounded by the Sierra Madre mountain range, rising air currents move pollutants up the valley to the wall of the mountains by day, only to return them with falling cooler currents of air at night. Thus pollutants are redistributed rather than removed.

1 Using large-scale atlas maps of Los Angeles and Mexico City, draw annotated sketch maps to illustrate the ways in which pollution builds up over time. Show the daytime and night-time air movements in different colours.

Urban pollution in Britain

Between 5 and 9 December 1952, a temperature inversion in London caused a dense fog with visibility of less than 10m for 48 consecutive hours. The concentration of smoke and sulphate particles from coal burning changed a natural atmospheric fog into a smog which resulted in 4000 deaths. Some 12 000 additional deaths were recorded between December 1952 and February 1953 compared with the same period twelve months previously. In 1956 the Clean Air Act was passed. The Act included measures to create smokeless zones, to encourage the use of smokeless fuels, and a move away from coal to natural gas and oil for space-heating. The rapid improvement in air quality following the introduction of the Act was dramatic, and there was a widespread belief that the 'old pea-soupers' were gone for ever.

However, the huge rise in vehicle ownership since 1960, and a change from rail to road as the means of distributing goods in the 1980s, has led to an increase in photochemical smogs during summer and winter anticyclones. In December 1991, levels of nitrogen dioxide reached 423ppb at a time when the Department of the Environment measure of 'very poor air' was 300ppb. In June 1994, thunderstorms during a hot summer spell led to levels that resulted in a record number of asthma attacks in London. A major nationwide survey of nitrogen dioxide concentrations for the Department of the Environment in 1992 found that 19 million people were living in areas where the level was greater than the EC 'annual mean guide value'. To quote Hansard (June 1993): 'Eight out of ten of the sites with the highest levels were in London, somewhat contradicting ministerial reassurances that air quality in London is "good or very good 97 per cent of the time".' Transport is estimated to produce 74–76 per cent of nitrogen oxides, 94–96 per cent of black smoke, and 99 per cent of carbon monoxides. Over the last decade there have been increases of 61 per cent in nitrogen oxides and a doubling of black smoke, whilst surface ozone is rising at 1 per cent per annum.

Monitoring air pollution in London – a decision-making exercise

You have been asked by the London Boroughs Association to produce a report on the levels of air quality in London. The London Boroughs Association, the Association of London Authorities, and the South East Institute of Public Health set up the London Air Quality Network (LAQN) in 1993 to monitor air quality across the capital. You should produce a report on the situation and suggest recommendations to be made to the Department of the Environment to improve air quality. Use Figures 6.23–6.31 to help you.

Your report should include the following:

a) An explanation of the causes of photochemical smog in cities and some definitions of poor air quality.

b) The differences between monitoring standards issued by the Department of the Environment, the World Health Organization (WHO), and the European Union (EU), and why these are important.

c) The use of graphic and cartographic techniques to illustrate levels of pollution across the capital, and a comment upon patterns that emerge.

d) An assessment of health risks associated with urban pollution, and which groups are most at risk.

e) Recommendations for the improvement of air quality in London.

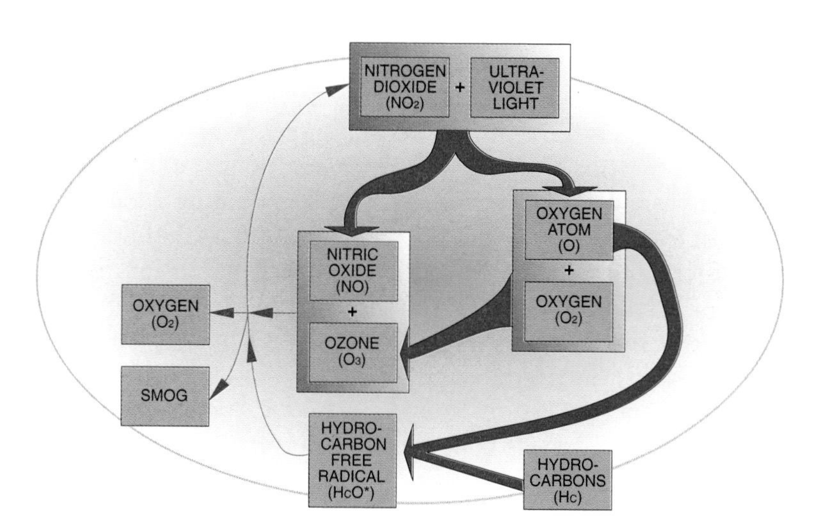

▶ **Figure 6.23** The causes of photochemical smog. Ultra-violet light causes natural nitrogen dioxide (NO_2) to break up into nitric oxide (NO) and free oxygen. Free oxygen (O) combines with natural oxygen (O_2) to form ozone. Ozone reacts with nitric oxide to produce more nitrogen dioxide, and oxygen. Photochemical smog results from this reaction and other chemical reactions of hydro-carbons from fossil fuels with free oxygen. This 'cocktail' is linked with increases in respiratory illnesses. The way in which the gases react with one another and with other pollutants is not fully known.

▶ **Figure 6.24** London's air pollution dome: note the marked inversion layer at the top of the haze.

POLLUTANT	Sulphur dioxide	Airborne particulates	Carbon monoxide	Ozone	Nitrogen dioxide
SOURCE	Combustion of sulphur-containing fossil fuels. Power stations contribute 73% of UK total. In cities levels can be boosted by smoke from diesel.	Coal burning used to cause most black smoke but vehicles are increasingly important, contributing nearly a third overall and up to 90% in urban areas.	Incomplete combustion of fuel. 85% from motor vehicles.	A secondary pollutant formed by a photochemical reaction: sunlight + nitrogen oxides + hydrocarbons. Highest levels occur when hot spells combine with heavy traffic.	Motor vehicles are the major source, providing 45% of total nitrogen oxides. Power stations emit a further 35%.
HEALTH	Breathing problems, e.g. bronchitis. Asthmatics are particularly badly affected. Most serious in combination with particulates which with moisture can form sulphuric acid in lungs. Long-term exposure can increase mortality from cardiorespiratory diseases.	Heavy metals and some complex organic compounds, such as polyaromatic hydrocarbons, carried deep into lungs on particulates. These potentially cause cancer.	Deprives body of oxygen by reacting with haemoglobin. Slows thought and reflexes, causes drowsiness and headaches. Increased pressure on heart. Fatal at high concentrations. Can retard foetal growth if inhaled by pregnant women.	Although beneficial in the ozone layer, it is harmful at ground level. High concentrations severely damage lung tissue and impair defence against infections. Lower concentrations cause coughing, impaired lung function, eye, nose and throat irritation and headaches, particularly in people who exercise. Aggravates asthma and bronchitis.	Increased susceptibility to viral infection, irritates lung tissue, increases risk of bronchitis and pneumonia.
ENVIRONMENTAL EFFECTS	The main constituent of acid rain, which damages aquatic life and increases concentrations of heavy metals in acidified water. Injury to plants, both vegetation and trees. Corrosion of buildings.	Soiling of buildings with associated cleaning costs. Reduced visibility. Odour.	Reacts with chemicals that would normally remove 'greenhouse gas' methane, and oxidizes to carbon dioxide, another contributor to global warming.	A major component of photochemical smog and a 'greenhouse gas'. Damages agricultural crops by reducing yields and implicated in forest damage. Damages plastics, rubber, and paints.	Responsible for about one-third of the acidity of rainfall. At low dosages, can stimulate plant growth, but also increases susceptibility to insect attack or frost damage.
AIR QUALITY STANDARDS	Smoke / SO_2 Median daily values for whole year: 80μ g/m^3 / $120\mu g/m^3$ smoke < $40\mu g/m^3$; $80\mu g/m^3$ smoke > $40\mu g/m^3$ Median daily values for winter: 130μ g/m^3 / $180\mu g/m^3$ smoke < $60\mu g/m^3$; $130\mu g/m^3$ smoke > $60\mu g/m^3$ Peak value for whole year: 250μ g/m^3 / $350\mu g/m^3$ smoke < $150\mu g/m^3$; $250\mu g/m^3$ smoke > $150\mu g/m^3$	*Notes* WHO 1–hour guideline for $SO_2 = \mu g/m^3$ WHO guidelines for combined exposure to SO_2 and particulate matter (black smoke measuring method) are: 24 hours $125\mu g/m^3$ 1 year $50\mu g/m^3$	There is no UK or EU legislation controlling levels of carbon monoxide in air. The WHO guideline values are: 1 hour 30 mg/m^3 8 hours 10 mg/m^3	There is no UK or EU legislation controlling levels of ozone in air. Ozone is usually measured in parts per billion. The WHO guidelines are: 1 hour 76–100 ppb 150–200 $\mu g/m^3$ 8 hours 50–60 ppb100–120 $\mu g/m^3$	The EU Directive agreed in 1985 is not very stringent. It sets a 98 percentile of hourly values throughout the year limit value of $200\mu g/m^3$. The WHO guideline values are: 1 hour 400 $\mu g/m^3$ 8 hours 150 $\mu g/m^3$
UK LEVELS	The EC Directive limits are not very stringent compared with those of the WHO. Some areas – mainly coal mining – have not complied with the standards. The 1-hour WHO guideline is regularly exceeded in London. WSL, the government's monitoring agency, believes it is exceeded fairly widely throughout Britain.	34 sites exceeded WHO guidelines and 10 sites exceeded EU limit value in 1985/86, mainly due to domestic coal burning.	The WHO 8-hour guideline was exceeded on 24 days between 1 October and 31 December 1988 at one London site. The highest reading was almost double the WHO guideline.	WHO guidelines were exceeded several times during the 1989 hot summer which saw some of the highest levels since 1976. Both rural and urban areas are affected. One Devon site reached 135 ppb.	WHO guidelines are likely to be exceeded at busy roadside locations, says WSL, the government's monitoring agency. WSL also concludes that under different meteorological conditions there is a risk of breaching the EU Directive as well. London Scientific Services data suggest the Directive is exceeded at its roadside station.

$\mu g/m^3$ = thousandths of a gram per cubic metre   ppb = parts per billion   mg/m^3 = milligrams per cubic metre

▲ **Figure 6.25** The main urban air pollutants.

▼ **Figure 6.26** Exceeding the European Union guidelines a) for nitrogen dioxide b) for ozone in some London boroughs.

a)	**Limit value**	**Guide value**	**Guide value**
Borough	EU guideline: calendar year 98%ile hourly means > 104.6ppb	EU guideline: calendar year 98%ile hourly means > 70.6ppb	EU guideline: calendar year 50%ile hourly means >26.2
Hounslow	No (74.33ppb)	Yes (74.33ppb)	No (17.5ppb)
City of London	No (57.32ppb)	No (57.32ppb)	Yes (29.85ppb)
Westminster	No (62.70ppb)	No (62.70ppb)	Yes (36.0ppb)
Barking & Dagenham	No (46.75ppb)	No (46.75ppb)	No (22.75ppb)
Bromley	No (36.6ppb)	No (36.6ppb)	No (14.25ppb)
Hackney	No (39.25ppb)	No (39.25ppb)	No (17.75ppb)
Southwark	No (54.0ppb)	No (54.0ppb)	Yes (26.83ppb)
Bexley	No (46.75ppb)	No (46.75ppb)	No (17.25ppb)

Entries indicate whether European Union guidelines have been exceeded.

b)	**Population information threshold**	**Population warning value**	**Health protection threshold**	**Vegetation protection threshold**	**Vegetation protection threshold**
Borough	1-hour means >90ppb	1-hour means >180ppb	Specified 8-hour means >55ppb	1-hour means >100ppb	24-hour means >32ppb
City of London	1	0	3	0	13
Westminster	0	0	0	0	0
Bromley	4	0	9	3	8
Hackney	0	0	0	0	0
Southwark	1	0	0	0	1
Bexley	11	0	13	4	14

Figures indicate number of times per year guidelines were exceeded. Compare with Figure 6.25.

▼ **Figure 6.27** Respiratory problems resulting from pollution, in increasing order of seriousness.

- Unpleasant odour
- Nose, eye, and throat irritation
- Respiratory tract infections
- Wheezing, coughing, and sputum requiring medical treatment
- Restriction of lung function
- Acute attacks in chronically ill patients
- Increased frequency of asthma attacks
- Increased incidence of lung cancer
- Increased mortality

▼ **Figure 6.28** Traffic congestion – levels of pollution rise dramatically at the kerbside, and young children are thus at increased risk.

▼ **Figure 6.29** Extract from Friends of the Earth, February 1994.

Public concern about the impacts of air pollution on health, in particular the prevalence of childhood asthma, is currently extremely high. There are an estimated 3 million asthmatics in the UK, with one in ten children affected.

There is now firm evidence, supported by recent UK research, that air pollution worsens the symptoms experienced by asthma suffers, and that high pollution episodes result in increased hospital admissions and higher use of medication. A direct causal link between air pollutants and asthma is currently lacking, but evidence of indirect effects, for example the finding that pollutants may act as a sensitizer, increasing susceptibility to allergens, is accumulating.

Carcinogenic emissions also merit attention, in particular benzene (known), 1,3 butadiene (probable) and diesel exhaust (recognized as a probable carcinogen by the WHO six years ago). There remains scandalously little public information about actual levels of carcinogens in ambient air in London, let alone a consensus about what levels can be regarded as 'safe'.

New Scientist has attributed 10 000 deaths a year in England and Wales to particulates, nearly 2000 in London.

7.0 SHORT-TERM RESPONSE: SMOG ALERTS

7.1 Smog alerts are important, not only to protect people with breathing difficulties, but to educate the public about the impact of the motor car on health, thereby building support for transport policies which restrain traffic growth.

7.2 The issuing of a strong statement by a Government minister urging motorists to leave their cars at home is a simple first step which, moreover, is cost free.

7.3 In numerous other countries formal 'smog alert' systems grade measures according to the severity of the pollution episode and, where appropriate, include powers to ban cars or require factories to cut production until an episode has passed. The German system is perhaps the most sophisticated, although designed to address emergency levels of SO_2 and NO_2 from the East, rather than modern-day photochemical smogs. During a smog alert in Hamburg in February 1987, hundreds of police blocked off the city centre with smog warning signs, 800 companies reduced their emissions by 40%, and 400 000 people switched from cars to public transport. Contingency plans – for example the laying on of extra buses and opening of first-class train carriages to all passengers – play an essential part in ensuring the exercise is successful.

7.4 A 50% reduction in car use would have ensured that nitrogen dioxide levels in London in December 1991 never entered the DoE 'poor' band. A greater reduction in car use could have alleviated the severity of the pollution episode. The public is likely to judge harshly Government unpreparedness in the event of such episodes.

8.0 TECHNOLOGICAL MEASURES

8.1 The UK Government should act swiftly to put into place pollution controls which can deliver significant short to medium-term cuts. This should include:

- review of the emission test for vehicles to require better cold-start performance;
- the fitting of Stage II petrol pump controls to reduce VOC (and benzene) emissions;
- tighter controls at EC levels for a range of air toxics from passenger cars;
- fiscal incentives to encourage the clean-up of the existing, and ageing bus fleet, as recommended by the Royal Commission on Environmental Pollution;
- powers to local authorities to fine drivers of smoky vehicles.

9.0 TRANSPORT POLICIES

9.1 It is now broadly accepted that further controls on vehicle emissions will fail to cut pollution to levels which protect health, unless complemented by transport policies which reduce car use.

9.2 'Sustainable Development, The UK Strategy' admits that although new emission standards will begin to reduce pollutant emissions per car over the next ten years, by 2012 overall emissions will begin to rise again as growth in traffic overtakes these gains. It concludes that if a progressive improvement in urban air quality, and the improvement in the quality of life in urban areas which it can bring, is to be achieved then '*contributions must also come from improvements . . . in traffic limitation . . .*'

9.3 Key policies should include:

- a requirement that local authorities ensure that air quality guidelines are met, and that this objective is reflected in transport and planning policies;
- adequate central Government funding for TPPs designed to meet local traffic reduction goals;
- a substantial increase in investment in public transport;
- the implementation of planning policies aimed at reducing car use;
- the use of funds raised from fuel price increases, and the diversion of funding from the national roads programme towards transport policies aimed at reducing traffic growth.

▲ **Figure 6.30** Ways forward: recommendations from Friends of the Earth, 1994.

Urgent action is needed to reduce air pollution in London and prevent deaths and respiratory illness, the London Boroughs Association said yesterday, publishing its first air quality survey.

The association said its findings were in sharp contrast to results given by the Department of Environment which, it claimed, failed to keep the public informed of the full danger to health from air pollution. John Gummer, the Environment Secretary, said the Government would be issuing a consultation document this year reviewing ways of dealing with the situation. But he appeared to rule out curtailing the use of private cars even at times when it was clear that air pollution had reached levels which were damaging public health. He said: 'I do not see a need for substantial traffic bans in our urban areas. We do not face problems on the scale of Athens and Rome. Nor is it clear that there is a need here to restrict movement by vehicles without catalysts during pollution episodes, although that is something we must continue to keep under review.'

Councillor Ronnie Barden, chairman of the LBA's environment committee said: 'The priority must be to persuade people out of their cars and on to buses and trains. Greater investment in public transport to make it more attractive and efficient is vital, along with other initiatives such as more bus priority measures, park and ride schemes and tougher parking controls.'

Fiona Weir, air pollution campaigner for the Friends of the Earth, said: 'Mr Gummer is making the same speech over and over again without doing anything. We have been promised targets and limits for pollution of certain substances since 1990. We are still waiting.'

The survey found high levels of nitrogen dioxide in Barking, the City of London, Westminster and Hounslow, and levels of carbon dioxide which breached WHO guidelines at Southwark and Hounslow.

High levels of ozone pollution were found in Bexley and Bromley last June, while sulphur dioxide was higher in Barking, Dagenham and Bexley than in central London. The air quality report showed that in the capital in December 1991 the problem could have been prevented if traffic had been cut by 50 per cent.

▲ **Figure 6.31** Extract from *The Guardian*, 14 July 1994.

Measuring nitrogen dioxide pollution levels

An excellent focus for an individual study at A-level would be to undertake an investigation of nitrogen dioxide levels in your local area. It is possible to monitor levels using simple, low-cost equipment called **diffusion tubes**. These are readily available from local scientific laboratories, which your local Environmental Health Office should be able to identify for you.

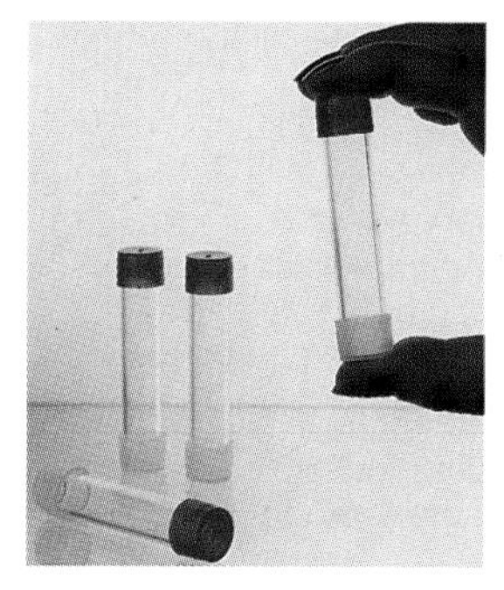

▲ **Figure 6.32** Diffusion tubes.

▲ **Figure 6.33** A typical local authority diffusion tube monitoring site.

The tubes are approximately 50mm long and require no power or other services. They can be mounted on the sides of buildings, or on street furniture such as lamp-posts or railings.

Brief guidelines for use

- Ideally you will want to site your tubes at a variety of places in your local area to get a detailed coverage. Some might be close to busy roads and others at distances of 10m, 20m, and 30m away. Others should be in urban background sites in residential areas, at least 50m from busy roads.
- Permission to site tubes may have to be obtained from landowners or the local council if street furniture is to be used.
- Maintain a standard monitoring period of one week or one month.
- Keep careful records of traffic flow and weather conditions to help you interpret your results.
- At the end of the monitoring period, send your carefully labelled tubes to your local contact analysis service. You will be charged a small fee. Keep a record of all the tubes sent, and their location.
- Think about the presentation of your data. You may wish to show small-scale variations using isoline maps and located bar charts, for example. Alternatively, graphs will enable you to show changes over short time-spans, e.g. one day, or trends over a week or month, using running means. The more sampling sites you have and the longer the sampling period, the more extensive your results will be.

Ideas for further study

1 Carry out a survey of pollution levels and air quality in a town or city known to you. Follow the guidelines in this chapter on page 144.

2 Investigate to what extent people of your and your parents' generation are prepared to change their lifestyles in order to see pollution levels and the threats posed by global warming reduced.

Summary

- Human activity has greatly modified the composition of the atmosphere. Quantities of naturally occurring greenhouse gases have increased with the growth of human activities which release them.
- Human activity may lead to an enhanced greenhouse effect which traps a greater percentage of the outgoing infra-red radiation below a thickening blanket of gases which include relatively recent pollutants such as CFCs. This change in the world's climate could produce global warming. This scenario is contested by some scientists and politicians who believe that other natural processes will offset any potential warming.
- The majority of scientific opinion is united in believing that adverse consequences for future generations are posed by global warming. This position is contested by opponents of global warming who feel that such scenarios are flawed, and that computers cannot model all the complexities of the atmosphere.
- The implications of a changing global climate and changes to weather systems, such as increasing storminess and rising sea-level, are so severe that a fierce debate is taking place on how to respond.
- How we repond will determine our ability to manage weather and climate in the future, and could have far-reaching effects on people's lives and activities. A variety of options is available, from radical changes in lifestyle to a 'wait and see' approach which places faith in the market and in technology to help us adapt to global warming if it ever occurs.
- Human activities modify the atmospheric characteristics of urban areas to create features such as permanent heat islands and short-term photo-chemical smog. Such changes can adversely affect the health of urban dwellers, especially those who already suffer from respiratory disorders.
- There can be wide variations in pollution levels across a city. There is not always a simple rural–urban divide.
- Reductions in pollution levels will only come about through changes in lifestyle and by radical shifts in central government policy in such areas as transport, and pollution warning systems for the general public.

References and further reading

Association of London Authorities, London Boroughs Association, South East Institute of Public Health, *Air Quality in London*, The First Report of the London Air Quality Network, 1994. (Second Report, 1995, now available.)

R. Balling and S. Idso, 'Historical temperature trends in the United States and the effect of urban population growth', *Journal of Geophysical Research*, 1992, No.94, pp.3, 359–63.

T. J. Chandler, *The Climate of London*, Hutchinson, 1965.

Friends of the Earth, *Climate Change: Our National Programme for Carbon Dioxide Emissions*, March 1993. (A detailed study of progress made from an environmentalist stand-point.)

R. Lindzen, 'Absence of scientific basis' in *Global Warming Debate*, A Scholarly Publication of The National Geographic Society, Spring 1993.

K. T. Pickering and L. A. Owen, *An Introduction to Global Environmental Issues*, Routledge, 1994.

People, weather, and climate: Summary

Key ideas	Explanation	Examples
1 Weather and climate play a key role in the working of the environment	Role of weather and climate as part of the global life support system.	• The influence of weather and climate on human survival, e.g. in excessively cold or hot regions
	Relationships between atmospheric processes and hydrological, geomorphological, and ecological processes.	• Links between weather and climate and other physical processes
	Weather and climate as part of natural systems.	• Weather and climate in tropical regions, e.g. in savanna regions, and how weather and climate vary across a region
2 There are important differences between weather and climate	Definitions of weather.	• Examples and definitions from the introduction
	Understanding what causes weather events – processes of condensation, precipitation, convection, and subsidence within the atmospheric machine.	• Forecasting and interpreting the 1990 storm in Britain on the basis of weather maps and satellite photographs
	Role of airmasses and their interaction along fronts.	• Different airmasses in the British Isles, and in tropical regions • Frontal systems in the north Atlantic, and the ITCZ
	Interpretation of weather systems.	• Depressions and anticyclones in the British Isles
	Definition of climate.	• Examples and definitions from the introduction to this section
	Relationship between weather and climate.	• Keeping your own weather records
3 Weather and climate have combined impacts on people's daily lives and activities	How day-to-day and seasonal variations of weather have shorter-term impacts on farming, business, health, leisure, and daily living and transport.	• Impacts of weather events, e.g. Hurricane Andrew • Contrasts in weather in tropical areas between rainy and dry seasons
	How the nature of climate has longer-term effects on the nature of economic activity such as housing, agriculture, energy provision, lifestyle.	• The Australian drought (Chapters 5 and 9) • The links between climate in savanna regions (Chapter 5) and the lifestyle of the Fulani people (Chapter 9) • The Sahel and the impact of drought over the long term
4 Management of weather and climate poses a continuing challenge for people	• Costs and benefits of weather forecasting. • Imperfections of weather forecasting, particularly of extreme events. • A study of the impact of extreme weather.	• Weather forecasting in the UK • Increased storminess and incidence of hurricanes • Hurricane Andrew
5 Human activities modify the atmosphere both in the short term and the long term, sometimes with adverse effects	• Impacts of urbanization, development, and industrialization on weather and climate. • Urban–rural contrasts, urban heat islands. • Impact of atmospheric pollution on air quality. • Impact of other activities such as afforestation and deforestation, water management, etc. • Long-term impact on climatic change; impact on ozone layer, and possible contribution to the greenhouse effect. • Possible effects – ozone and health issues; climate change caused by global warming.	• Urban climate in London and other cities • The impact of industrialization on pollution and climate • Examples of global climate change (Chapter 6) • Effects of and debate about global warming (Chapter 6) • The potential for long-term change in human activity caused by global warming • El Niño

Ecosystems and human activity

The Russian oil spill, October 1994

Tundra regions are found in the most northern latitudes of the world. The tundra environment appears at first to be a barren landscape supporting little life. In fact, many plants and animals are adapted to survive in the difficult environment found in higher latitudes of the northern hemisphere. Figure 1 shows that northern Russia including Siberia, parts of Scandinavia, northern Canada, and Alaska are all tundra areas. All systems have a balance which, once upset, has implications for different parts of it. This balance is a particularly delicate one in the Arctic.

▼ **Figure 1** Tundra regions of the world.

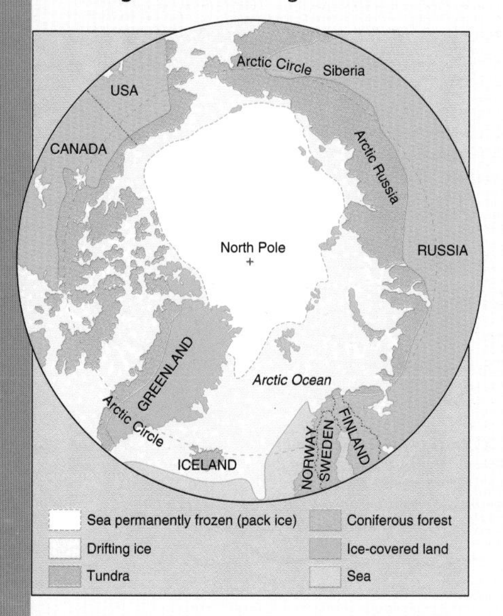

Arctic threatened by oil spill in northern Russia

By George Graham, John Thornhill, and Anthony Robinson

A burning oil slick spilling from a pipeline in the northern Russian province of Komi is threatening to create an environmental disaster in the fragile Arctic.

Komi officials asked Moscow for help in dealing with the spill yesterday, estimating that 200 000 barrels of oil had escaped. But Mr Bill White, the US deputy energy secretary, said a US company at the site had estimated the spill at 2m barrels, nearly eight times the size of the *Exxon Valdez* disaster in Alaska.

Another Western company had measured the slick at 1 metre deep and 12 metres wide, stretching for 10km.

'It is a significant spill, whether it be 100 000 barrels or 2m or somewhere in between. The fact is the Arctic environment heals a lot more slowly than other environments,' Mr White said.

The oil slick is reported to have built up from leaks in a 47km pipeline run by Komineft, a Russian oil production association. The leak is believed to have begun in February, but had been retained by an earth dam. After heavy rains early this month, the dam burst, spilling oil into the Kolva River, which runs indirectly into the Pechora River and thence to the Barents Sea. US officials say the slick was probably set on fire in an attempt to contain it.

▲ **Figure 2** The Russian oil spill, as reported in the *Financial Times*, 26 October 1994.

In October 1994, a massive oil leak in Arctic Russia focused international attention on the damage inflicted on fragile environments by oil spills. The oil spill was from a leaking pipeline in Komi. In terms of the volume released, the spill was worse than either the *Exxon Valdez* disaster in Alaska or the wreck of the *Braer* off the Shetland Islands. Each of these events had also caused significant damage to the environment. The Russian spill started in February 1994. The spilt oil was trapped by the construction of an earth dam, but heavy rains in October caused the dam to burst, carrying the oil along rivers and into the sea. Over a period of eight months, an enormous quantity of oil had accumulated. The newspaper article in Figure 2 tells the story of the spill. The oil company, Komineft, tried to clean up the oil using vacuum pumps and then set the oil slick alight, in the hope of reducing damage.

Kominefft very much played down the size and significance of the spill, claiming that environmental groups had overreacted to the accident. Figure 4 shows how local people reacted. They were largely resigned to the situation; most had seen oil spills in the area many times before and knew of the damage. Others relied either directly or indirectly upon Komineft for their jobs. They feared that tighter restrictions on the oil company would mean fewer jobs in the future.

Even on a smaller scale, the escape of oil is likely to have serious effects on the environment. When a spill occurs in a fragile and sensitive environment such as the Arctic, its effects can be devastating. Why should this be the case? Why does the Arctic ecosystem take so long to recover? Is the damage permanent?

1 Using Figures 2, 3, and an atlas, draw a sketch map to show the location of the Kolva river in Russia. Label it fully to show:
   a) when and where the oil escaped
   b) the estimated amount of leaked oil
   c) why the oil leaked
   d) steps taken to clean up the spill.

2 From your atlas, shade in the extent of the area of northern Russia covered by tundra.

3 Study Figure 4. In pairs, categorize the views expressed by people in the area. To what extent do views conflict? What reasons does each person have for expressing the views shown?

▲ **Figure 3** The location of the oil spill.

▼ **Figure 4** How people reacted to the oil spill.

Effect of oil on ecosystems

The term 'ecosystem' refers to a community of plants and animals together with the environment in which they live. It includes both living parts – plants, animals, and micro-organisms – and non-living parts such as rocks, soil, and climate. Where conditions are harsh, as in the Arctic, only a small number of species are found. A balance exists within all ecosystems which, once upset, affects other parts of the ecosystem. If large tracts of land or water are inundated with oil, and plants and animals destroyed, the whole ecosystem is threatened.

When oil escapes from a pipeline, it is in an unprocessed state, known as 'crude oil'. While warm, it is relatively liquid and flows easily. Once it is exposed to the atmosphere, many of its component chemicals evaporate, reducing it to a thick sludge. Both the liquids that evaporate readily, such as benzene, and more stable compounds have the potential for environmental destruction. Sulphur and other chemicals in the oil are toxic to both plants and animals. When oil escapes into water, it can poison fish and aquatic species. Birds and mammals which feed on fish are affected by the oil when feeding. The oil on a bird's plumage destroys its buoyancy, which it needs in order to float without drowning. Many of the chemical compounds within crude oil also have a toxic effect on birds, and are consumed when birds preen.

There are few ready solutions to a spill of this kind. Two methods are commonly applied. First, the use of detergents helps to disperse the oil, but does not destroy it. Detergents may do more damage to the skin of an animal. Second, setting fire to the oil destroys it, but not without releasing toxic fumes and smoke into the environment, causing environmental destruction of plants and animals. In the longer term, toxic compounds may remain in soil for several years.

What questions are raised?

The main issue raised by such an incident is the conflict between economic need and environmental protection. Oil is essential to provide standards of living that people in developed countries are now accustomed to, and demand is likely to increase as more countries develop economically. Yet costs to the environment can be enormous. Accidents threaten environments, as does lack of care in the use of resources, and the deliberate actions of people or organizations whose concern is economic gain with no regard for the environment.

Whether it is oil in the Arctic, timber from tropical rainforests, or the demands of people to be able to enjoy the countryside, the environmental costs of resource exploitation are far from fully realized. Section 3 aims to help you understand the structure of ecosystems and how they function, and to consider how people can manage the environment. You will then understand more about the conflicts between economic benefits and environmental costs which lie behind many news stories.

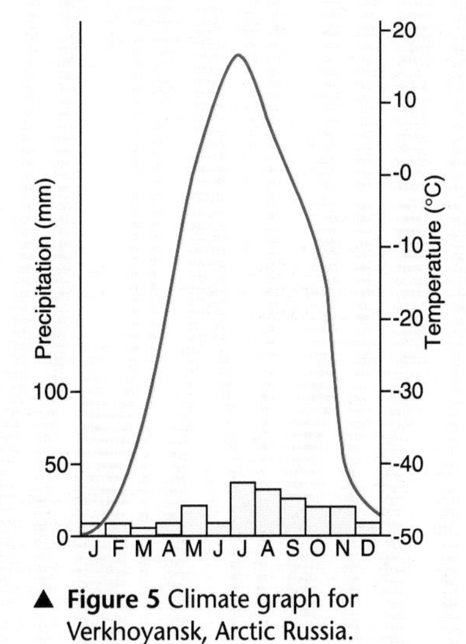

▲ **Figure 5** Climate graph for Verkhoyansk, Arctic Russia.

1 Study Figure 2. Reports of the oil spill claimed that 'the effects of the oil spill will be felt far away from the point where the spill took place'. What was meant by this?

2 Study Figure 5, which is a typical graph for a tundra climate. What difficulties for cleaning up the oil did the climate present:

a) in the month of the spill

b) in the following two months (time of the clean-up)

c) in April–May when the final winter thaw takes place?

3 Explain why the Russian oil spill may have serious implications for this Arctic region. How might the implications be different if the same incident occurred in the UK?

4 a) Make a list of ecosystems which you feel are threatened by their use or abuse by people.

b) In each case, identify the economic benefits and the environmental costs.

7 Small-scale ecosystems

▲ **Figure 7.1** Summer in the tundra at Taimyr, Arctic Russia. Shallow soils lie on permanently frozen subsoil called permafrost. Temperatures only rise above freezing for about three months of the year, and in winter may fall to –40 °C or less. These are the cold deserts with air so cold and dry that there is little precipitation; annual amounts are often below 250mm.

Introduction to ecosystems

In order to understand the ways in which human activities affect ecosystems, this chapter deals first with two questions:

- What is an ecosystem?
- How do ecosystems function?

What is an ecosystem?

An ecosystem is a community of plants and animals together with the environment in which they live. It has two main components:

- abiotic or non-living elements, such as rocks, soils, and water
- biotic or living parts, such as plants, animals, and micro-organisms.

Ecosystems are found within the biosphere, the part of the Earth and its atmosphere that supports life. The biosphere includes oceans, land, and the lower atmosphere. Life within the biosphere depends on the exchange of materials with the hydrosphere, the lithosphere, and the atmosphere. These are defined in Section 2.

Boundaries of ecosystems are difficult to identify. Plants may not have fixed boundaries, so that the edge of a woodland may be indistinct from the beginning of grassland. Inorganic components may move from one ecosystem to another. An ecosystem is therefore an *open* system because it is affected by factors outside it.

Ecosystems exist at a range of scales, from a puddle or a rotting log, to a lake, a woodland, an ocean or forest. Ecosystems also exist at a global scale, where they are referred to as biomes. Biomes are plant and animal communities like those characterized on world vegetation maps – tropical rainforests, desert, or tundra, for example. The introduction to this section has outlined some of the effects of the Russian oil spill on the tundra ecosystem. The following part of this chapter uses the example of the tundra to explain how ecosystems function.

How do ecosystems function?

To study the structure of the Arctic tundra or any ecosystem, we must look at both biotic and abiotic elements of the system. Living components are plants, animals, and micro-organisms which are interdependent and survive together in this hostile area. Living things also depend on the non-living world, and adapt to survive in the harsh physical conditions.

How an ecosystem functions

Figure 7.1 shows a tundra scene in summer. Although daylight lasts for 24 hours at this time of year, the tundra is a hostile environment for plants and animals, with as few as three months in the year when temperatures exceed the 6 °C minimum needed for plant growth.

Mosses and lichens may colonize bare rock surfaces. As soil development takes place, dwarf shrubs, hardy grasses, sedges, and flowers such as buttercups and anemones grow in well-drained and sheltered places. Flowering plants and grasses die back in winter but seeds survive and germinate the following summer. Waterlogged areas, common where permafrost lies under the surface, provide habitats for mosses, cotton grass, and sedges. No trees survive as soils are too shallow above the permafrost and the climate too harsh.

Small mammals such as lemmings, voles, and shrews survive throughout the year on tundra vegetation. Caribou, musk-ox, and other large herbivores migrate into these areas during the summer. Many birds spend the summer months here feeding on the large insect population.

1 Copy the food web shown in Figure 7.2.

2 Construct a diagram, and write 300 words to explain the impact of an oil spill on this food web. The introduction to this section should help you.

▼ **Figure 7.2** The tundra food web.

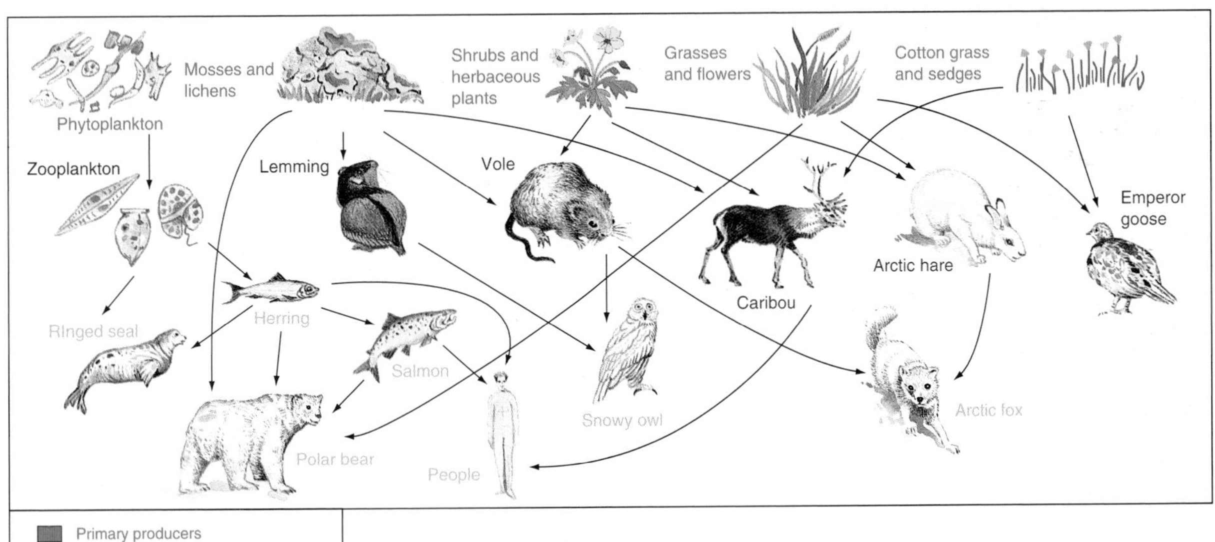

	Biomass (kg/m^{-2})	Biomass ratio above-ground : below-ground
Tropical rainforest (Amazonia, Brazil)	45	4 : 1
Temperate deciduous forest (oak woodland, UK)	30	3 : 1
Temperate grassland (Prairies, North America)	1.6	1 : 5 – 1 : 9
Arctic tundra (Alaska)	0.6	1 : 4 – 1 : 5

▲ **Figure 7.3** Biomass in four selected ecosystems (from G. O'Hare, *Soils, Vegetation, Ecosystems*, Longman).

Biomass and ecosystems

Both biomass and dead organic matter are part of the structure of ecosystems. Biomass means the total amount of plant and animal matter present. It is expressed as the dry weight of tissue in an area, either as tonnes per hectare (t/ha) or kilograms per square metre (kg/m^{-2}). Biomass also occurs below the surface, for example as roots, animals, and micro-organisms, and above the surface as stems, leaves, and animals. In tropical rainforests, four times as much vegetation biomass exists above the surface as below it; in tundra regions four or five times as much living plant matter exists below the ground as above it.

3 Explain the differences in biomass shown in Figure 7.3. Remember that most animal biomass is found below the surface, for example earthworms and insects.

Dead organic matter (DOM) is also important. Dead and decaying plant matter, consisting of surface litter and soil humus, may be more than living biomass by weight. In tundra regions, DOM may exceed living biomass by three times. Biomass is the product of complex systems of energy and nutrient flow within ecosystems. Tundra ecosystems, like all ecosystems, function in two ways, which are closely linked. These are:

- energy flow
- the cycling of nutrients (sometimes called biological cycling).

The following theory boxes explain these.

Energy flow

Energy is essential for all processes within an ecosystem. Virtually all energy is derived from the Sun, although a small amount comes from geothermal energy. Solar radiation, or **insolation**, reaches the ground in varying amounts depending on latitude, altitude, season, and time of day. Green plants use some of this light energy in photosynthesis, where inorganic carbon, oxygen, and water are converted into organic compounds, particularly sugars. Oxygen is released during this process.

$6CO_2 + 6H_2O$ + energy from sunlight
$= C_6H_{12}O_6 + 6O_2$

The rate at which energy is converted into organic matter through photosynthesis is called **gross primary productivity (GPP)**. GPP is measured in kilograms per square metre per year ($g/m^2/yr$). GPP varies enormously between world ecosystems because rates of photosynthesis differ. Tropical rainforest, hot deserts, and tundra regions each have different GPP rates, because different amounts of light energy and moisture are available for photosynthesis.

Energy fixed by plants during photosynthesis can be:

- used by the plant for life processes, such as respiration
- stored as plant material or animal tissue, known as biomass
- passed though the ecosystem along food chains or webs.

Plants use energy during respiration, and energy is then re-converted back to the atmosphere. Energy fixed by photosynthesis (GPP) minus that lost through respiration is called **net primary productivity (NPP)**. Like GPP, it is measured in grammes per square metre per year. Figure 7.4 shows how NPP varies between different biomes.

▼ **Figure 7.4** Average NPP and the range of NPPs for selected world ecosystems.

	NPP per unit area $gm^{-2}y^{-1}$		World NPP billion tonnes per year
	Range	Mean	
Tropical rainforest	1000–3500	2200	37.4
Cool temperate deciduous forest	600–2500	1200	8.4
Tropical grassland (savanna)	200–2000	900	13.5
Temperate grassland	200–1500	600	5.4
Boreal forest	400–2000	800	9.6
Woodland	250–1200	700	6.0
Tundra and alpine	10–400	140	1.1
Desert and scrub	10–250	90	1.6

1 On a world map devise ways to show the information in Figure 7.4. What spatial patterns emerge?

2 Which natural factors determine NPP?

3 Why is it necessary to consider both mean NPP and the range of NPP for different ecosystems?

4 What does the information on your map reveal about the potential for each biome for food production?

5 How might human activity alter NPP in a tundra ecosystem?

Energy is stored in plant tissue, but may be passed from plants to animals through a food chain. All animals obtain food either directly or indirectly from plants. Figure 7.2 shows a simplified food web for the tundra ecosystem. Primary producers (plants) are consumed by primary consumers (herbivores) such as lemmings, shrews, and Arctic hares. These are in turn consumed by secondary and tertiary consumers (carnivores), such as Arctic foxes and snowy owls. When plants and animals die, they are consumed by organisms such as bacteria and fungi. Each stage in the food chain is known as a **trophic level**. Producers are known as **autotrophs**, and consumers as **heterotrophs**. Energy is passed through the ecosystem in this way, but at each trophic level a number of things happens.

- Energy is used by the animal for life processes such as moving, eating, and respiring.
- Energy is lost through animal waste.
- Energy is stored as body tissue, such as muscle, and becomes part of the biomass.

Whenever energy passes from one trophic level to the next, about 90 per cent is lost and only 10 per cent is passed on. This huge loss is common to all ecosystems, and helps to explain why there are rarely more than five trophic levels in any ecosystem. It also explains why many vegetarians point to the inefficient use of world food resources in the production of meat.

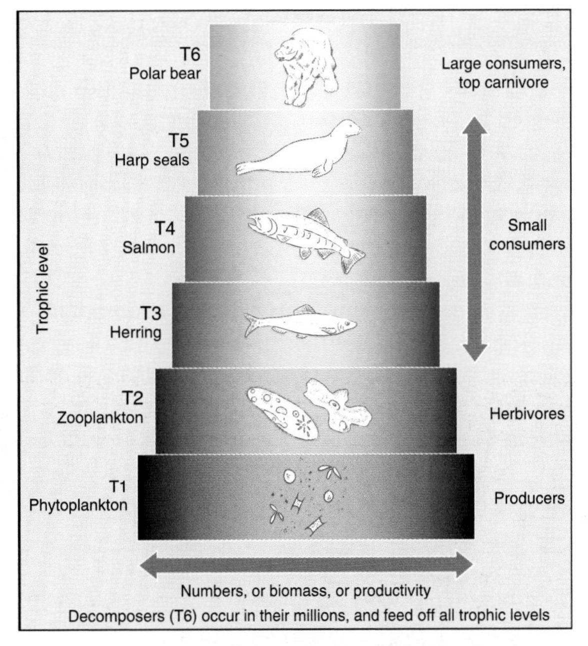

▲ **Figure 7.6** Trophic levels for tundra regions. In general, smaller numbers of living organisms exist at each successive trophic level, and thus there are larger numbers of primary consumers than secondary ones. This is shown here. In tundra regions, the decrease in numbers at successive trophic levels is marked because there are few animal species. The ecological pyramid may also be constructed using biomass, or energy produced by each trophic level. If decomposers T5 are included, the pyramidal shape alters, as decomposers are by far the largest group.

▼ **Figure 7.5** Energy flows in an ecosystem.

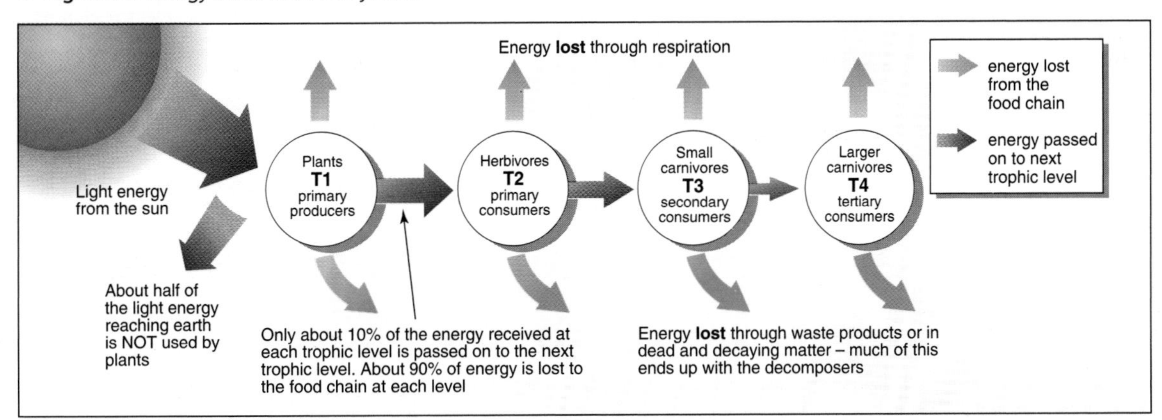

The cycling of nutrients within ecosystems

Nutrients are chemical elements and compounds needed by plants for growth. Unlike energy which enters and flows *through* the ecosystem, nutrients are cycled *within* it. They may be introduced or lost either naturally or by human actions. Essential plant nutrients include carbon, nitrogen, oxygen, hydrogen, calcium, phosphorus, and potassium.

Ecosystems obtain nutrients from the atmosphere, lithosphere, and hydrosphere, which are the three main nutrient pools. Oxygen and carbon dioxide enter and leave the system as gases, but others such as nitrogen are absorbed by plants in solution from the soil. The process of obtaining nutrients and recycling them depends entirely on the flow of energy. Nutrients are used by plants to grow and are converted into tissue. When organisms die and decompose, nutrients are released and are used again. Thus nutrients are continuosly recycled.

Nutrients may be introduced into an ecosystem by visiting birds and animals, or wind, and may be lost in surface runoff or leaching of soils. Changes can occur as a result of human activities, such as deforestation or because fertilizer finds its way into water courses.

Nutrient cycling within the tundra ecosystem is shown in Figures 7.7 and 7.8. It is slow because there are long periods of sub-zero temperatures and low precipitation. Tundra has low species diversity and a small, fluctuating animal population. The surface layer of soil which thaws out in the summer is often less than 50cm deep. The release of nutrients from the bedrock is limited by permafrost, which in turn prevents leaching. Soils become saturated during summer because of poor drainage, which leads to peat accumulation. Peat decomposes slowly, so the release of minerals is slow, and nutrients remain stored in peat. Nutrient input from the atmosphere is negligible because precipitation is both clean and low.

▼ **Figure 7.8** Nutrient cycling.

(a) Nutrient circulation, input, and loss.

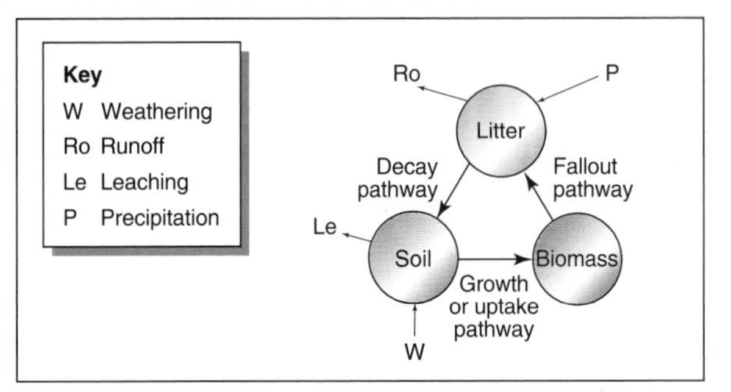

(b) Nutrient cycling in tundra. Nutrients are stored in biomass, soil, or litter. They are cycled between the stores by growth, fallout, and decay. Nutrients are added or removed by weathering, runoff, leaching, and precipitation.

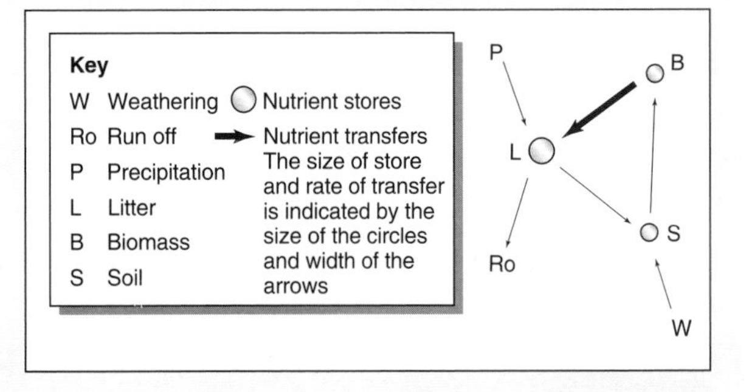

▼ **Figure 7.7** Typical nutrient flows in the tundra ecosystem.

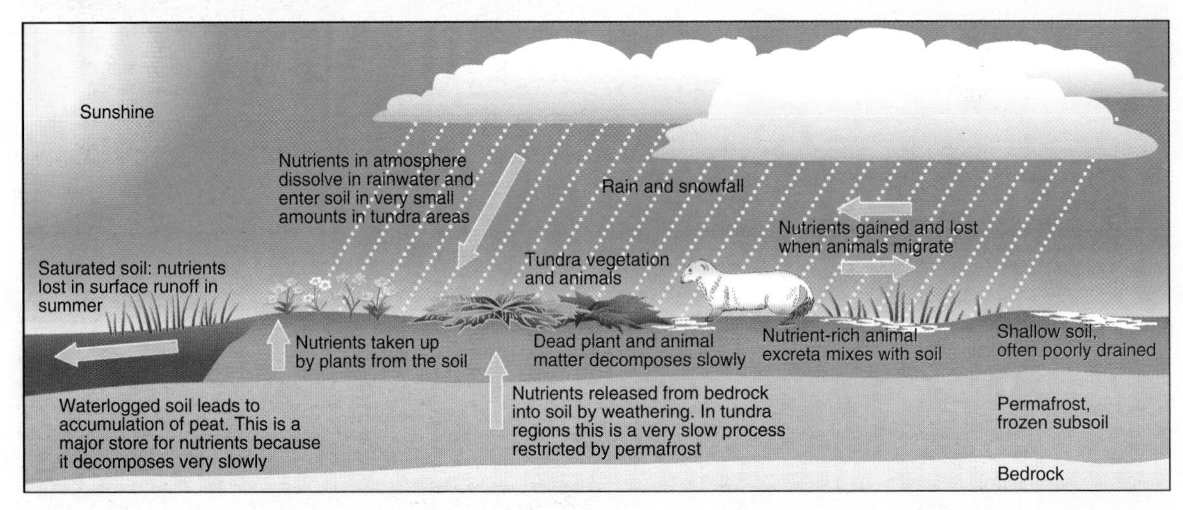

The slow rate of nutrient cycling, together with the small number of plant and animal species, make the tundra a particularly fragile ecosystem. Exploitation of these regions for oil and other minerals can lead to large-scale destruction of soil, plants, and animals. Ecosystems can recover only very slowly, if at all.

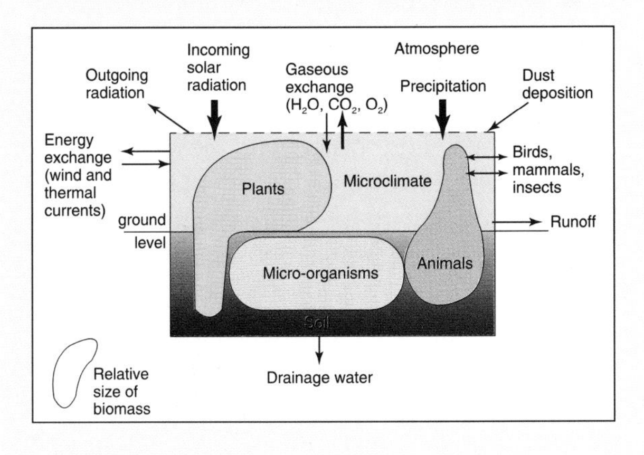

1 Study Figure 7.8a. Describe the processes involved in the cycling of nutrients between:
   a) soil and biomass stores
   b) biomass and litter
   c) litter and soil.

2 How effectively does Figure 7.8a show how nutrients are added or removed from an ecosystem?

3 Study Figure 7.8b. Describe and explain the low rate of nutrient storage and circulation shown.

4 What do you understand by 'structure' and 'functioning' of an ecosystem?

5 How effective is Figure 7.9 in showing the structure and functioning of an ecosystem?

◀ **Figure 7.9** Plan of an ecosystem. This shows the structural components of an ecosystem, and exchanges of energy and materials with surroundings.

Studland: a sand dune ecosystem

Studland Beach stretches for about 6km south-east from the entrance to Poole Harbour in Dorset. It is owned and managed by the National Trust. Extensive sand dunes have formed at the back of the beach, and they are part of the Studland Heath National Nature Reserve. This area is owned by the National Trust, but managed by English Nature. In summer it is popular with visitors. Most come to enjoy the long sandy beach but some explore the nature reserve and follow nature trails. Others pass through walking the South Western Coastal Footpath.

▼ **Figure 7.10** Studland Heath National Nature Reserve. See Figure 7.12 for section through A–B.

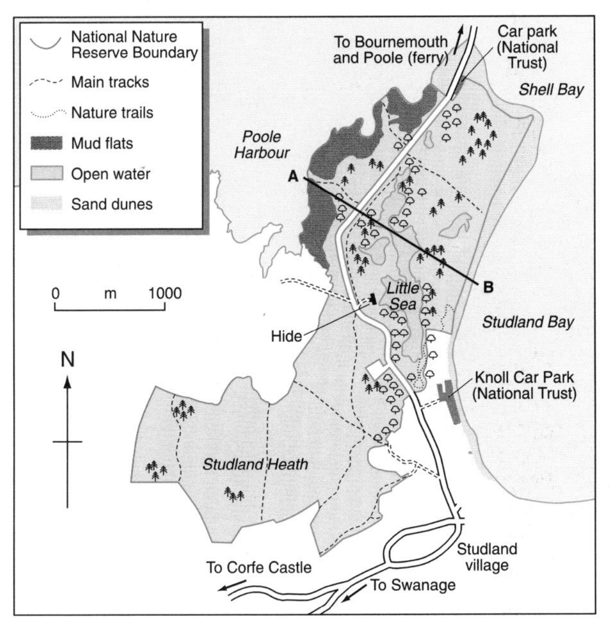

▼ **Figure 7.11** Studland sand dunes.

Plant succession on the Studland Bay dunes

Sand dunes provide an excellent place for studying plant succession. It is surprising that plants ever grow on sand at all; it is often dry, and moisture is salty. Dry sand is mobile, and begins to form dunes along the strand line. This process is described fully in Chapter 2. Small embryo dunes form and are easily destroyed, unless the sand is sufficiently moist and stable to be colonized by plants. At Studland, embryo dunes are colonized by sea couch grass and sea lyme grass, which are salt-resistant. Their spreading roots help to bind the sand, allowing the embryo dune to grow. Marram grass becomes established once the dune is about a metre high. Marram has long roots, enabling it to obtain water. It grows with the dune. It can withstand dry weather but not salt water. Once established, it helps to trap sand more quickly and dunes can grow by up to a metre a year.

Thus embryo dunes grow until they are large enough to be called mobile or yellow dunes. At Studland, the mobile dune is named Zero Ridge. It has been established for about 50 years and is colonized by red fescue, sand sedge, and marram grass. Few animals are found on the embryo dunes, but Zero Ridge supports snails and insects as well as visiting rabbits, lizards, and sea birds. The air behind sand dunes is calmer, allowing hollows known as dune slacks to develop.

When marram dies and decays, nutrients are added, enabling the dunes to support other plants which can survive in dry sandy conditions. As more organic matter is added, the soil increases in depth and its moisture content rises, which encourages different species to invade and colonize. Thus mobile dunes become more stable and are known as semi-fixed and, later, fixed dunes. Studland First Ridge is colonized by dandelions, sea bindweed, and other wild flowers, together with heather, moss, and lichen. Marram grass is still common but there is little bare sand except where it has been trampled. Animals too are more varied. Butterflies, meadow pipits, grass snakes, and lizards are found here. Second Ridge is covered with heather and gorse, but there is little marram grass. A thin black line of humus is found here, showing the first stage of soil formation. This allows more varied plant and animal species to inhabit the dunes.

Inland from Second Ridge, some parts of the dune slacks contain stretches of water, the largest of which is Little Sea. Small trees, such as birch, are found as well as water-loving plants such as bog myrtle, and aquatic animals, ducks, and other waterbirds. Inland, gorse and heather give way to birch and hazel trees. This is known as the climax community of Studland Heath, and would take over the area if the trees were not deliberately cut to maintain the diversity of habitats.

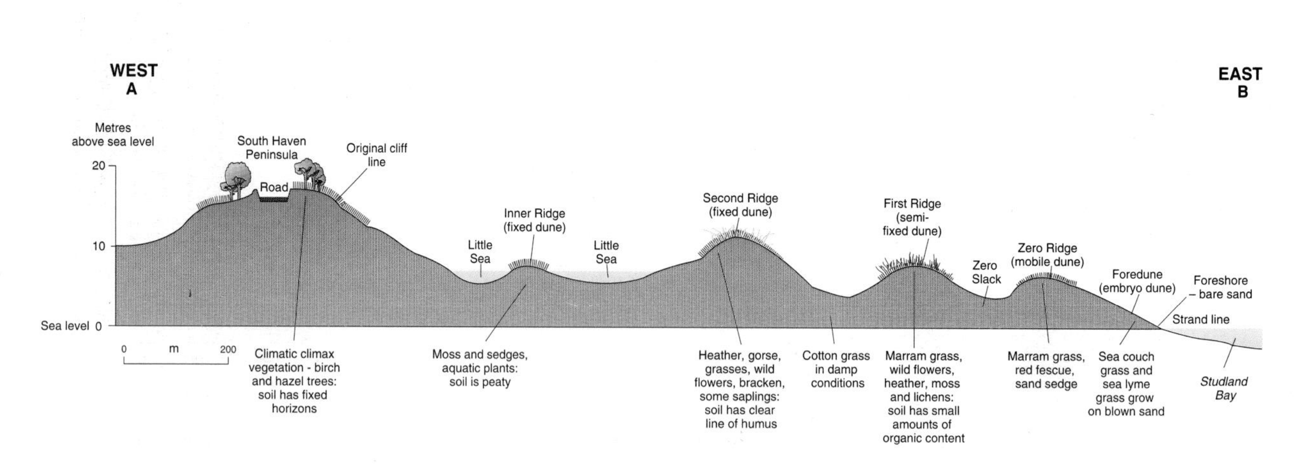

▲ **Figure 7.12** Sand dune succession at Studland Heath. See transect line A–B on Figure 7.10.

Plant succession

Bare ground such as exposed rock or wasteland in a city does not remain bare for long. Similarly, sand dunes or shallow water at the edge of a pond are gradually colonized by plants that are able to survive in these conditions. These are pioneer species and are the beginning of the process of plant succession, whereby other plants invade and take over until an equilibrium is reached. If bare ground has never previously been colonized, this process is known as **primary succession**, and may take hundreds of years to complete. If the land has been exposed by human activities then the process is known as **secondary succession** and may be more rapid.

Pioneer species survive on exposed sites without competition. These begin to modify the environment by forming and binding soil, and adding nutrients when they die and decay. Creeping plants or those with leaf cover help the soil to retain moisture. These changes allow other species to colonize, at the expense of pioneers which can no longer compete. The invaders in turn modify the environment by providing shade as well as improving soil. Birds and insects also start to find food and shelter.

Further modifications attract different species until stability is reached. If there are no limiting factors, the final community will be adjusted to the climatic conditions of the area and is known as the **climatic climax community**. Each stage in the process of succession is known as a **seral stage**.

Often plant succession does not reach the climatic climax, because something prevents full succession taking place. This may be soil conditions, relief, or drainage characteristics, or the human management of the land for agricultural purposes. Such factors are known as **arresting factors**, and result in the development of a **subclimax community**.

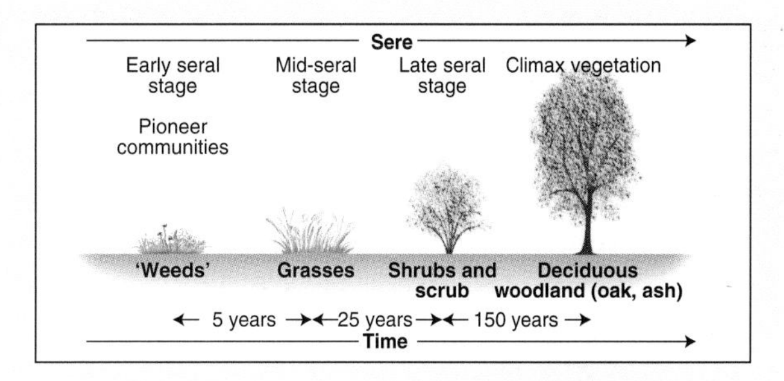

▲ **Figure 7.13** Plant succession in an abandoned field.

Managing pressures on Studland Heath and Studland Bay

On a busy day in summer, up to 25 000 people visit Studland. The vast majority come for the beach but some come to visit the sand dunes and nature reserve. The number of visitors is increasing each year, and problems for both beach and dunes are also increasing.

The main problems for the area are the following.

- Dune erosion caused by people walking through the dunes from car parks to the beach or for shelter. This threatens plant and animal species.
- Traffic congestion in car parks and roads leading to the area.
- Visitors leave over 12 tonnes of litter per week. Unless it is put in the bins provided, it is dangerous to small animals and birds.
- At least once a year, heath fires destroy plants and animals. The most common cause is a discarded cigarette end. Lizards and snakes can escape by burrowing, but may not escape predators once the vegetation cover has gone.

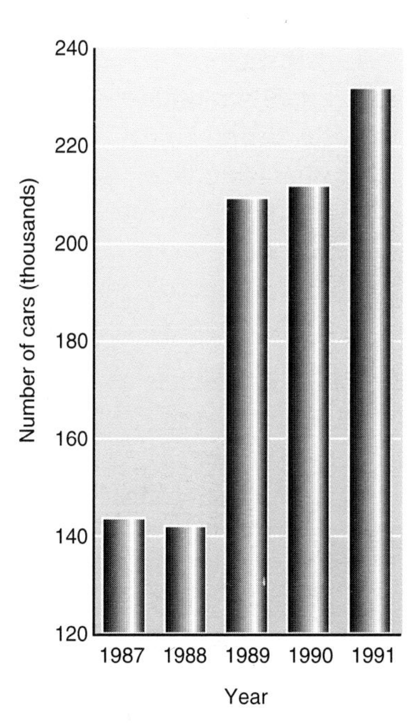

▲ **Figure 7.14** The use of car parks in Studland.

Management strategies at Studland

Both the National Trust and English Nature have introduced schemes to try to protect Studland Heath and the dunes, while allowing the public to visit the area. Since 1982 the National Trust has:

- enlarged the four main car parks to increase capacity by 800

Options for Studland Heath and Bay

1 Form groups of two or three people. Consider the following options for Studland Heath and Bay. Decide on points in favour of and points against each option.
   a) Restrict access to a maximum of 15 000 people on any day.
   b) Charge admission to the Heath and Bay.
   c) Remove all management strategies, and leave the Heath and Bay to the public.
   d) Close all access to the heath and dunes except by means of boardwalks and laid paths.
   e) Establish Studland Heath and Bay as a Site of Special Scientific Interest, and allow access only for special study.

2 Which option is preferred
   a) by you, and
   b) by the rest of the group?

- built a visitor centre with shop, café, and information point
- increased the number of toilets, and provided facilities for the disabled
- closed some paths and fenced off parts of the sand dunes
- planted marram grass
- placed litter bins on paths and at the back of the beach
- placed fire beaters and water hydrants on the heather and gorse heath, and made fire breaks
- erected information boards and provided leaflets to educate people about the area.

The National Trust and English Nature differ in their approach to management. The beach is managed by the National Trust and the Heath and nature reserve by English Nature. The National Trust is a charity, funded by subscription. The Trust is keen for people to visit the beach and to use the facilities it has provided. These facilities bring about £0.5 million a year to the Trust, which is used for improvements to the beach and its facilities, and for conservation projects in this part of Dorset. However, English Nature, a government agency, prefers to restrict access to the nature reserve to specialist groups, for example, to protect and conserve this environment. The use of the beach by the National Trust encourages more people to visit the nature reserve, and this can cause problems. A policy that combines conservation with public access has inherent difficulties but there are probably few alternatives in this case.

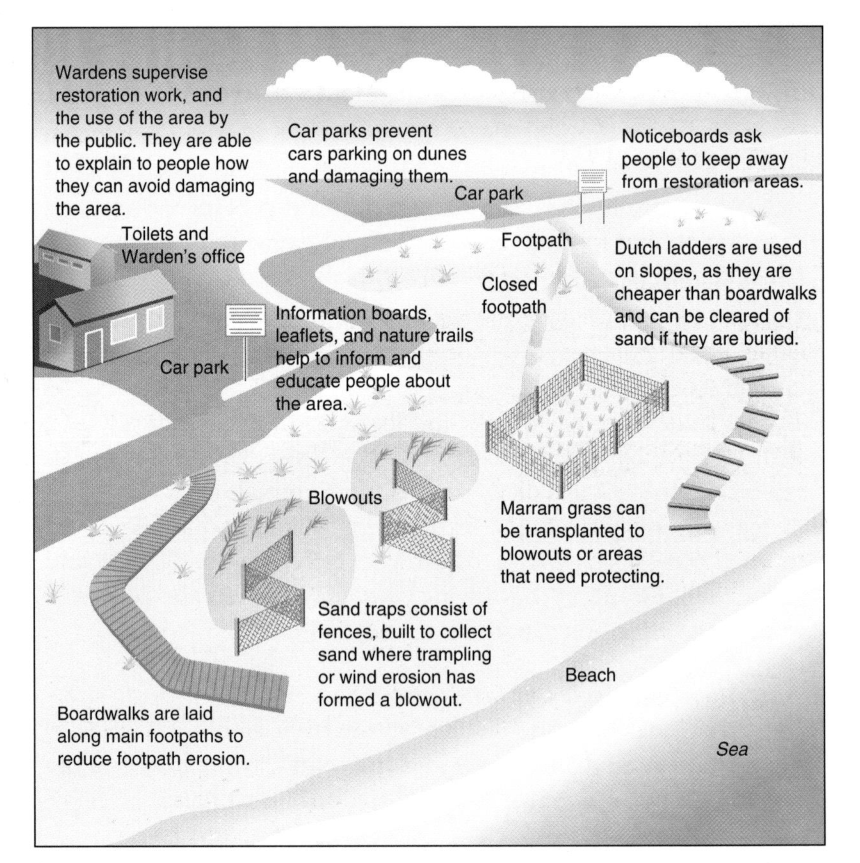

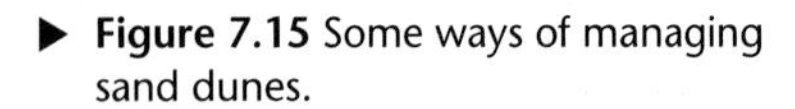
▶ **Figure 7.15** Some ways of managing sand dunes.

Transect through sand dunes

Taking a transect through the sand dunes from the strand line to the fixed or semi-fixed dunes is a useful way of collecting data. A transect of 300–500m will probably take you three or four hours.

You cannot record information along the whole transect, so choose a method of sampling. Two methods are shown here. A third method, **stratified sampling**, is described in Chapter 2, page 37.

Systematic sampling

The easiest method is to sample systematically, which means sampling along the transect at regular intervals. If you choose 50m intervals a transect of 350m would mean selecting seven sites at which to collect data.

Random sampling

Random sampling means that points are chosen at random along the transect, using random number tables – see Figure 7.16. Before selecting sample points, decide how many sites you can deal with in the time available to give you the information you need.

Once you have chosen a transect, you need to decide what data to collect and how to do it. Some ideas for this are presented in the box below. Before choosing methods of data collection, think about what data you need for your study, and how you are going to record the information you collect.

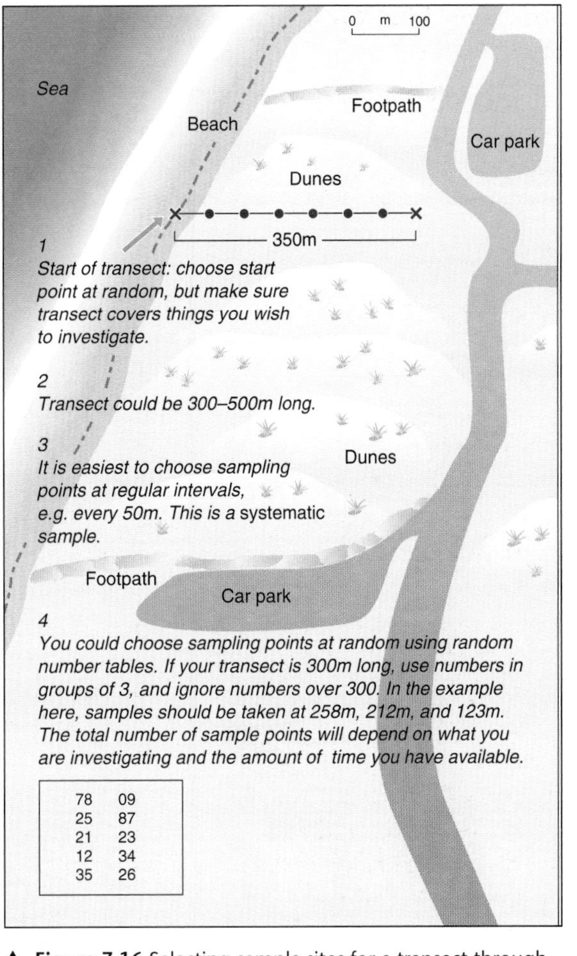

▲ **Figure 7.16** Selecting sample sites for a transect through a sand dune.

Collecting data in sand dunes

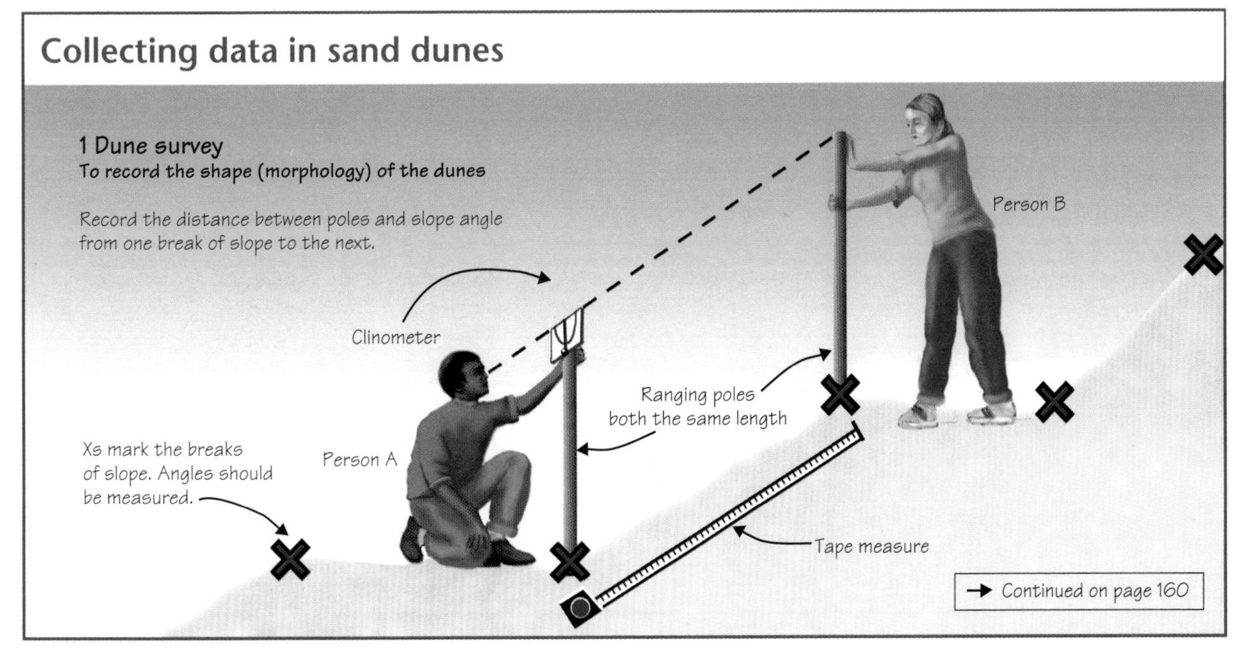

→ Continued on page 160

▶ Continued from page 159

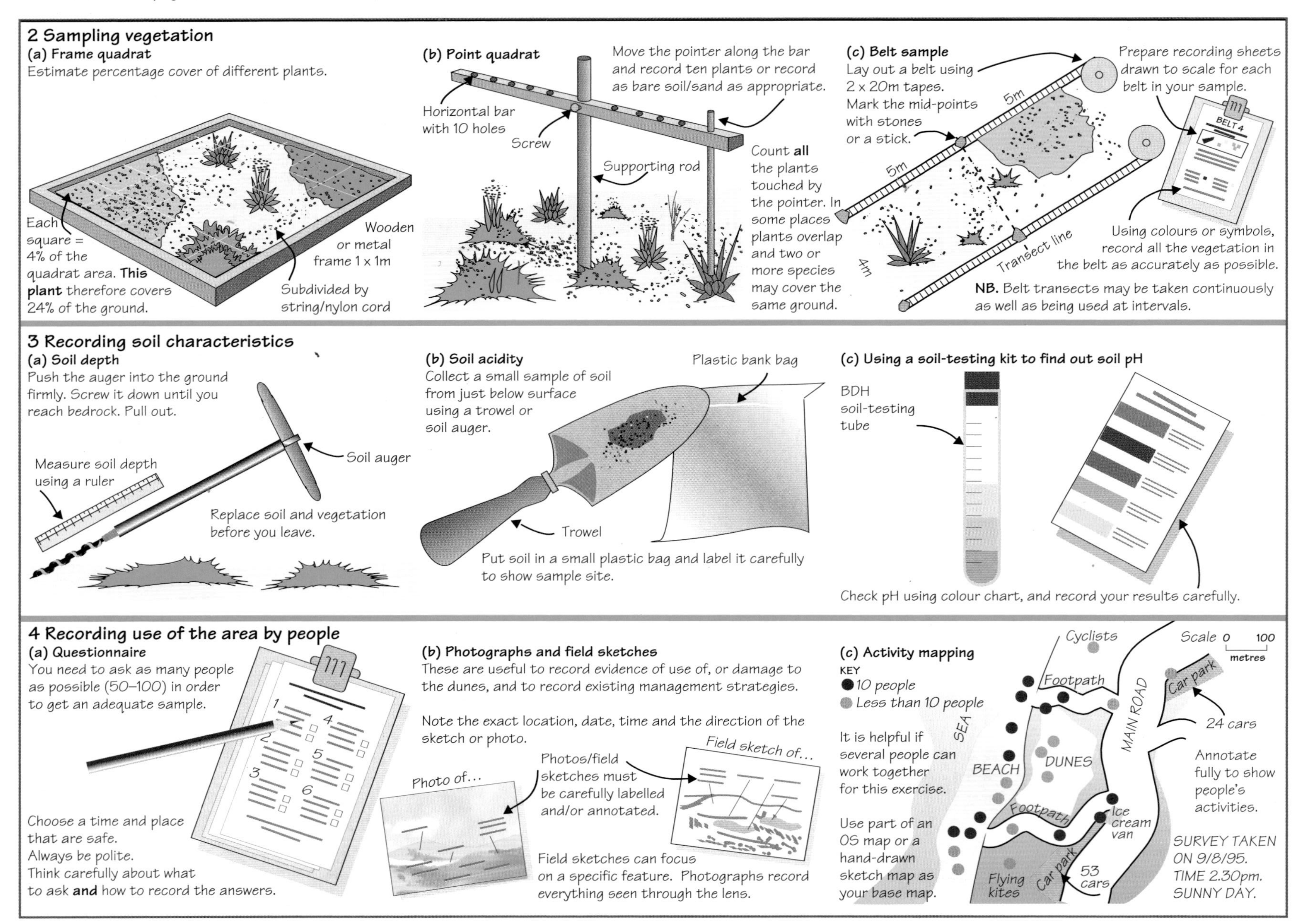

Collecting field data: Ladies Spring Wood, Sheffield

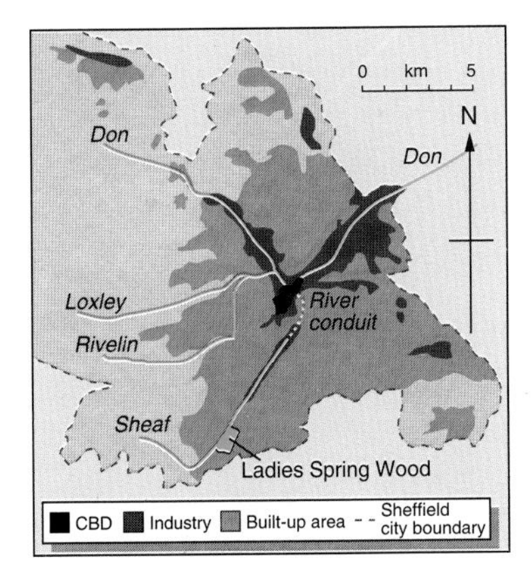

▲ **Figure 7.17** The location of Ladies Spring Wood.

The climatic climax vegetation of much of Europe is temperate deciduous forest. Much has been cleared, but pockets of woodland remain. These may provide places to study a woodland ecosystem on a small scale. Ladies Spring Wood on the outskirts of Sheffield lies on a steep slope which leads down to the river Sheaf. Although surrounded by housing, much of the woodland is too steep to be built on. It is valued by local people for recreation. There are opportunities to study not only a woodland, but also some of the issues which arise from the pressures of visitors.

The following shows how you can carry out an investigation of a woodland ecosystem. You will need to obtain a location map of your study area, like the one of Ladies Spring Wood in Figure 7.17.

When you have located your study area, you will need to select a transect, such as the one shown for Ladies Spring Wood in Figure 7.18a, and decide on a method of sampling. For instance, Figure 7.18a shows a diagrammatic transect between two sites along the slope. You can combine your vegetation survey with one of soil characteristics. Figure 7.18b shows how to record soil profiles, using soil colour and depth to help you to identify horizons.

You should then decide what equipment you will need. Figure 7.19 shows vegetation height, which reveals that Ladies Spring Wood, like most natural woodland areas, has a layered structure. This information has been obtained by measuring the height of trees, using a clinometer.

▼ **Figure 7.18(a)** A transect through Ladies Spring Wood.

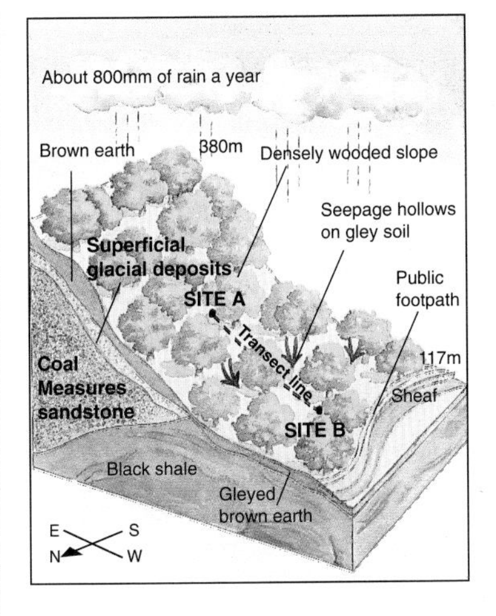

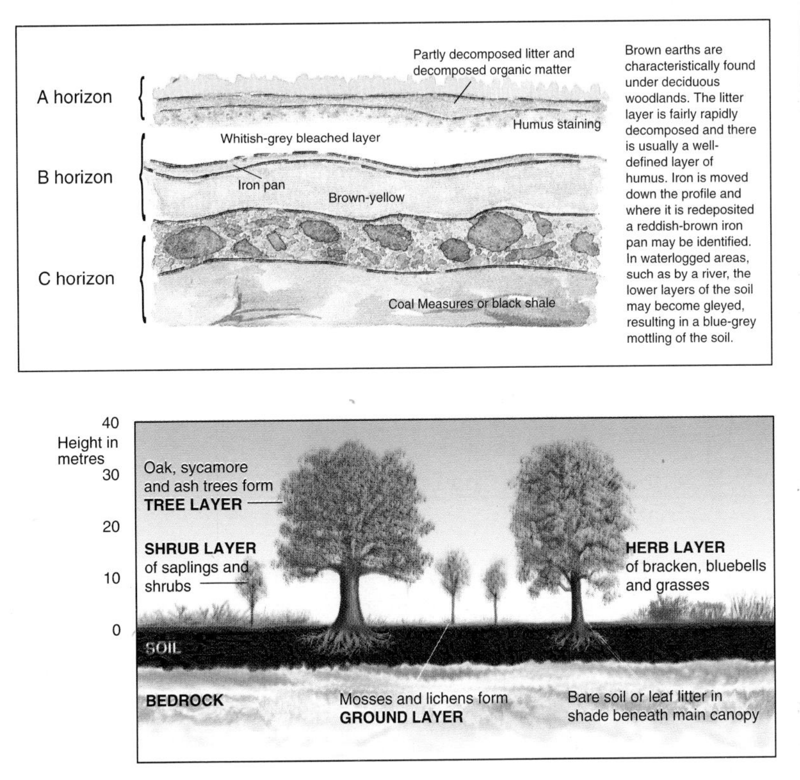

▶ **Figure 7.18(b)** A soil profile diagram. It is usual to dig to a depth of about 60–100cm to produce a profile such as this. Always replace all soil afterwards.

▶ **Figure 7.19** Stratification in Ladies Spring Wood.

Having collected your data, you will need to consider how to present it. Although vegetation may be shown in map form, the use of kite diagrams, shown in Figure 7.20b, is helpful in showing plant communities and a range of different species.

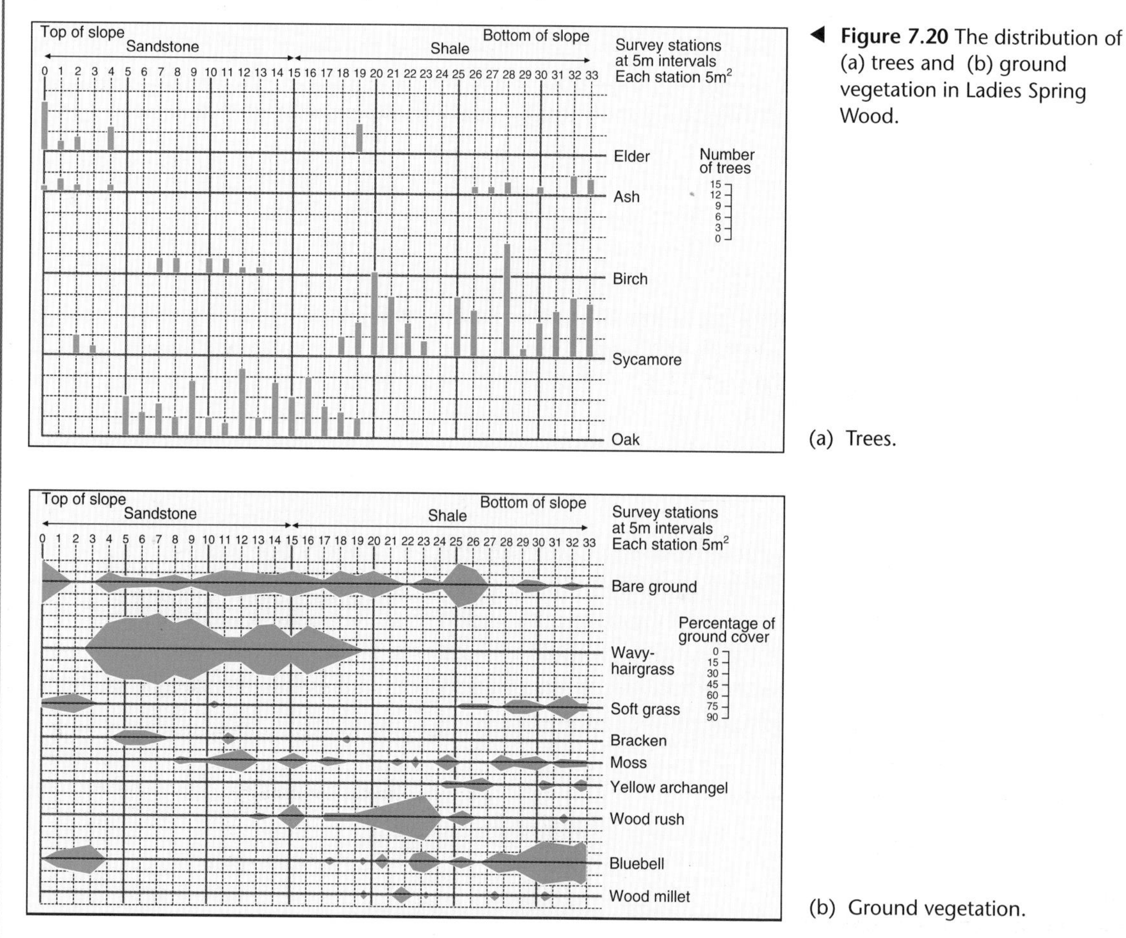

◀ **Figure 7.20** The distribution of (a) trees and (b) ground vegetation in Ladies Spring Wood.

(a) Trees.

(b) Ground vegetation.

1 Suggest other information not recorded on Figure 7.20, which could have given information about:
a) soils
b) microclimate
c) animals within the wood.
Describe possible methods of collecting this data.

2 In Figures 7.18 and 7.20, a block diagram and kite graphs respectively have been used to present information. What are the advantages and drawbacks of using these two methods to show data?

3 Use the information in Figures 7.18 and 7.20 to:
a) describe the differences in the woodland ecosystem between the upper slopes (sites 1–15) and the lower slopes (sites 16–33)
b) suggest reasons for the differences you have noted.

4 A well-used public footpath runs through this woodland close to the river Sheaf. How might access by people to this woodland affect the ecosystem?

Measuring footpath erosion in the field

Fieldwork can easily be carried out on a footpath to measure its width and depth and to assess damage caused by trampling boots or by other forms of erosion.

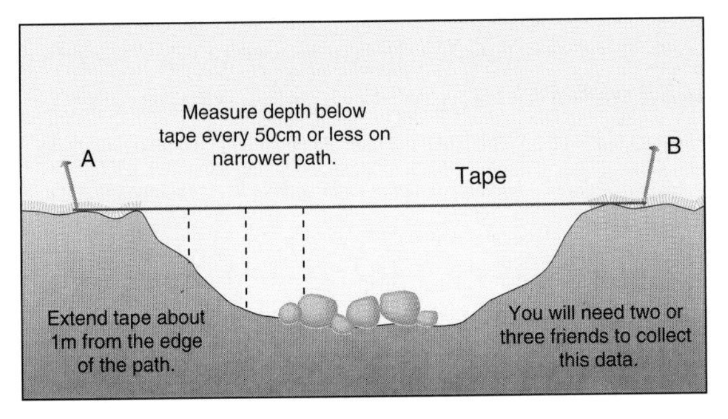

▲ **Figure 7.21** Measuring footpath erosion.

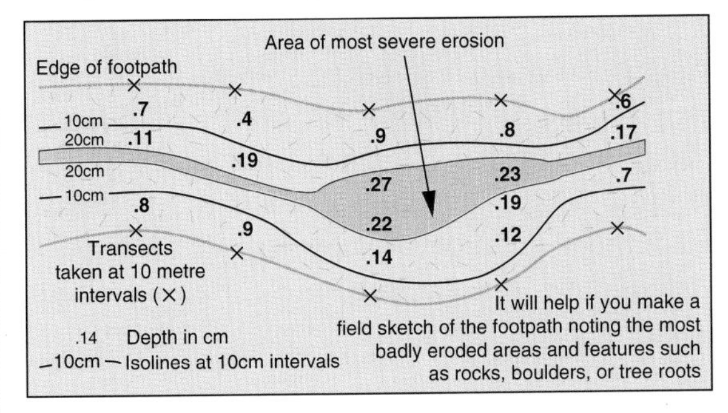

▲ **Figure 7.22** Isoline map of an eroded footpath.

1. Stretch the tape across the footpath.
2. Extend the tape at least one metre from the edge of the path at each side in order to record untrampled vegetation.
3. Record the width of the path (A to B).
4. Record the depth of the path at point A, then at regular intervals, for example every 0.5m, and at point B. Present information as cross-sections drawn to scale.
5. Repeat the measurements at other places on the footpath. You could select sites systematically, for example every 10m, or randomly using random numbers.
6. Present your data as an isoline map to show depth of the footpath – see Figure 7.22.

If the footpath is worn with patches of bare soil, but not badly eroded, you can use a quadrat as shown in Figure 7.23. You should record data about bare soil both for an eroded area and for an undamaged area away from the main area of trampling.

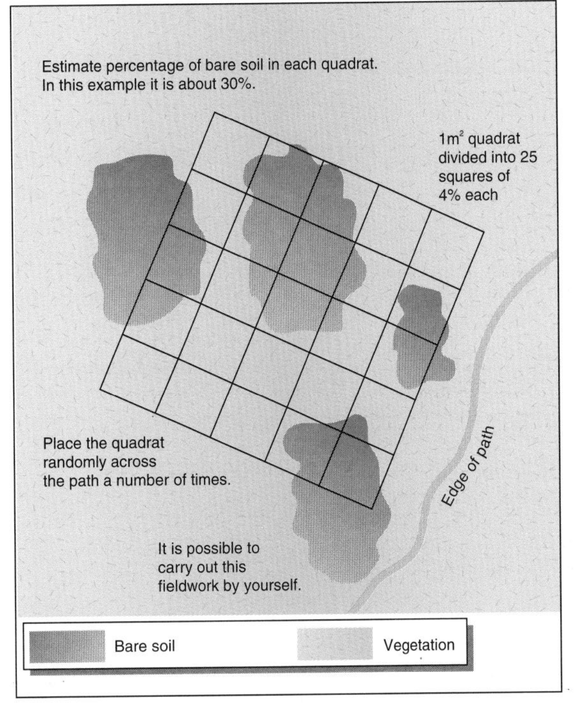

▲ **Figure 7.23** Measuring damage on a worn footpath.

Organizations or individuals who own woodlands are responsible for managing them. Ladies Spring Wood is owned and managed by Sheffield City Council.

1. Describe some of the management strategies which might be used in deciduous woodlands:
   a) to protect and encourage the growth of trees and other natural vegetation
   b) to enable the use of the woodland for recreational purposes without causing damage to the ecosystem.
2. What specific measures could be taken to protect footpaths in Ladies Spring Wood?

1. Using an atlas and information from this section, draw a map to show the location of Kinder Scout. Annotate it to show information about rock type, soils, and climate.
2. Draw a sketch of the photograph in Figure 7.26. Annotate it to show the plant communities on Kinder Scout.

Kinder Scout in the Peak District

Kinder Scout is one of the highest peaks in the southern Pennines. It is between Manchester and Sheffield, and lies in the Peak District National Park, Britain's most visited national park, and the most under pressure from visitors. Its moorland ecosystem is increasingly threatened.

Kinder Scout lies on the first stage of the Pennine Way from Edale to Kirk Yetholm on the Scottish border. It is known as the Dark Peak, with wild, open moorland overlying grits, shales, sandstones, and mudstones. Kinder Scout is featureless, and at 636m above sea-level its summit is hardly noticeable. Gritstone edges are found around the plateau.

In summer, the purple slopes of the heather moorland are stunning, but in winter much of the vegetation dies down, leaving a barren landscape. Poor weather can make this a cold, bleak, desolate, and dangerous place. Annual precipitation is about 1550mm. In winter this may fall as snow, and the snow may lie for more than 70 days. Average temperatures are only 11 °C in July and fall to 0 °C or below in winter. Strong winds make it feel even colder.

Kinder Scout is part of the Dark Peak Site of Special Scientific Interest (SSSI), and is owned by the National Trust which protects and manages it. The Dark Peak is an SSSI because it is an extensive area of semi-natural upland vegetation, it supports rare bird species, and it is of geological interest. The geology and features of Kinder Scout are shown in Figure 7.25.

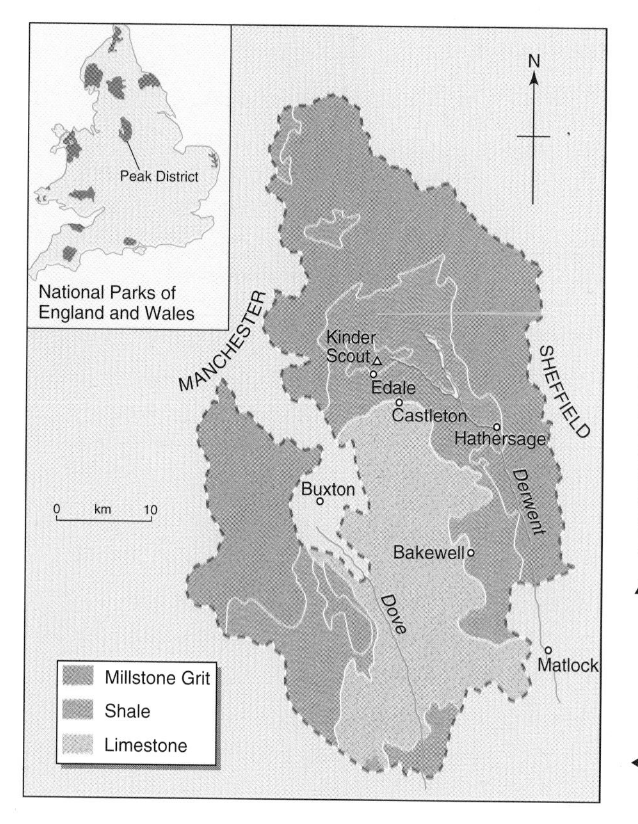

▲ **Figure 7.24** Small rock stacks and pedestals on the Kinder plateau.

◀ **Figure 7.25** The Peak District, Kinder Scout, and the underlying geology.

Plant and animal communities on Kinder Scout

Blanket peat covers the whole of this area. The peat bogs were formed as a result of clearance of natural oak woodland which impoverished the soil. As the climate became colder and wetter from 600 BC, tree growth was further inhibited and peat began to form. Peat formation has been more substantial in this area than anywhere else in Britain, probably because deforestation took place earlier here than elsewhere.

Although Kinder Scout appears to be an homogenous environment, a variety of plants and animals live in different communities. The moorland divides physically into two – the plateau where the peat is deepest, and surrounding hillsides. The food web in Figure 7.27 shows that many plant species are found in the two areas, though dominant types differ. Cotton grass, cowberry, and bilberry dominate the blanket bog, although heather is also found. In some places, mosses such as sphagnum occur, but high levels of atmospheric pollution have drastically reduced the numbers and variety of bog mosses since the early 19th century.

▲ **Figure 7.26** Blanket bog with cotton grass on the Kinder plateau.

1 Study Figure 7.27.
 a) Give examples of primary producers, and primary, secondary, and tertiary consumers. Identify the consumers as herbivores, carnivores, or omnivores.
 b) Where are the main energy flows likely to be found within the Kinder Scout ecosystem? Is there evidence of a trophic pyramid?

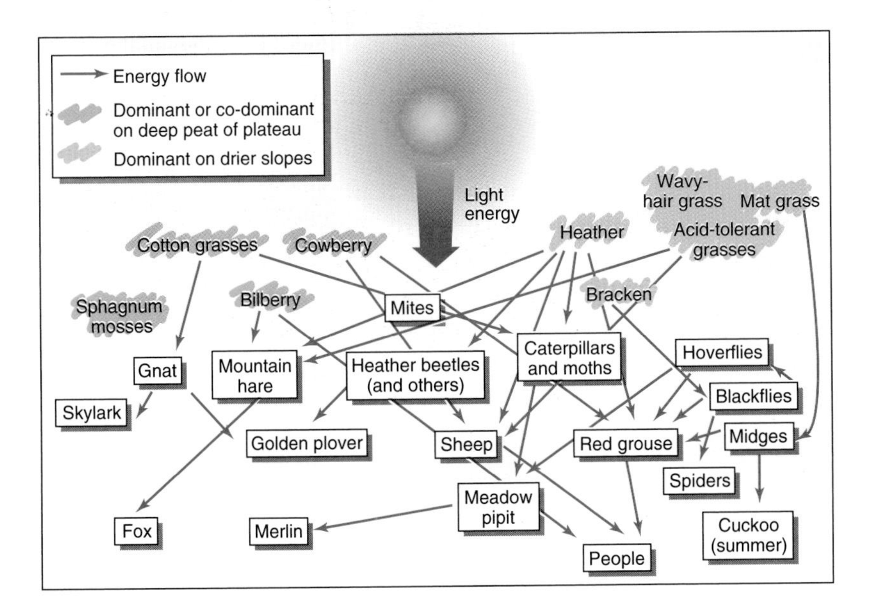

◀ **Figure 7.27** Food web for Kinder Scout.

On the drier slopes below the watershed, heather dominates with more grasses growing here than on the plateau. Many birds and animals living on Kinder Scout find their food from these two plant communities.

Moorland vegetation must be adapted to live in acid, wet peat and survive in wet, cold conditions at relatively high altitudes. Because the vegetation is specialized in this way, it supports a limited number of animal species which are also able to tolerate these conditions. There are limited numbers of soil fauna, and an absence of earthworms because wet soils contain little oxygen. Snails and other molluscs are rare because they need lime to make their shells. Butterflies too are rare, although moths, flies, and mosquitoes are to be found in substantial numbers.

Peat accumulation

High rainfall and poor drainage on Kinder Scout mean the process of decay is slowed down because of reduced supplies of oxygen, high soil acidity, and low temperatures. As plants die, the remains only partly rot and begin to accumulate, forming peat. Sphagnum moss is the most important plant in the formation of peat because it absorbs a lot of water and nutrients.

Little sphagnum moss is now found on the Kinder Plateu. Investigate the nature of sphagnum moss and those influences which reduce its growth. The information you find will help your evaluation of the National Trust's management strategies, considered on page 172.

Soil formation

Soils are divided into types on the basis of horizons. These are visible when a soil pit is dug. Horizons develop as a result of the soil formation processes, such as weathering of bedrock, oxidation of minerals including iron, leaching, and the accumulation of humus. The A horizon contains mixed mineral and organic material, but organic material is fully incorporated in the soil. It may lie below a layer of organic material such as leaf litter or in this case peat. The B horizon is below the surface and is a mineral layer with little organic material. The C horizon is known as the **parent material** horizon.

The soils forming on upland plateaus are usually **podzols**. The name 'podzol' is given to the pale grey layer found at the top of the B horizon, from which most iron has been leached, or washed out beyond the reach of most plant roots. The iron is re-deposited lower in the soil, forming a reddish iron pan.

Peaty gleyed podzols are found in many parts of upland Britain where there is high rainfall. This encourages **leaching** – the movement of nutrients from the upper to the lower layers of the soil by water – and **gleying**, which gives soils a characteristic green-blue tinge and red mottled patches resulting from iron oxidation. Gleying occurs where waterlogged anaerobic (airless) soils cause iron to be reduced, giving the soil its green-blue colour. Where air or fresh oxygenated water is able to penetrate the soil in cracks or root channels, iron is oxidized resulting in reddish-brown colouring. These soils are acidic and infertile and support only a limited range of plants that are able to tolerate these conditions.

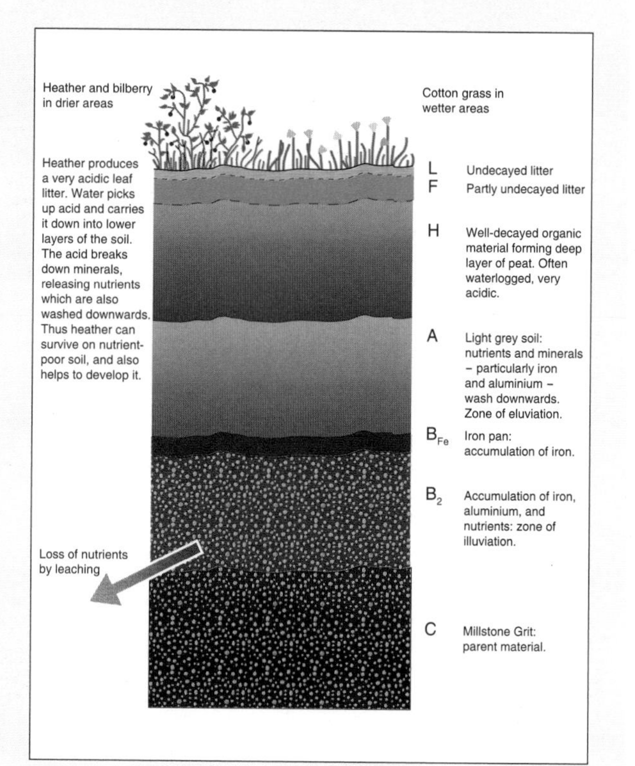

▶ **Figure 7.28** Soil profile showing a peaty gleyed podzol.

How is heather moorland managed?

Heather survives on acidic, nutrient-poor soil and produces acidic leaf litter. Soil acidity reduces availability of nutrients for other plants. Few soil animals survive in the organic matter produced beneath heather, and there are no earthworms. As a result, decomposition of organic matter is slow, leading to layers of acid peat on the soil surface. Heather is semi-natural vegetation maintained by grazing and burning. This allows it to regenerate rapidly with little competition from other plants. If it is left without management for fifteen years or more it degenerates, plants become woody, and the branches fall outwards allowing other species to take over. Without management, the area would revert to scrub, and at lower altitudes oak woodland would establish itself through plant succession.

▼ **Figure 7.29** Life cycle of heather.

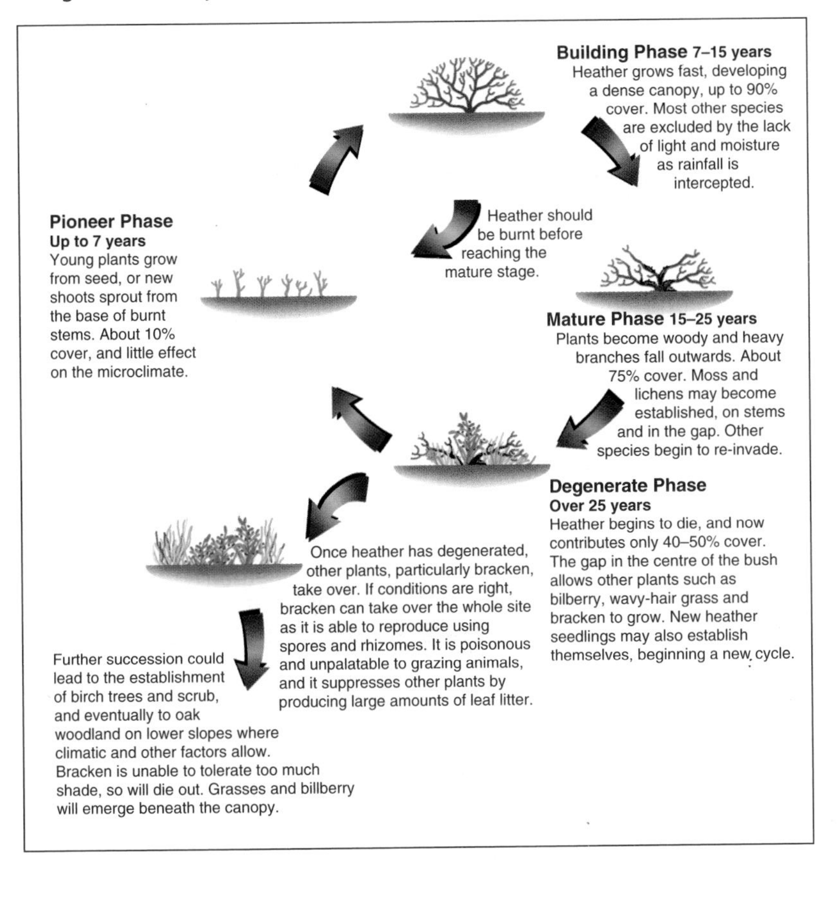

1 If heather moorland remains unmanaged, it reverts to scrubland. Complete succession to woodland on Kinder Scout is unlikely. Why?

2 Study Figure 7.30. Describe and attempt to explain the changes in plant species shown. Suggest what would happen to the plant community if controlled burning were carried out:
   a) after 20 years
   b) after 30 years
   c) never.

3 The data in 7.30 were collected over ten years on six different sites using quadrats. Suggest why only five species were chosen to be identified. Describe some of the problems of collecting data in this way.

▼ **Figure 7.30** Changes in the plant community on the Quantock Hills, Somerset. Data collected by Ackworth School.

Years after burning

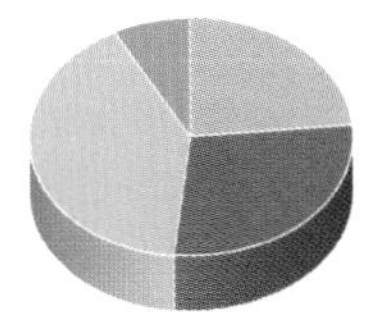

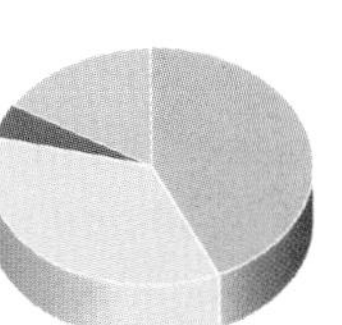

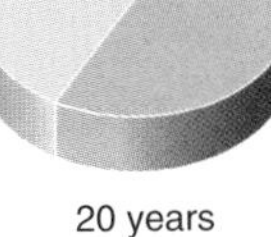

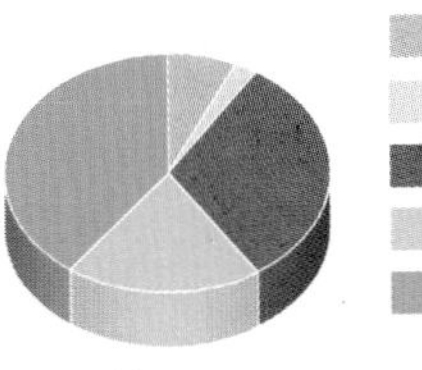

Managing Kinder Scout – the problems and solutions

Kinder Scout's location as a feature along one of the country's most used footpaths is bound to create pressure. This section explores the key issues threatening moorland ecosystems on Kinder Scout. It examines the changes taking place, the processes responsible for the change, and the influence of human activities.

The evidence of an ecosystem under stress

There has long been concern in the Peak District about peat erosion. This led the Moorland Erosion Study, set up by the Peak District Planning Board (PDPB) in 1979, to examine the increase in moorland erosion in the national park, and to try to find the causes of it. The results showed how serious the problem had become, particularly on Kinder Scout.

The extent of peat erosion is most obvious at the triangulation point. When it was erected in the 1920s the concrete base was level with the peat surface. By 1981 over a metre of peat had been removed. Peat erosion is caused by two processes:

- the destruction of vegetation
- the removal of exposed peat by running water, wind, or gravity.

The blanket peats of Kinder, over 4m deep in places, are now severely eroded. Up to 60mm of peat a year is being lost, and gullies over 2m deep are found on the plateau. Where vegetation and peat are lost, the balance of the ecosystem is upset. The survival of birds and animals is at risk from the reduction in number and variety of primary producers, and the loss of plants and soils upsets nutrient cycling. The rapid rate of erosion of the peat on Kinder Scout is due to a number of factors – pollution, sheep grazing, and climate.

Pollution

Kinder Scout lies between Manchester and Sheffield and suffers severe atmospheric pollution. Sphagnum moss and other bog plants do not tolerate high levels of sulphur dioxide or nitrous oxides, yet these are brought to the moorlands as acid rain. Sphagnum moss acts like a sponge and protects the peat surface. Pollution levels are too high to allow it to re-establish.

Sheep grazing

The number of sheep on Kinder and Bleaklow increased from 17 000 in 1914 to 60 000 in the 1970s. This was due to government subsidies which paid farmers on the basis of how many sheep they kept. When the National Trust bought Kinder Scout in 1983, there were often 2000 sheep grazing there. Some belonged to the owners of the moor, but many were 'trespass' sheep from local farms. The lack of fences on Kinder and low levels of shepherding meant that the sheep could wander freely. Most sheep were breeding ewes which are selective in their grazing, choosing heather and bilberry rather than other plants. This has increased the spread of moorland grasses at the expense of more valuable shrubs. The 30 per cent loss of heather in the Peak District this century can be attributed to overgrazing by sheep. In places, grazing has led to extensive areas of bare ground broken only by coarse grass. The loss of food and cover has led to a decline in bird species.

Walkers

As many as 10 000 people annually walk the length of the Pennine Way. More than 150 000 visit Kinder Scout, exceeding the capacity of every major path, and most of the minor paths in the Kinder area. Between the mid-1970s and mid-1980s, the average bare width of peat on the Pennine Way increased from 1.45m to 3.53m. Plants such as cotton grass, mosses, bilberry, and heather are easily damaged by trampling, particularly on slopes and in wet areas. Once plants are removed, peat is churned up by boots and becomes waterlogged. People avoid boggy areas, but widen the footpath by damaging vegetation at the edges. A sponsored walk can cause as much damage to a path in one day as a year's normal use. Walkers go on to Kinder in all weathers, and at all times of the year, giving vegetation no time to recover.

▲ **Figure 7.31** Walkers on the Pennine Way, Kinder Scout.

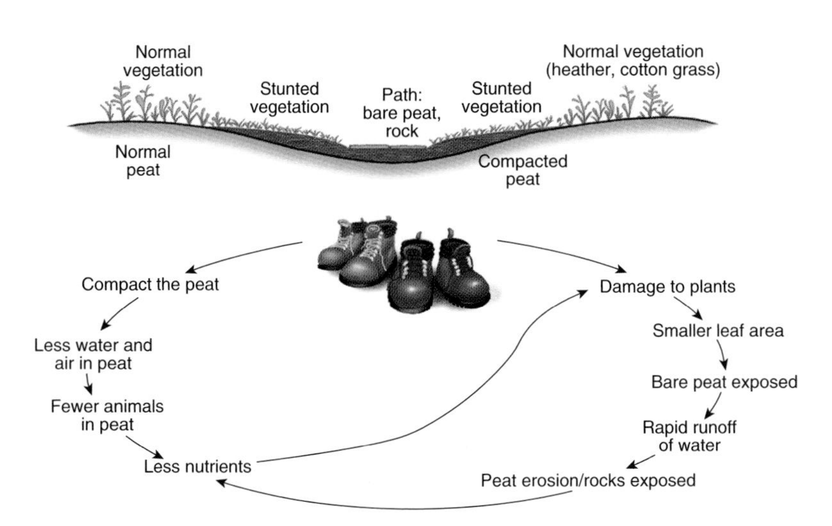

◀ **Figure 7.32** Damage by walkers to footpaths.

Moorland fires and tourism

Heather moorland fires are beneficial, removing old stands of heather and returning nutrients to the soil. Accidental fires in summer, however, can have devastating effects, especially on dry peat. They tend to burn longer and deeper, destroying all vegetation, dormant seeds in the peat, and sometimes the peat itself. Insects, birds, and animals may also be killed. These areas become exposed to erosion and some areas never recover. Summer fires are almost always associated with visitors. Most start at the weekend, and 60 per cent start on or near a footpath or road. Once vegetation cover has gone, peat is easily removed by rain and wind. The high annual rainfall, high-intensity rainstorms, and strong winds cause great damage.

Climate

Heavy and frequent rainfall means that peat remains saturated most of the time. It has a low infiltration capacity, so much of the water runs away as surface runoff. This carries peat in suspension in rivers and streams, and some is deposited in the reservoirs surrounding Kinder. Long periods of low winter temperatures inhibit plant growth which would stabilize the peat.

1 What evidence is there of severe peat erosion on Kinder Scout? How does this compare with average surface lowering of bare peat, which is thought to be 5–10mm per year?

2 The first stage in peat erosion is the removal of vegetation. Discuss the causes of vegetation loss.

3 How significant is the role of climate in peat erosion?

4 In what ways is the ecosystem affected by erosion?

5 Suggest reasons why reservoir infill by peat is greater in the southern Pennines than for other upland areas of Britain. For whom is this a problem?

6 Copy and complete Figure 7.34, which summarizes peat erosion. With reference to the key, use two colours to highlight aspects which you think could be tackled in attempts to reduce peat erosion on Kinder, and those which would be difficult to tackle.

7 Why are some causes of peat erosion difficult to address?

8 Form groups of three or four people. Decide how you would tackle each problem shown in your diagram. Present your plan to the group. Discuss difficulties you might face.

▼ **Figure 7.33** Bower's classification of peat erosion. Peat erodes in two ways.

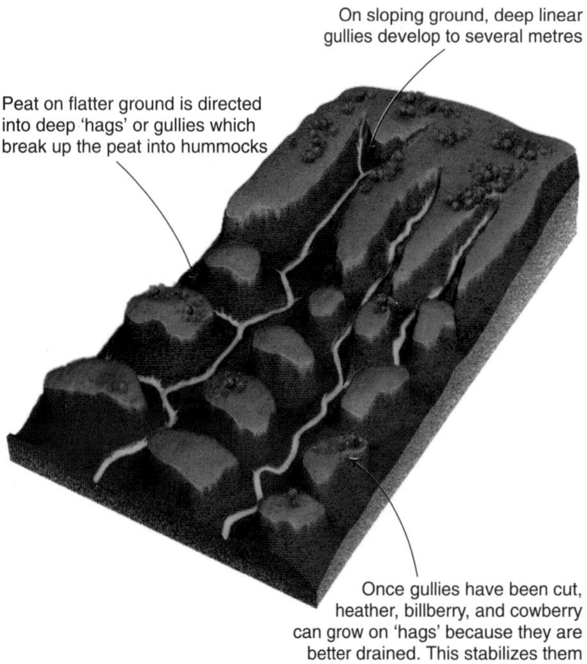

On sloping ground, deep linear gullies develop to several metres

Peat on flatter ground is directed into deep 'hags' or gullies which break up the peat into hummocks

Once gullies have been cut, heather, billberry, and cowberry can grow on 'hags' because they are better drained. This stabilizes them

▼ **Figure 7.34** Summary of processes of peat erosion.

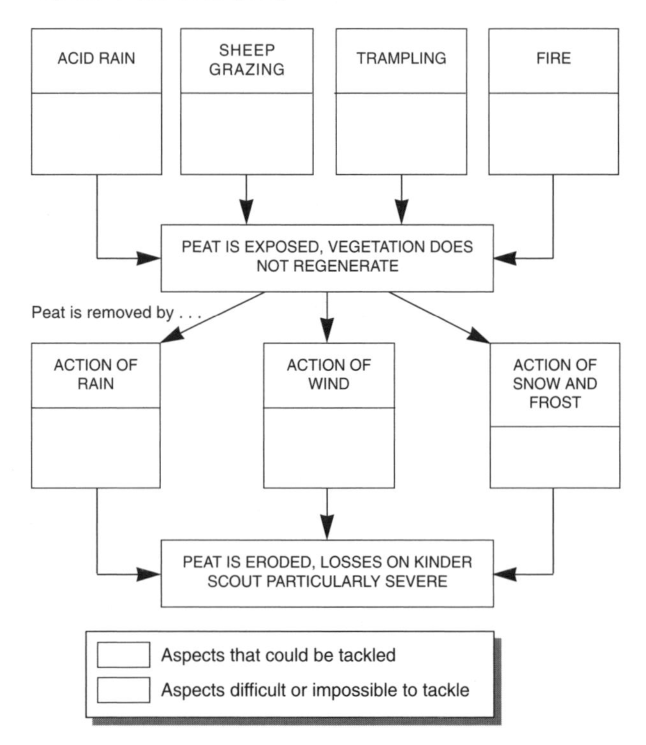

Management proposals

Management strategies

In 1982 Kinder Scout was bought by the National Trust. It was suffering from severe erosion. In consultation with others, the Trust drew up a plan which had three main objectives:

- to preserve and enhance the natural beauty of the area
- to halt and, if possible, to reverse the moorland erosion
- to continue to provide access for the public.

The main management strategies employed are outlined below.

Sheep grazing controls

The National Trust's first action (in 1983) was to ban all sheep grazing on the Kinder moors. Fences and drystone walls help to keep the sheep out. It is not a permanent ban though, and when the vegetation is fully re-established, controlled grazing will be allowed.

Re-vegetation

The National Trust and the University of Sheffield have been investigating how bare peat can be re-vegetated. The best methods appear to be the following.

- Apply lime and fertilizer to reduce soil acidity and add nutrients. This process aids growth and germination. Now it is applied on a large scale using helicopters. It is extremely expensive – about £600 for 3ha (hectares). Kinder has about 900ha of open moorland.
- Plant native species such as cotton grass, and introduce seeds or cuttings of heather.

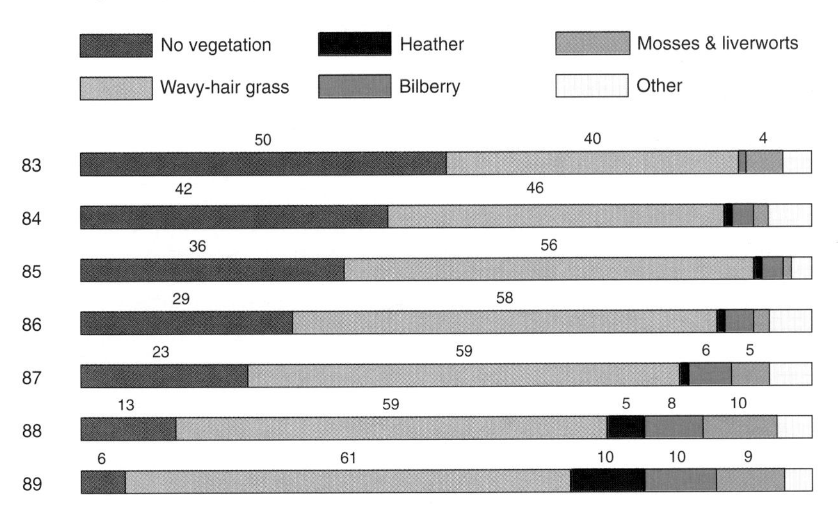

▲ **Figure 7.35** Changes in ground cover, 1983–89. The application of lime and fertilizer encourages the growth of wavy-hair grass which soon seeds itself on surrounding bare peat. As the grass develops it protects the peat and allows other plants to develop, such as heather.

▲ **Figure 7.36** Footpath repair on Kinder Scout. Footpath repair and maintenance are extremely costly, particularly as paths may be a long way from roads. Materials may have to be airlifted in, and people may have to walk for over an hour to the site.

Footpath repair

Paths across Kinder are particularly affected by trampling. The National Trust is anxious that this should not threaten the natural landscape. In some places drainage channels or the relocation of boulders and turf are enough to protect a path from erosion but elsewhere it has been necessary to lay new surfaces. Several types have been tried.

- Two major routes have been 'stone-pitched'. Large blocks of local gritstone have been embedded in the ground, creating a hard-wearing path.
- Floating paths have been laid in badly drained places. These involve laying fibre matting and covering it with either aggregate or boulders.
- Gritstone flagstones have been laid on gentle slopes to provide paths across deep or waterlogged peat.

Fires

The damage caused every year by accidental fires has been tackled by the National Trust by:

- providing specialized fire-fighting equipment in strategic locations on Kinder, which wardens are trained to use
- close co-operation with the Fire Service
- better education of the public, using posters, information leaflets, and the 'No Moor Fires' campaign.

How successful are the National Trust's management strategies?

1 Look back through this section on management strategies employed by the National Trust on Kinder Scout, and the other information in this chapter. Summarize the benefits and drawbacks of each strategy: the sheep grazing ban, re-vegetation, footpath repair, and fire fighting.

2 Form groups of three or four people. Select *one* of the following situations, then write a paper to justify a management strategy for that situation.
- A meeting with the local Ramblers Group who are upset about the expense involved in footpath repair, and the increasing number of surfaced paths, which they consider to be unnatural. They want you to look at alternative strategies.
- A meeting with local farmers concerned about sheep grazing bans on Kinder. They are upset about the loss of grazing which had been available in the past, and about difficulties caused when their sheep are cleared from Kinder Scout by National Trust wardens. They want you to allow sheep to graze on Kinder.
- A meeting with the Finance Committee of the National Trust, who wish you to justify the cost involved in managing Kinder Scout. The Committee suggest that you reduce costs by concentrating on a smaller number of projects.

3 In your opinion, is the National Trust justified in the policies it has chosen? Which – if any – of the projects might be abandoned? How else might erosion be reduced on Kinder Scout?

Ideas for further study

There are lots of opportunities for fieldwork in sand dunes, woodlands, and moorland ecosystems. Recording changes in vegetation, soils, and microclimate can provide you with information for a local case study of your own. Each provides an excellent focus for individual enquiries. Studies could focus on:

- the characteristics of the ecosystem
- differences in soils or microclimate in different parts of the ecosystem, such as embryo dunes compared with mobile or fixed dunes
- a comparison of vegetation along parts of a transect
- the impact of people and any damage being caused
- an evaluation of strategies to manage those impacts that are already in place, or are planned
- a presentation of your solutions for resolving current problems.

Summary

- An ecosystem is a community of plants, animals, and micro-organisms, with the non-living environment in which they live.
- Ecosystems function using a system of webs and flows of energy and nutrients. The most fundamental source of energy is sunlight which is converted, through photosynthesis, into sugars.
- Energy flows through an ecosystem, but is removed at each succeeding trophic level. By contrast, nutrients are cycled. They come from the atmosphere, hydrosphere, and lithosphere.
- People are an increasingly important component of ecosystems. Human activities may significantly alter ecosystems. This may be unintentional (because of ignorance), or for economic gain.
- Ecosystems develop and change over time, until they reach equilibrium. This can be upset, by natural or human changes.
- Studies of plant colonization and ecological succession enable us to understand the components of ecosystems more easily.
- Ecosystem management is likely to be successful where damage is repaired and strategies taken to prevent further damage. To be successful, the structure and functioning of the ecosystem must be fully understood.

References and further reading

The National Trust, *Kinder Scout Ten Years On*, 1994.

G. O'Hare, *Soils, Vegetation and Ecosystems*, Oliver and Boyd, 1988.

RSPB, *Ecosystems and Human Activity*, Collins Educational, 1994.

T. Stott, J. Hindson and R. Crump, *Sand Dunes: A practical coursework guide*, Field Studies Council Publication No. 25, 1994.

8 Woodland and forest ecosystems

Tropical rainforest ecosystems

Myths about rainforests

Forests once covered 75 per cent of the Earth's land surface. The figure is now less than 30 per cent. Tropical forests alone once covered 14 per cent of the land; now they cover less than 7 per cent. The rate of tropical forest destruction, or deforestation, has increased by 90 per cent since the mid-1980s. Currently, 14 million hectares are being destroyed annually, and a further 15 million hectares degraded. If these rates continue, tropical forests will have practically vanished in your lifetime. If tropical forests are threatened, what are the consequences for biodiversity, world climate, and ecosystems?

The study of the rainforest ecosystems and the effects of human activity upon them are affected by our ideas, perceptions, prejudices, and the messages we receive. This chapter explores rainforests, their ecosystems, and the way in which human interference threatens not just the existence of rainforests but the ecology of the Earth itself.

▲ **Figure 8.1** Playing with figures.

Perceptions about rainforests

Look at Figure 8.1. Work in pairs.

1 What message lies behind the cartoon?

2 Is there any factual basis for the cartoon?

3 Are there links between the message and your own lifestyle?

4 To what extent do you agree with the message being conveyed?

Myths and reality about the Amazon rainforest

The Amazon rainforest is probably the best known rainforest in the world. Many of our beliefs, attitudes, and values about tropical rainforests have been shaped by images and messages we have received from Amazonia. Recently, more has been written about tropical rainforests than about any other ecosystem. This is because of their importance in maintaining biodiversity and world climate, and because of the threats to the ecosystem.

	Assertions (1 = strongly agree, 5 = strongly disagree)					
1	The Amazon is the lungs of the world and if the forest disappears less oxygen will be produced.	1	2	3	4	5
2	The Amazon could be a major source of world food.	1	2	3	4	5
3	Logging is to blame for the forest's destruction.	1	2	3	4	5
4	Deforestation in the Amazon has occurred because fast-food chains in the USA need cheap beef.	1	2	3	4	5
5	Small settlers and their unsustainable farming practices are to blame for the disappearing rainforest.	1	2	3	4	5
6	The Amazon rainforest has almost disappeared.	1	2	3	4	5
7	The Amazon basin has extensive mineral deposits.	1	2	3	4	5

▲ **Figure 8.2** Statements on the Amazon rainforest.

Figure 8.2 shows seven statements about the Amazon rainforest. Score on a scale of 1 to 5 whether you agree or disagree with each statement. Discuss your views with a neighbour. Use the following guidelines to help you form your opinions.

1 a) What is the link between rainforest destruction and additions of carbon dioxide to the atmosphere?
b) What link is there to the 'greenhouse effect'?

2 a) What do you know about rainforest soils?
b) Can a rainforest sustain a large human population?
c) What types of agricultural system are best suited to rainforest areas?

3 Why is deforestation taking place?

4 a) Is there a link between fast-food chains and deforestation?
b) What propaganda have you heard?
c) What do you believe?

5 a) What are the differences in the roles of small farmers and large landowners?
b) What forms of agriculture need most land? Why?

6 a) What is your perception of the problem?
b) When did the rate of destruction begin to accelerate?
c) How much rainforest is left?

7 a) What links are there between mineral exploitation and deforestation?
b) What are the pros and cons of exploiting minerals here?

Where are the rainforests, and how are they used?

Figure 8.3 shows the location of the world's tropical rainforests. However, this distribution is changing, as more and more forest is destroyed, particularly in Africa and Latin America.

▶ **Figure 8.3** World distribution of tropical forests.

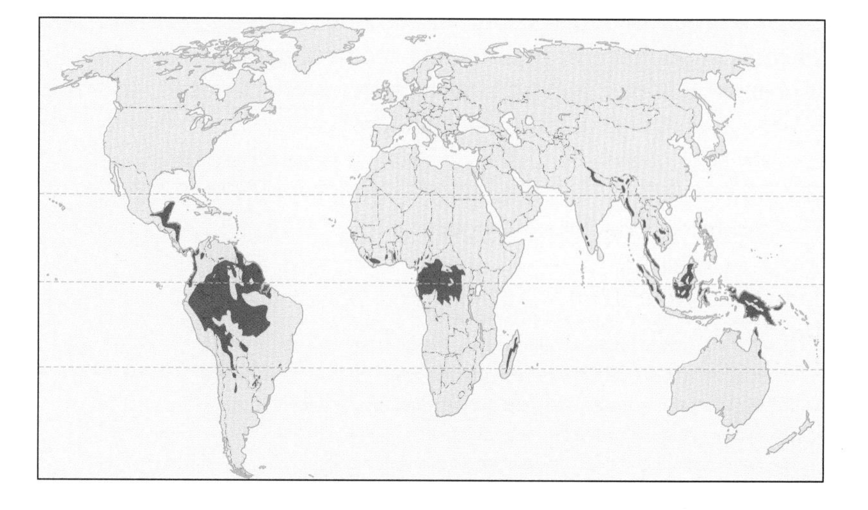

▼ **Figure 8.4** Distribution of tropical forests in Central and South America. In Latin America, the total tree cover is 895.7 million hectares in 23 tropical countries. These too are being destroyed. The main causes are felling and burning to create small farms and cattle ranches, and opening up of areas for colonization and development schemes.

▼ **Figure 8.5** Distribution of tropical forests in Africa where the total tree cover has been reduced to 703.1 million hectares in seven tropical countries alone. The main causes of destruction are cutting for fuelwood, and to clear the land for small-scale and shifting agriculture.

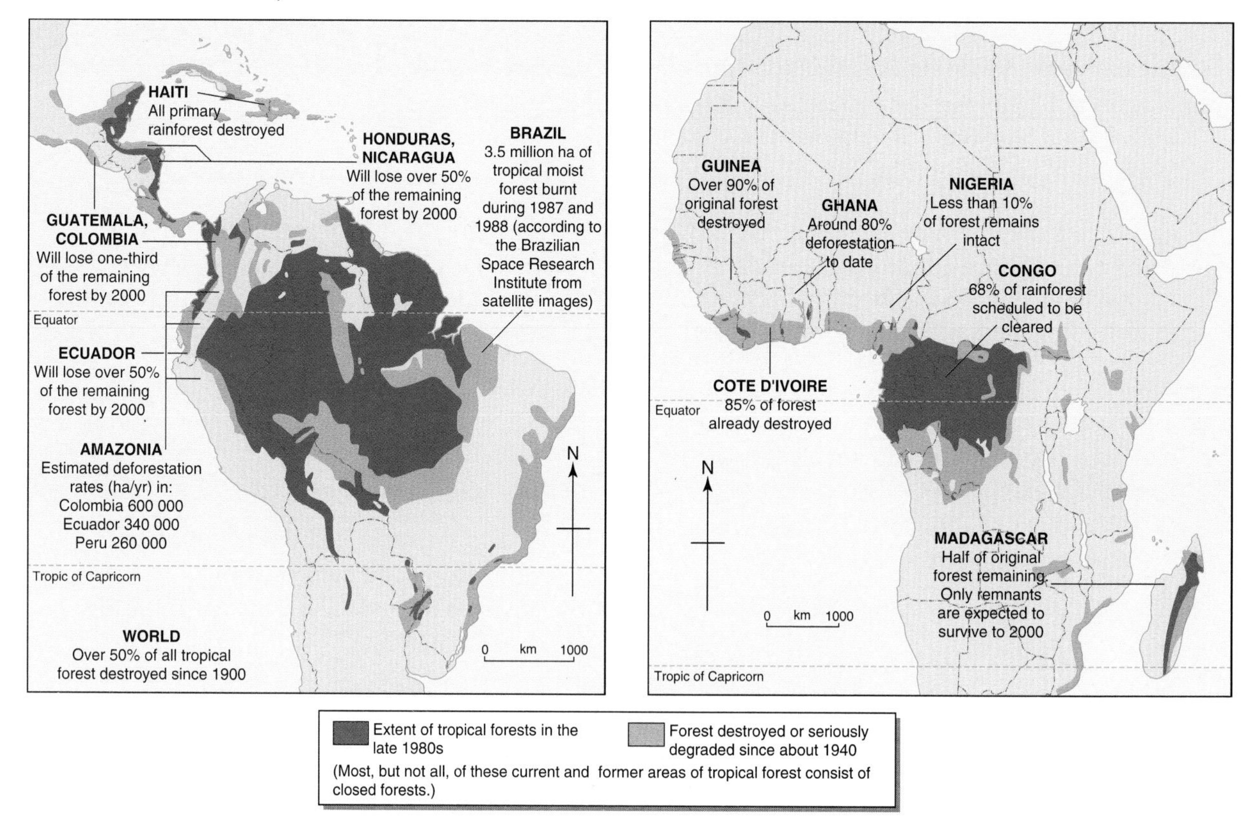

Biodiversity of tropical forests

Medical benefits from tropical forest plants

There is a huge range and number of plant and animal species in rainforests. This is called 'biodiversity'. Biodiversity exists even within very small areas of rainforest. For example, 700 tree species were found in ten 1-ha plots in Borneo. Genes from rainforests are used to improve the resistance of crops to pests or disease. A high proportion of the world's bird and primate life is found in these areas. One-fifth of all bird species are found in Amazon forests, and 90 per cent of all primates are found only in tropical regions of Latin America, Africa, and Asia. This biodiversity is under threat. Scientists believe that 10 per cent of the world's species will become extinct by the year 2000, and 25 per cent by 2009. Indigenous peoples are also in decline. Ninety different Amazon tribes are thought to have disappeared this century.

Indigenous peoples appreciate the medicinal qualities of rainforest habitats and have known about them for generations. Forest Indians of north-west Amazonia use more than 1300 plant species as medicines, and worldwide over 3000 different species are used by indigenous peoples to control fertility.

To date, less than 1 per cent of the 255 000 known rainforest plants have been screened for their potential use in developing life-saving drugs. There have been some notable discoveries. For example, a substance called diosgenin, from the wild yam species in Mexico and Guatemala, enabled the contraceptive pill to be developed. Tubocurarine, which is made from curare from a member of the Amazonian liana species, is used as a muscle relaxant during surgery. Forty per cent of our drugs come from the 'wild' areas such as rainforests, and the trade is worth US $40 billion per year world wide! Yet the Indians in the rainforest have up to now received nothing for this. Debates are now raging about the means by which communities can be paid for medical 'discoveries', which are later developed by multinational pharmaceutical companies for financial gain.

Tropical timber: where does it go?

The degree of management and interference in rainforest ecosystems varies from country to country, depending on the degree and nature of forest exploitation. Some countries have timber resources, but do not exploit them. Others have them and exploit them. Of those that do exploit them, some do so for fuelwood, and others for timber sales overseas. These are complex patterns – see Figures 8.6 and 8.7.

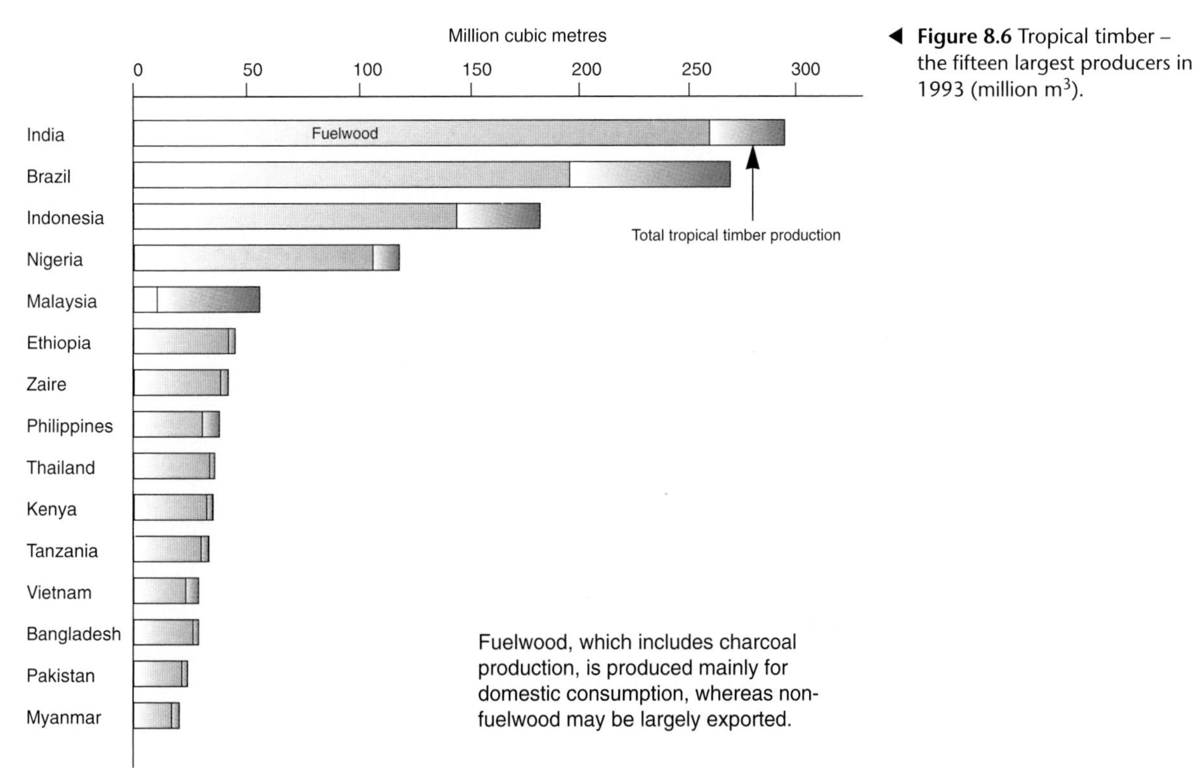

◀ **Figure 8.6** Tropical timber – the fifteen largest producers in 1993 (million m³).

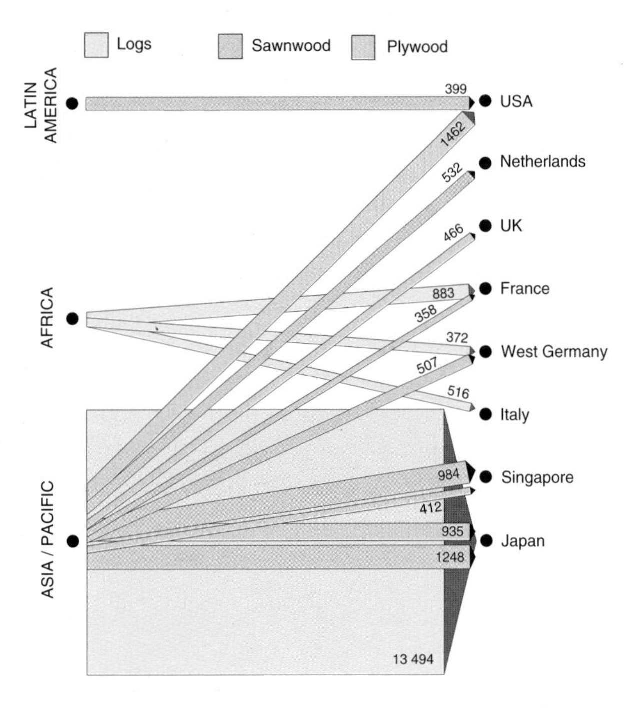

▲ **Figure 8.7** The trade in tropical timber (thousands m^3).

Look at Figures 8.6 and 8.7, and at Figure 8.3.

1 On a world map shade in and label the fifteen largest tropical timber producers. Show by proportional circles the amount used as fuelwood.

2 Using Figure 8.7, show on your map the direction and volume of trade between producers and consumers.

3 **a)** Which countries have significant reserves but do not feature as major producers?
  b) How do you think this can be explained?

4 Explain the pattern of trade.

The dynamics of the rainforest ecosystem

The first known written description of a tropical rainforest was by Christopher Columbus. He wrote the following to King Ferdinand and Queen Isabella on 28 October 1492: 'Never [have I] beheld so fair a thing; trees so beautiful and green, and different from ours, with flowers and fruits each according to their kind; many birds, and little birds which sing very sweetly.' When he realized he had not found the Indies, he decided he had discovered the Garden of Eden.

Rainforest structure

The tropical rainforest has been referred to as the 'jungle', the 'primitive' or 'primeval rainforest', 'equatorial rainforest', and 'tropical wet broad-leaved evergreen forest'. The tallest trees rise to about 60m, and average about 37m. Below these are about five tiers of straight-trunked trees. The upper layer, difficult to see from the forest floor, consists of emergent trees, rising above the canopy. The bases of many of the medium and larger trees send out flying buttresses from heights of 6–9m up the trunk. The heads of the trees are bound together by lianas, or woody vines, some approaching 250m in length. Below the canopy, visibility is good for about 15m and often more. Light levels are low on the forest floor, with less than 1 per cent of light on the canopy reaching the ground, and as a result vegetation is sparse at ground level.

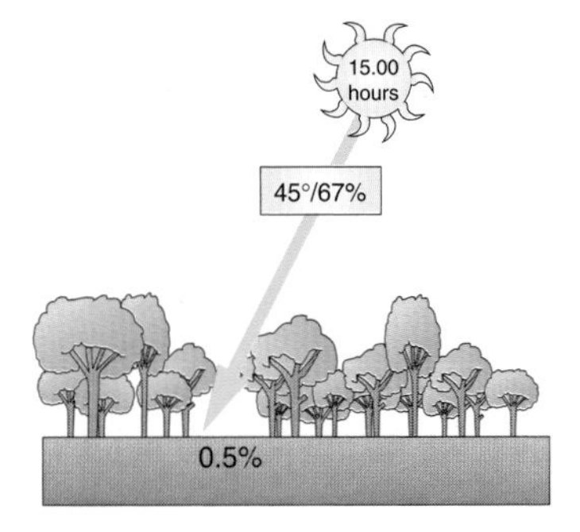

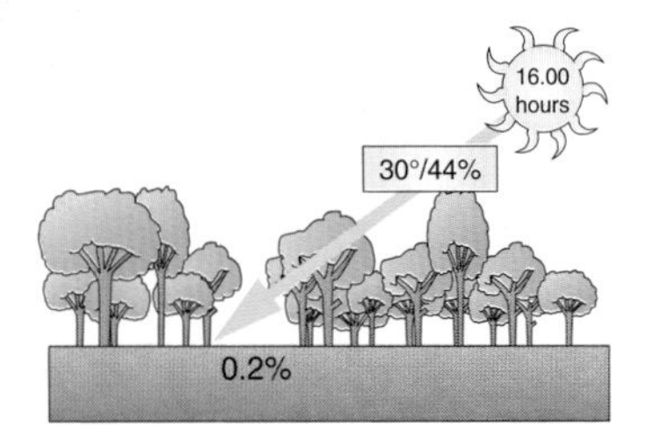

◀ **Figure 8.8** The passage of light through tropical rainforest. Light is absorbed with great effectiveness. If we use noon as a 100 per cent value for light intensity above the canopy, by 4pm (16.00 hours) on a cloudless day only 0.2 per cent of light will reach the forest floor. The light of a full moon is not even perceived here.

▲ **Figure 8.9** The multi-tiered rainforest.

▲ **Figure 8.10** The multi-tiered closed canopy. The four-storey forest is a useful way of dividing the forest, as each overlapping layer has its own distinct communities.

Describe the characteristics of tropical rainforest climate. What are the key differences between Cochin and Singapore climates?

Tropical rainforest climates

The driving force behind rainforest ecosystems is the rainforest climate. The pattern of high temperatures which are maintained in each month of the year, low daily temperature range, and temperatures over the 6°C minimum required for plant growth, means that plant growth is continuous throughout the year. Rainfall varies more than temperature, with a regime of rainfall in each month, but with a maximum season which corresponds with the passage of the overhead Sun. This is described fully in Chapter 5.

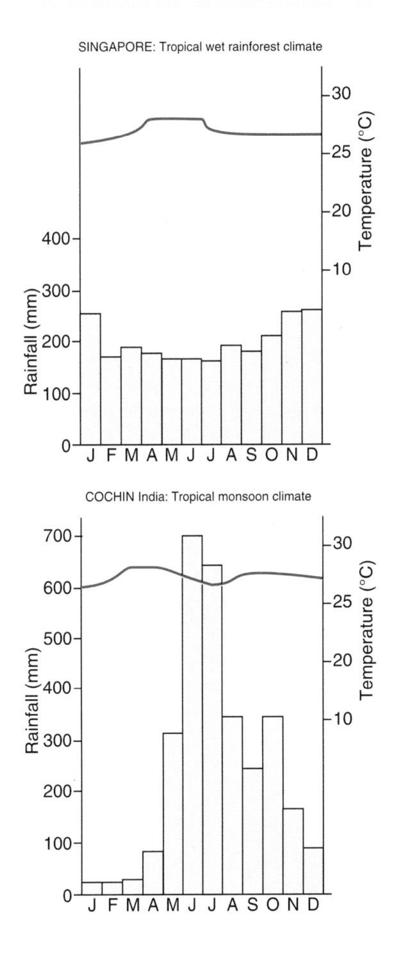

▶ **Figure 8.11** Climate graphs.

▲ **Figure 8.12** A recycler: termite mounds in the Korup rainforest, Cameroon.

Amazonia: a fragile ecosystem

Decaying vegetable matter in rainforests is rapidly consumed and recycled by insects, bacteria, fungi, and soil humus. Plants and trees take most of their nutrients from rain and wind-borne dust, rather than through their roots. Many plants, because they have no contact with the soil, use their roots for stabilization. Rain falling on the forest canopy has more nutrients than soil moisture on the forest floor.

The recycling process is so effective because fairly continuous annual rainfall (2000–3000mm per annum), and high temperatures which cause rapid evaporation, enable a wide variety of organisms to survive. Death is the source of new life. Rainforests are dominated by trees (see Figure 8.14). Each tree is a complicated chemical factory, combining the Sun's energy, rainwater, and carbon dioxide from the atmosphere, in the process of photosynthesis. The result is an energy product which, as well as fuelling the trees' growth, is used for the manufacture of a range of biochemicals – sugars, amino acids, proteins, cellulose, tannins, lignins, fats, waxes, hormones, drugs, and fragrances, to name a few. On death, all of the biochemicals and components are recycled by decomposers. These include:

- bacteria and blue-green algae
- fungi, including mushrooms, toadstools, yeasts, moulds, and slime moulds
- green plants such as algae
- animals, from single-celled protists to mammals.

The rainforest is a balanced, self-sufficient system which depends on recycling for its existence.

"Biological resources constitute a capital asset with great potential for yielding sustainable benefits. Urgent and decisive action is needed to conserve and maintain genes, species and ecosystems, with a view to their sustainable management and use of biological resources."

▲ **Figure 8.13** Extract from The Earth Summit, Agenda 21: The United Nations Programme of Action from Rio, UN 1992.

▶ **Figure 8.14** The Amazon river.

Characteristics of rainforest plants

In rainforest climates a young leaf may grow so rapidly that it outstrips the capacity of the system to strengthen new tissue, so it hangs limp. It may be filled with red pigment to protect the chlorophyll against too much sunlight. A leaf growing on a myrmecophyte – a tree that lives in a state of symbiosis with colonies of ants – is protected from insect attack. Leaves that survive insect attack often develop a thin cuticle to minimize water loss, and a drip-tip to channel water quickly from the latima.

Some leaves have motor cells at their base, allowing them to turn green solar cells towards the Sun. High in the canopy, the leaf could overheat and die, so it turns its edge away. But if the Sun is obscured by cloud, it faces upwards again. Its ability to adapt to photosynthesis shows that leaf structure, anatomy, and life chemistry are geared to minimizing stress and damage. The daily intake and uptake of carbon dioxide are in equilibrium. Rainforests neither recharge the atmosphere with oxygen, nor deplete its carbon dioxide. They do, however, hold much carbon in store. When burned, carbon becomes carbon dioxide, a potent greenhouse gas. Hence the burning of huge areas of rainforest is an environmental issue of global importance. This huge carbon store is secured by living roots which hold both vegetation and water in place. Some of the water is used as a raw material for photosynthesis. The rest evaporates from the floor and insulated by the canopy, it condenses to produce an atmosphere of mist and humidity.

Fungi, the key to rainforest recycling, do not photosynthesize, but feed by digesting dead organic matter with enzymes. In the process, they release nutrients for other plants. The mycorrhizal fungi are essential to the rainforest ecosystem. They weave threads around and inside the roots of canopy trees, feeding from substances within that are rich in sugar and nitrogen – hence the symbiotic relationship.

Agricultural impact on the rainforest

The massive forest destruction of the Amazon rainforest is well documented. The destruction has been so vast that satellite images show change yearly, and even provide a photographic record of the fires that rage, and the effects of smoke on the atmosphere. Figure 8.15 shows these effects.

The key to recognizing agriculture as a cause of deforestation in the tropics is to realize that soils and rainfall here are different from those in most other regions, where agriculture is stable. A remote-sensing team at the Space Research Institute of Brazil (INPE) used LANDSAT images to assess rates of tropical deforestation in Amazonia. What they found in the 1980s contradicted the image of rapid deforestation due to logging and agriculture. The LANDSAT images by 1989 suggested a total of 5 per cent vegetation loss due to deforestation, and that the rate was steady. It contradicts a figure of 47 per cent calculated seven years earlier.

How do you react to the INPE findings described above? Who might be interested in the new findings? Should we accept the figures without question?

◀ **Figure 8.15** LANDSAT image showing inroads into the Amazon rainforest in Brazil. The colours have been chosen to highlight the destruction: dark green of the natural forest contrasts with the pale green and pinks of the levelled forest.

'Slash and burn'

Unsustainable farming practices in the tropics are known as 'slash and burn'. The entire biomass of a future farm plot is felled and burnt in preparation for planting. Some nutrient matter in the ash remains on the surface, and seed is sown in this enriched soil. Without the protective layers of tree cover, rains erode the soil, flushing away nutrients. Only two or three crops may be harvested before the farmers move on to fell more forest. Within five years, soil fertility levels for agriculture are marginal.

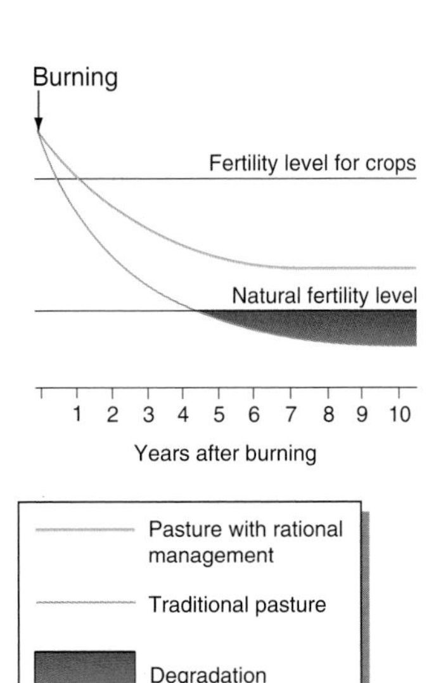

▲ **Figure 8.16** Declining soil fertility. Following forest clearance and burning, peak fertility levels fall sharply.

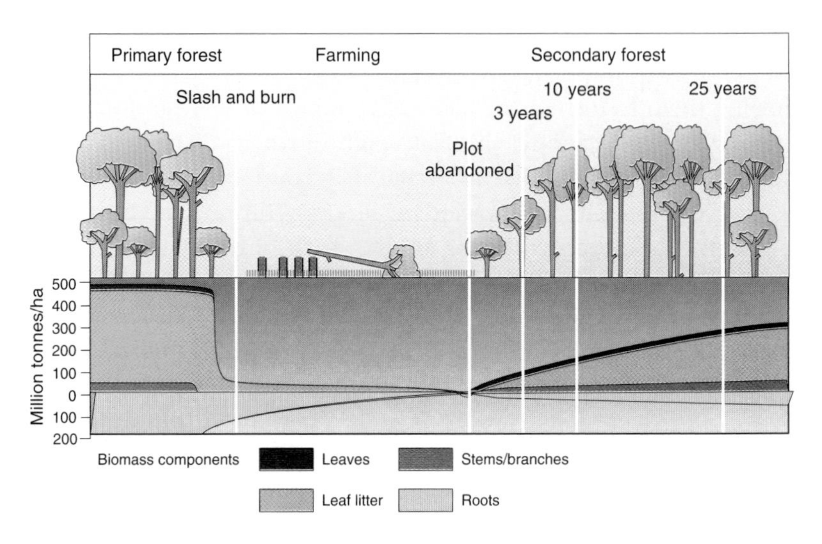

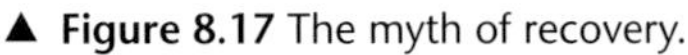

▲ **Figure 8.17** The myth of recovery.

When primary forest is cut down, used for agriculture, and then abandoned, there are marked changes in the biomass. Nutrients are stored in living biomass and forest floor litter. When the forest is cut and burnt, its mass is reduced to almost nothing and its nutrients are exploited before the area is abandoned. Thirty years later, re-growth bears little resemblance to the original forest – it is poor in total biomass and in species diversity.

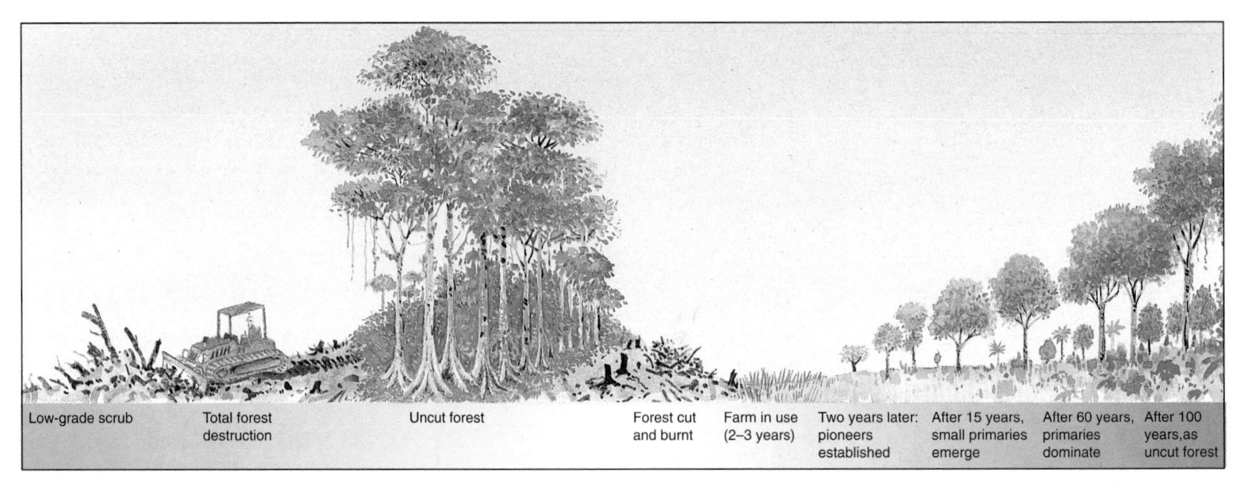

▲ **Figure 8.18** The pattern of forest regeneration.

Forest regeneration

When large gaps appear in the forest canopy, the microclimate of the mature forest disappears. The forest floor is exposed to direct sunlight, air and soil become dry, and temperatures fluctuate widely between day and night. Some pioneer species take advantage. For most, growth is fast and life is short, unlike an undisturbed primary forest. Second-wave pioneers also demand plenty of light; they grow rapidly but live longer, perhaps for a century. These trees fix large amounts of nitrogen, replenishing the nutrients and setting the stage for the return of mature-phase species. If vegetation is bulldozed for logging or agriculture, soils may become irreversibly poor, and once-spectacular forest may be replaced by low-grade scrub.

▲ **Figure 8.19** The result of destruction.

The Korup Project

Tropical evergreen rainforest is the most species-rich ecological community on Earth. The South-West Province of Cameroon still has some rainforest, as at Korup, in pristine condition. It was recognized that rainforest was rapidly being destroyed in the rest of Africa. In 1986 the Cameroon government declared Korup a national park, and signed an agreement in 1988 for help in establishing and managing the park. The management objectives for the park are to benefit science, education, and tourism. The Korup Project aims to protect one of the oldest and most diverse forests. At the same time it will protect water resources for local people, without whom development could not succeed.

1 Study Figure 8.20. Produce a table with two columns, outlining:
   a) the reasons for the project advanced by Benjamin Itoe, Minister of Tourism
   b) your own reasons why tropical rainforests might need to be conserved.

2 Divide these reasons into categories – economic, social, environmental, and political. Is this categorization difficult? Why?

▼ **Figure 8.20** Extract from the Preface to the Korup Project – the plan for developing the Korup and its support zone.

The Republic of Cameroon has been identified by the World Bank, the IUCN and the WWF Biodiversity Task force, as one of 13 mega-diversity countries in the world.

The Cameroon government places a very high priority on the conservation and sustainable development of our natural resources. Our national policy is to conserve at least 20% of our surface area under the National Parks and Forest Reserves.

It is because of the importance which he attaches to the conservation of our very rich and natural ecosystems and the role which these resources will have to assume in preserving the environment, in fostering scientific research as well as in enhancing tourist development, that His Excellency Paul BIYA, President of the Republic of Cameroon, on the 13th April 1989, and for the first time in the political history of our country, created a Ministry of Tourism charged precisely with the responsibility, amongst others, of preserving wildlife, the environment and its biodiversity, and of patterning education programmes to foster these objectives.

Korup in the South West Province of Cameroon is Africa's oldest and most diverse forest. It is our first Tropical Rainforest National Park. This Master Plan is the first of its type to be prepared by our government. We are very proud to hear that it is already being used as a model in other countries.

With the backing of friendly countries and the availability of abundant renewable natural resources within our frontiers, coupled with a highly educated population that is backed by a stable government, the Republic of Cameroon is determined and strategically well placed to make conservation of tropical rainforests and their wildlife a success that should serve as an example to emulate.

Certainly, with the cooperation of our friends around the globe, the implementation of the Korup Plan will before long become a living reality in the service of nature and mankind the world over.

Dated this 15th day of December, 1989, Benjamin ITOE, Minister of Tourism

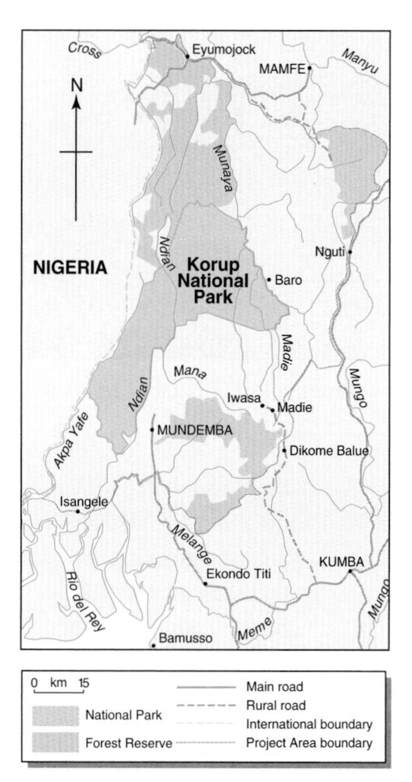

▲ **Figure 8.21** The Korup project area.

Discuss the priorities for the park, with particular reference to the possible effects on ecosystems.

1 How are the aims suited to the forest?

2 How well are they likely to be accepted by local people?

3 Which aims would you see as
   a) easiest to achieve
   b) most difficult to achieve?

4 Try to produce a rank order to show which aims you would most like to see achieved.

What is the project about?

The project is as much about people as it is about rainforests. The survival of the rainforest in Cameroon depends on how people use it. The project's priorities are to harmonize the impact of people on the rainforest. It has different purposes according to which people are being targeted. Priorities are to:

- educate indigenous people whose hunting, trapping, and agricultural activities contradict the aims of the park
- pursue rural development in areas surrounding the park
- involve people of the park in the process of change
- move villages from within the park, and provide roads, health, education, and water supplies
- create alternative sources of revenue for those who depend on hunting
- develop sustainable agriculture, including access to markets for cash crops
- construct an education centre
- increase access points, nature trails, campsites, and guard posts
- increase educational opportunities for teachers
- attract tourists, especially scientific visitors.

What is the Korup region like?

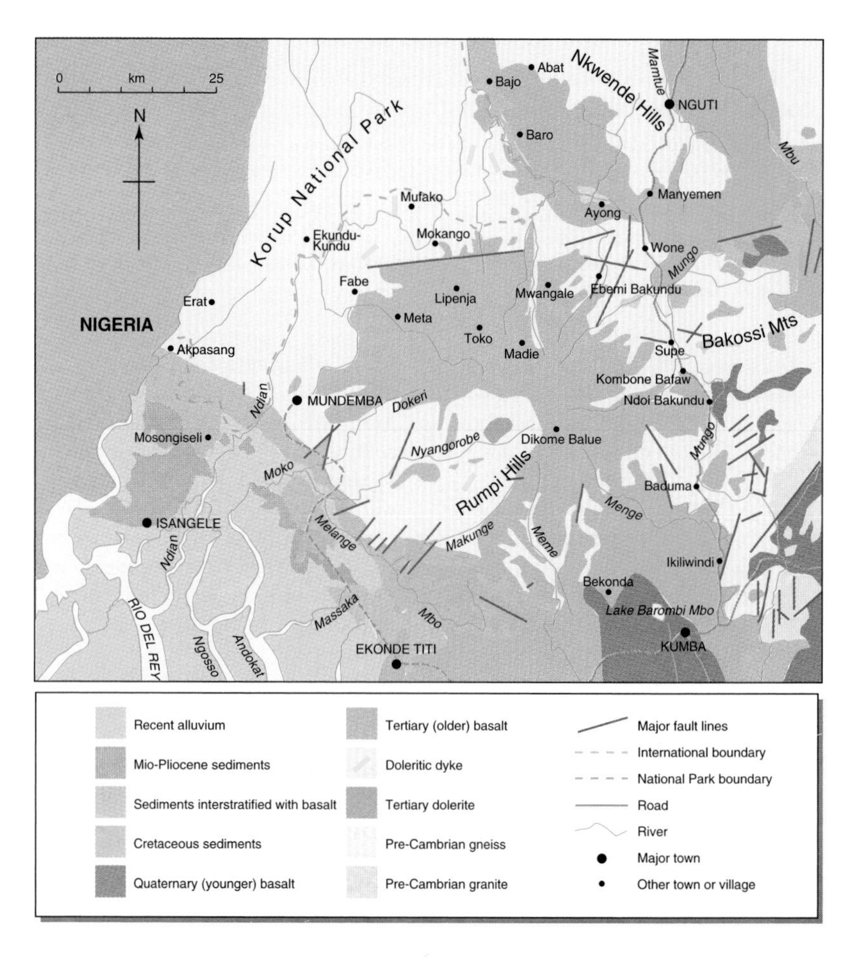

▶ **Figure 8.22** Known geology of the Korup area.

The physical background

Korup is a dense lowland forest. The western forest is unsuitable for agriculture, because soils here are poor, and so is relatively untouched. The eastern forest, with soils of volcanic origin, is attractive for farming and has been cleared in parts. The western forest has many plant species for which Korup is the only known locality in the world.

Animal life is also diverse and rare. There are 143 known passerine (perching) bird species – 49 per cent of Africa's total number of passerines. Also found here are 25 per cent of Africa's primate species and 52 mammal species, belonging to 20 families. These include the water-shrew, red colobus monkey, chimpanzee, golden cat, forest leopard, and forest elephant.

In the south, soils are acid and poor in nutrients. Farmers avoid them, though they have been used for commercial oil palm production. In the east, volcanic soils are reddish in colour, fine in texture, fairly clayey, and sticky when wet. They are important for coffee and cocoa cash crops, as well as for food crops. In the west, soils are granitic, coarse-grained, gritty, and poor for agriculture. Forestry is unimportant, because the area is isolated, the terrain rugged, and there are few commercially desirable species.

1 Construct a sketch map, using Figures 8.21 and 8.22, and the text, and annotate it with information about the ecology and geology of Korup.

2 Does the information on ecology and geology change your order of priorities for Korup? Re-order them if necessary, outlining the reasons for your choices.

Location	Village	Population
Within project area	Erat (Ekon II)	278
	Ekundu-Kundu (Ekabo Old Town)	92
	Esukutang	155
	Bera	55
	Ikenge	146
	Bareka-Batanga	24
Villages to the south of the park	Akpasang (Ekundu-Kundu II)	302
	Monsongiseli	325 *
	Mundemba Town	6628 *
	Ndian Town	1072 *
	Ikassa	81 *
	Ndian Estate	2000 *
	Last Bush	210 *
	Last Banana	34 *
	Bekoko	n.a.
	Ngumu	n.a.
	Okabo	n.a.
Villages to the north of the park	Ngenye	82
	Mopako	68
	Lobe	28
	Babiabanga	40
	Tombel	18
	Banyu	31
	Baro	120
	Bajo	64
	Basu	31
	Ekogate	26
	Bakut	166
	Mbofong	39
	Akwa	81
	Ekoneman Ojong	43
	Nguru	3
	Ekon I	258

n.a. = not available * = estimated figure

▲ **Figure 8.23** Villages of Korup with population of the park and surrounding areas.

▼ **Figure 8.24** Villages of the Korup.

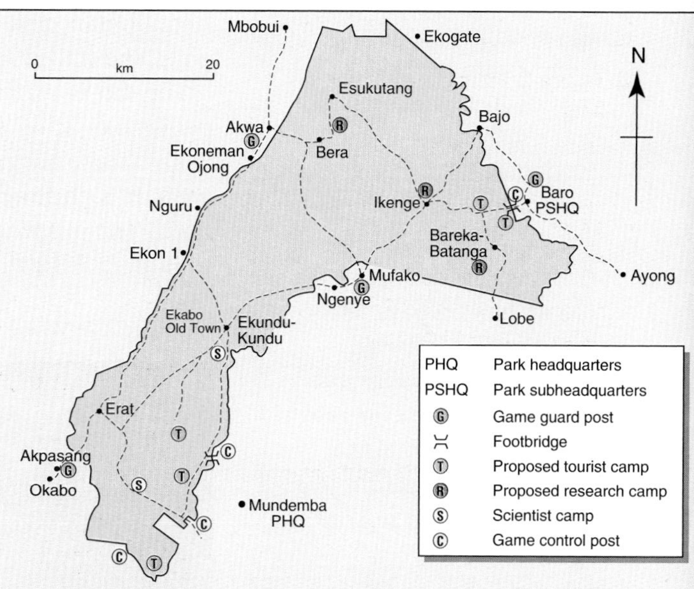

The population of Korup

Of the total population of Korup park, 51 per cent are under the age of 15. There is a declining birth rate, caused by an exodus of people of reproductive age. With reference to Figure 8.23:

1 Calculate the total population of the villages within the park.

2 Compare this figure with populations to the south and north of the park, and annotate your sketch map of Korup.

▲ **Figure 8.25** Production of palm wine, an illicit activity.

The project management team placed as a priority the resettling of villages from within the park to outside its boundaries. Do you think this is desirable:

a) socially
b) ecologically
c) economically?

Use the subheadings in the theory box on page 187 to organize your answer. Discuss this in pairs, and share results with the rest of the group. Produce a chart showing reasons *for* resettlement and reasons *against* it.

The economy of Korup

Economically, the area is a mixed subsistence and cash crop economy. Subsistence crops include cassava, bananas, plantains, cocoyams, maize, and yams. Cash crops are cocoa and coffee. Together, Korup villages have 18.9 per cent of coffee bushes, and 18.7 per cent of cocoa trees found in or near the park, but 61 per cent of the population. Other economic activities must, therefore, be important. Fishing is a seasonal occupation in all but the wettest months of July and August. Freshwater crayfish have important potential as a cash crop. Poisoning of streams with vegetable toxins is a traditional practice, now being replaced by insecticides intended for cocoa trees. This damages the sustainability of fishing by threatening fish populations, and endangering human health. Most adult men have firearms for hunting, which is illegal. They hunt at night for meat to be traded in surrounding villages.

In the margins of the forest, snare traps are common. These comprise wire fence, hundreds of metres long, with gaps carrying wire nooses and a trap sprung with a powerful sapling. This is efficient hunting if traps are monitored regularly, but between one-third and one-half of animals trapped are found in a decomposing state. Rattan cane and saplings of other trees are used for building poles and traded for other goods.

The sap of raffia palm is collected and used as a beverage. This is perishable, so is only traded locally, though several villages have 'stills'. These are guarded and palm wine is distilled in an illicit gin or *afofo*. It is cheaper than imported spirit, and is a major item of local trade. The plantation 'labour camps' of PAMOL at Mundemba are an important market.

▼ **Figure 8.26** Village in the Korup, with electrification.

How sustainable is a rainforest?

The sustainability of a rainforest needs to be measured against the economic services it provides. Korup can be measured economically against certain criteria.

Gene pool

Forests contain a diversity of species, habitats, and genes, which are probably their most valuable asset. They provide a gene pool that protects commercial plants against pests and can provide the raw material for high-yielding strains of seed. Wild strains of avocado, banana, cashew, cacao, cinnamon, coconut, coffee, grapefruit, lemon, paprika, oil palm, rubber, and vanilla – exports worth more than US $20 billion in 1991 – are found in tropical rainforests.

Water

Forests absorb rainwater and release it slowly into streams, preventing flooding and extending water availability in dry months. Forty per cent of farmers in the developing world depend on forest watersheds. In India, forests provide water regulation and flood control valued at US $72 billion a year.

Watershed

Forests help prevent soil erosion. Siltation of reservoirs costs the world economy about US $6 billion in lost hydro-electric power and irrigation water.

Fisheries

Forests protect fisheries in lakes, estuaries, and coastal waters. 75 per cent of fish sold in the markets of Manaus, Brazil, are nurtured in seasonally flooded forests.

Climate

Forests stabilize the climate. Tropical deforestation releases the greenhouse gases carbon dioxide, methane, and nitrous oxide, and accounts for 25 per cent of the net effect of all greenhouse gases. Replacing the storage function of tropical forests would cost an estimated US $3.7 trillion – equivalent to Japan's total GNP.

Project planning and implementation for Korup

Who should benefit from the project?

A sustainable future is one that does not use up a resource for future generations. But who should the Korup plan be for?

- For indigenous peoples? They will be resettled beyond the margins of the project area.
- For the people of Cameroon? Many are unaware of the Korup's existence, as communications are poor. Most Cameroonians regard the rainforest as 'somewhere else', rather than something to be prized.
- For the rest of the world? Why? So that it can be kept like a zoo in a fossilized state, never to develop?

To make sustainable planning decisions, a range of data is needed.

A WWF team have been mapping and surveying the project area to record the level of its infrastructure. They found that, prior to project development:

- there was no adequate road network
- the only place with electricity was Mundemba
- water was drawn from streams and wells, except in Mundemba and Nguti
- there were schools only in the three main centres
- hospitals were only in the main urban centres, with health centres scattered through the project area.

Consider each of the project's priorities. You are trying to win a place on the Operation Raleigh scheme to help with environmental improvements in this area. Before your interview, you must produce a report on the likely effects of human interference on the ecosystem, if these priorities are implemented. Write a report of 750 words on those aspects of the project which you feel will have an impact on the ecosystem of Korup.

The team identified a number of problems which, because of the lack of infrastructure, were affecting people's quality of life.

- Farmers cannot take produce to market easily, and farm inputs are costly to bring in.
- Lack of daylight hours is a problem.
- The disease filaris causes villages to be sited high above rivers where the filaria fly breeds. Women spend most of their day fetching water.
- Many villages have no access to schools, thus reducing literacy levels.
- In rural areas there are few medical facilities with laboratory analysis services.

The most sensitive issue facing the project concerns six villages lying within the park. Cameroon law prevents villages being within a national park. The government was concerned about the effect of this law and therefore set up socio-economic surveys. These found that villages were on poor soils, and that crop yields were low. There was no means of transporting the crops out. The main source of income came from hunting and exporting dried animals. This conflicts with laws on hunting within a national park, carrying weapons without a permit, and exporting wildlife without a permit. Village populations are declining as people move to the towns.

The project had a number of aims for the sustainable development of the Korup: resettlement, hunting zones, tourism, roads, agriculture, water supplies, health facilities, and schools. Figure 8.27 shows how these priorities were to be phased in and action taken.

Activity	1989–90	1990–91	1991–92	1992–93	1993–94	1994–95
Establish Park infrastructure	—	—	—	—	—	—
Air photography/ satellite imagery		—				
Vegetation mapping			—			
Soil/land use surveys			—			
Mundemba–Toko road						
– design		—				
– construction			—	—		
Mundemba–Isangele–Akwa road						
– design		—				
– construction			—	—		
Rural road rehabilitation		—	—	—	—	—
Agricultural, livestock, fisheries development	—	—	—	—	—	—
Provision of water supply, health, and education facilities		—	—	—	—	—
Resettlement						
– 5 villages (excl. Erat)						
– planning		—	—			
– relocation			—	—	—	
Resettlement – Erat						
– planning			—	—		
– relocation				—	—	—

▲ **Figure 8.27** Time-lines for implementing project priorities.

◀ **Figure 8.28** Gateway to the Korup park, near Mundemba. This bridge was built by participants of Operation Raleigh.

Resettlement

The proposal was to relocate six villages within the park – Erat, Bareka-Batanga, Ikenge, Esukutang, Bera, and Ekundu-Kundu – see Figure 8.24. Sites would be chosen and people relocated after the construction of the Mundemba–Toko road. Research suggested that people would move willingly if it meant social and economic advancement. The other option was to keep villages where they are and restrict hunting to a subsistence level. The government's decision was:

- to encourage villages to relocate on fertile soils beyond the park – villages will be in the same ecological zone and in their tribal groups
- to allow villagers to hunt and gather forestry products with which they are familiar.

Hunting zones

People need to be taught to manage their own resources, including animals. Wildlife laws will be reviewed to allow villages to extend hunting beyond the park, in return for stopping it within it. Each village will be given a quota of animals to hunt, and taught to identify protected species.

Tourism

Tourism in Korup has the potential to contribute to the local and national economy through foreign exchange, employment, and business opportunities. The project insists that local people must benefit from such development, but tourism must be kept in check. Tourists will only be able to experience the rainforest on foot. The sighting of large animals is infrequent here – the rainforest is not like the African savanna. Rainforests currently have popular appeal to international tourists, and specialist appeal to smaller numbers of people – adventurers, and scientific tourists who are interested in flora or fauna. They have particular appeal to butterfly, bird, and botanical groups. There are constraints, however. Depending on the season Mundemba is at the end of a dusty or muddy road from Kumba. The park entrance is another 10km further on. Facilities are poor. Unless tourists are on a circuit that includes a boat trip from Limbe, they will have to backtrack.

Roads

The main developments are shown on Figure 8.29. These include the following.

- Construction of the Mundemba to Toko road. This needs to be completed early, within two dry seasons, to promote rural development.
- The Iwasa–Dikome Balue road serves the area east of Toko. Although built recently, its gradients are often steep, and it is impassable in all but the driest months. It needs to be redesigned and rebuilt.
- From the Kumba to Mamfe road, there is a rural road to Banyu, with a spur off to Baro. A track to Osirayib is being extended by the logging company working in the area. All of these pass through areas that are potentially rich for agriculture. It is proposed to make them accessible throughout the year.
- The Cameroon government plans to construct a road through Toko connecting Mundemba with the Kumba–Mamfe road.

▼ **Figure 8.29** Roads in the project area. The road programme is part of a plan to create a 'buffer zone' around the park, to develop the area around the park rather than within it.

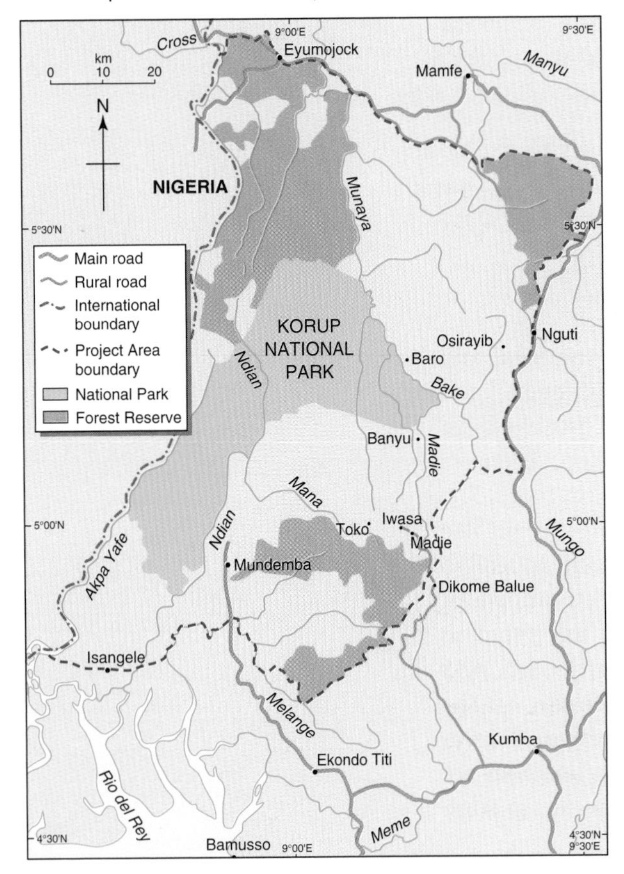

The park headquarters are to be located at Mundemba. The Trans-African Highway is to pass to the north of the national park. The park subheadquarters at Baro will be able to handle limited numbers of tourists, and villages will be encouraged to provide basic accommodation. The bridge across the Bake river on the Nguti–Baro road will give all-season access to northern Korup.

Agriculture, agroforestry, livestock, and fishing

The Korup project has placed great emphasis on sustainable agriculture. Project authorities know that sustainable agriculture is possible, given the population density of the area. To achieve this, a series of trial demonstration centres will be set up in the Toko/Baro area and the rural development zone. These will act as centres for:

- demonstrating techniques in sustainable agriculture and fishing
- testing new crop varieties
- promoting new crops
- breeding 'improved' livestock
- liaison with community groups and schools
- training local farmers, and seminars on sustainable development
- a multipurpose co-operative for food marketing
- an agricultural credit-in-kind scheme for small farmers, to provide improved plants, livestock breeds, and agricultural equipment.

Water supplies, health facilities, and schools

No detailed plans have been finalized, but plans have been put forward for the construction of improved water supplies, schools, and health facilities in existing villages in the Toko/Baro area.

Evaluating the project

You have been asked to evaluate the Korup Project.

1. What would you look for, to see whether the Korup Project had been a success? Who would you interview? What other data would you collect?
2. Is the Korup Project a model you would like to see adopted elsewhere in the world? Why?/Why not?

Montane forests

The Kilum Mountain Forest Project

Montane, or mountain, forests are different from rainforests. The Kilum mountain forests in north-west Cameroon are the last remnant of a unique and endangered ecosystem. The area has a temperate climate because of its relief. At 3011m, Mount Oku, or Kilum, is the second highest point in mainland West Africa. The Bamenda Highlands montane forest ecosystem, of which the Kilum forests are the largest part, has a unique flora and fauna.

However, population growth has put pressure on the land. This is a highly populated area, by Cameroon standards. Over 100 000 people depend on the forest for a variety of products, and for their water supply in the dry season. Communities have problems: the impact on the ecosystem is leading to unsustainable practices. Immediate action is needed, which is why the Kilum Mountain Forest Project (KMFP) was set up.

In 1987/88 the KMFP set out to help local people to improve the management of their resources. Priorities have been identified, and the project completed its second phase of development in 1994. The objectives of the Kilum Mountain Forest Project are to help conserve the Kilum mountain forests and to promote the sustainable use of natural resources in the Kilum area. The project has a social as well as an ecological function. The project is supported and funded by the International Council for Bird Preservation (ICBP), the World Wide Fund for Nature (WWF), and the Overseas Development Administration (UK government).

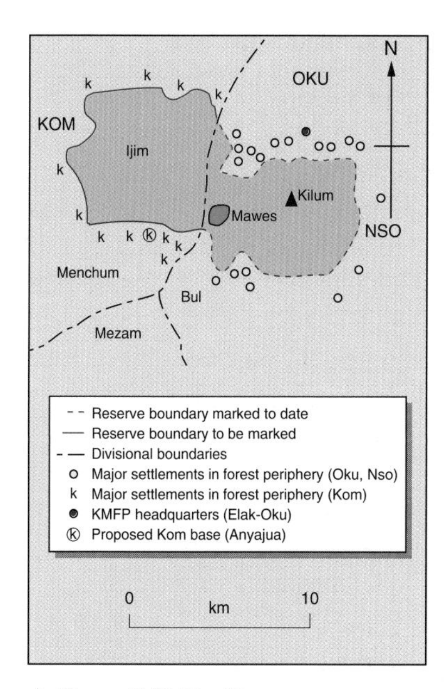

▲ **Figure 8.30** The Kilum area.

What are the priorities?

It is estimated that about 200 million hectares of new trees must be planted worldwide during the next decade if the less developed countries are to meet their people's needs for tree products. If the planting was undertaken commercially, at £375/ha, the investment needed would be about £75 billion. Most of this is required in countries that are least able to afford it. Governments of these countries cannot finance tree planting, even with overseas support. Most must therefore be done by the rural people themselves. This is being tackled by social forestry policies.

Social forestry

To develop local participation, new approaches through social forestry programmes are essential for foresters to stimulate and guide tree-growing by rural people. What makes social forestry different from government or industrial forestry is the involvement of local people in growing trees for their own use. Social forestry rarely involves large forests; instead, small numbers of trees are planted here and there, on a small village plot, along the road, or interspersed with crops in fields. Small-scale activities by millions of tree-planters can be significant.

▶ **Figure 8.31** Agricultural improvement – KMFP staff at work with farmers.

Socio-economic importance of the forest

Rights to the forest area are traditionally held by three ethnic groups: the Oku, Nso, and Kom peoples. Over 100 000 people depend on the forest for firewood, building materials, honey, medicinal plants, and bushmeat. The forest holds cultural significance for the people here, and plays a crucial role in regulating the water supply. Fuel consumption in Oku's cool climate is high, and domestic firewood is collected from the forest. Many building materials, including pole bamboo, fibres, and thatching grass, are harvested from the forest and montane grassland. Because of its biodiversity, Oku has become a centre for traditional medicine. There is a tradition of beekeeping in the area, and honey is a constituent of many herbal remedies. The bark of the African cherry prunus, sometimes known as red stinkwood, is used for its medicinal properties.

What are the causes of the problems?

Forest destruction

Between 1963 and 1983, half the forests in this area were destroyed or badly degraded. The principal cause has been the practice of 'slash and burn' for creating new farmland. Thousands of goats and sheep have been brought into the forest, preventing regeneration and causing erosion on the steep, thin-soiled slopes. Beehives have been destroyed by fire. More than 80 per cent of mature prunus trees have been felled. Pressure on land to provide food for people has led to poor soil management. In the dry season, livestock wander uncontrolled through farmland.

Physical features

Until the problems of unsustainable agriculture are addressed, forest conservation is unlikely to succeed. Kilum's agricultural economy is poor in crop diversity, market access, and soil and livestock management. There is serious soil erosion and impoverishment here. This has been due to:

- the hilly nature of the terrain
- often torrential seasonal rainfall

- friable soils
- a growing cash economy which has forced more people into cultivation.

Draw a flow diagram to show how the present management of the ecosystem is unsustainable.

Population pressures
Much new farmland is steep marginal land. Population pressure means that people need more cash, so they turn to cash crops to pay for medical services and for education for their children. With so many people growing cash crops, market prices fall and everybody is worse off. The traditional fallow period has been abandoned, so soil has no chance to recover. Even on steep slopes, most of the permanent vegetation is cleared. Trees and stumps are burnt. As land deteriorates, farmers look to the forest for new productive farmland.

Project management
The results of poor management are evident. As forests have receded, firewood collection has become increasingly difficult. Dry-season water supplies have dwindled. Soil losses are heavy, landslides common, and crop yields often poor. Thus the priorities are:

- to educate villagers about sustainable farming methods
- to ban deliberate burning and free grazing of stock
- to control the harvesting of the bark of the red stinkwood tree
- to conserve soils
- to establish tree nurseries in order to restock woodland.

Education of villagers
Environmental education is the most important of the project's activities. It is aimed at schools, farmers, women's groups, traditional leaders, agricultural staff, and government administrators. School visits include tape/slide presentations, portable display, a demonstration of soil conservation, nature walks, and teachers' workshops. Many schools have set up tree nurseries and soil conservation trials on their own farms. The *FEN* magazine now attracts regular letters – see Figure 8.33.

◀ **Figure 8.32** In-service training for teachers at a high school.

From Nsakse Frederick, Legal Department, Kumbo

When I read in the fourth issue about the firewood crisis in Kenya, it reminded me of something that many young people, especially those who leave Oku, do not know about - the part that firewood plays in Oku culture.

In the Oku culture, firewood stands out as the most precious gift one can give to the Fon, father or Sub Chief. In the early days, groups of young children carrying small branches were led to the palace by a Juju known as 'Nkom'. Whenever they came, everything else was suspended. The Fon saw to it that the children were well entertained. This aspect of culture was instilled in their minds.

In Oku, there is no fixed bride-price for a wife. What is most important for a son-in-law is firewood. When his wife becomes pregnant, he must begin to pack firewood which is taken to his parents-in-law when the child is delivered.

Whenever a Fon is crowned, a Sub Chief has to arrange for wood to be taken to the Palace. When a son has wronged an elder, the Oku culture calls for a fine of five logs of firewood.

From Chung Peter Kongnyuy, CBC, Elak

Kilum Mountain, a special hill
Holding life with trees
From there one can hear
The forlorn cry of 'Fen'
A bird for time, a bird that keeps
tradition alive.
Kilum Mountain, a hill with dense forest
The trees are rich and deciduous
The deep black soil holds rain and mist
Which sink slowly into the ground
Feeding the streams that serve mankind
Too many goats, too many bush fires
Threaten the natural forest
The forest for us, the forest for
our children
Keep it, guard it, care for it
For it keeps men, guards men, cares
for men
Destroy it and man is destroyed.

But the rich green hill breaks down
Gradually changing nature
For it now grows red and bare
It can no longer hold rain and mist
Leaving the streams dry and man is
suffering
The great red hill stands desolate.
The earth has come away like flesh.

The lightning flashes over them
Clouds pour down on them
The dead rivers come to life
Full of the red blood of the earth
Down in the valleys, women scratch the soil
that is left
Maize hardly reaches the height of a man
These are valleys of old men and women
Young men are away, young girls are away
The soil cannot keep them anymore
Today, man has come to live
And tomorrow, man will die away.

▲ **Figure 8.33** Two letters to the editor of the *FEN* newsletter.

Forest conservation

Forest conservation is an important aim of the KMFP. Before the project, 3 per cent of the forest was lost annually. Clarifying the forest boundary has been a top priority, in collaboration with local communities. Boundaries have been agreed in 19 of the 40 villages settled on the edge of the forest.

Burning and free grazing

Fire has been the most destructive agent in Kilum every dry season. Fire is used to clear forest, to manage grassland, to harvest wild honey, and to dispose of farm refuse. Forest fires are usually the result of fires getting out of control. The KMFP has issued fire-fighting equipment to the village representatives.

The practice of keeping goats and sheep in the forest is recent. It is forbidden, but the ban has not been enforced until recently. Most sheep and goats have now been removed from the forest, but they graze on the surrounding grassland, where they cause erosion. In the dry season, many move into the forest, and their grazing prevents plants from regenerating.

Most households in Kilum keep chickens, and a few goats or sheep. Cattle are reared by Fulani herdsmen and usually graze on Kilum's grasslands in the rainy season before descending to the Ndop plain for the dry season. Goats and sheep are tethered throughout the rainy season but during the dry season they are allowed to roam free, which is when they cause damage. The project has started trials to see if goats and sheep can be reared in controlled conditions. One problem is the high cost of fencing materials. Experiments to create hedges are being made with several different species of tree and shrub. Guatemala grass has been introduced to farmers for dry-season fodder, and the project is experimenting with making silage.

Harvesting Pygeum *bark*

In the past, the harvest of *Pygeum* bark was carried out on a sustainable basis, with bark cut every seven years so that the tree had time to regenerate between harvests. The commercial exploitation of *Pygeum Africana* bark began at Kilum in 1976. Controls over its exploitation are inadequate, and 80 per cent of the mature *Pygeum* trees here have died as a result of poor harvesting. Illegal harvesting is still going on in the Ijim-Jung forests on a large scale. The project is now making considerable efforts to curb this exploitation.

Soil conservation

Farmers are aware of the problem of erosion – the Oku language has words for 'soil-wash', 'gullies', and 'landslides'. But its scale is unprecedented. Farmers have built soil ridges following the contours (rather than up and down the slope) since the 1940s. However, ridges often stray as much as 10° or more off the line of the contour.

Soil conservation measures have been discussed with farmers since the beginning of the project, and 40 farms are now used as demonstration farms. Trials suggest that the most effective conservation measure on the farm is the use of a contour hedge of permanent vegetation at intervals. These are spaced every 2–3m up the slope. Hedges act as a barrier against soil-wash, they increase soil fertility by fixing nitrogen, and they raise nutrients in the soil that are normally beyond the reach of farm crops. When a farm is ready for planting, contours are marked out and seed is sown in a drill at intervals of 1–2cm.

Tree nurseries

Tree nurseries play a crucial role in ecosystem management. Nurseries supply planting materials for soil conservation, agroforestry, fuelwood plantations, live fencing, windbreaks, animal fodder, demarcation of the forest boundary, and planting in degraded areas.

For each difficulty identified in Figure 8.34, explain:

a) the effects on the ecosystem

b) steps needed in ecosystem management to relieve the difficulty.

An experimental tree nursery, with a capacity of about 20 000 seedlings per annum, has been established since 1988. Farmers, women's groups, and schools have all been encouraged to establish their own tree nurseries.

The economy of Kilum once depended heavily on the forest. Hunting, wood-carving, traditional medicine, beekeeping, and manufacturing still bring in a fair amount of income. The project supports the development of medicine, beekeeping, and wood-carving. The income-generating capacity of the forest is a powerful argument for its conservation.

Trees and forests provide us with many things: firewood, oxygen, building and carving materials, medicines, food, and paper. Forests hold the rainwater which is so important for our water supply. Trees help to prevent soil erosion.

As our population grows, we cut down many trees for our own use. We cut down forests to make room for our farms and homes. Our natural forest is disappearing. Who is planting trees for tomorrow?

As we know, water is a limited resource. Rain that falls during the rainy season has to provide us with water during the dry season. Good, clean water is needed by everyone for cooking, bathing, drinking, and for growing our crops.

We are losing much of our water because of deforestation. Often our water is too dirty and polluted to drink, and it brings disease. Sometimes our streams dry up and there is not enough water for our growing population.

Land is one of our most important resources. We farm the land and get our food and energy from it. We build our houses and live on the land.

As our population grows, our need for land becomes greater. Will we have enough land for everyone? We must care for our land and use it wisely.

▲ **Figure 8.34** The project view of unsustainable lifestyles. These cartoons are from *FEN*, the KMFP magazine. They highlight the difficulties faced by the project and Kilum communities.

The KMFP team directors change every three years. You have to advise the next KMFP team on the best ways forward to manage an ecosystem under threat. They need to be briefed in London before they depart for Cameroon. You must advise them on the following aspects of ecosystem management:

- social forestry
- agroforestry
- environmental education.

Using all the material in this study, prepare a report of about 1000 words summarizing the work of the current KMFP team. Identify what you feel still needs to be done.

Upland temperate forests

Britain's landscape has changed dramatically during this century, with the continuing afforestation of some of its uplands with coniferous plantations, and the decline of the traditional lowland deciduous woodlands. Why is this? Forests are a major source of income and require as much careful management as any agricultural system. The conifer plantations shown in Figure 8.35 generate income and make use of otherwise marginal land.

Britain's forests continue to make the news. Upland conifer plantations make significant changes to the landscape, and increasing areas are in private ownership. This is likely to change if the Forestry Commission is privatized. The Forestry Commission was established in 1919, shortly after the end of the First World War. It was felt that Britain should in future be self-sufficient in timber supplies. The Commission had powers to buy land, create its own plantations, and establish them. In Britain today about 800 000ha are devoted to softwood plantations, divided equally between the Forestry Commission and private landowners.

The original reason for creating forests (self-sufficiency during wartime) is now, of course, a thing of the past. The justification for retaining these woodlands has had to shift. The recreational use of woodland in an age of increased leisure time is one reason why they are kept. If the area of land under forest increases today it will be because the return on marginal farmland makes it more worth while to turn it over to forestry. Fieldwork techniques can be used to explore the potential return on forested land.

▼ **Figure 8.35** Coniferous plantations around Ladybower reservoir, Derbyshire.

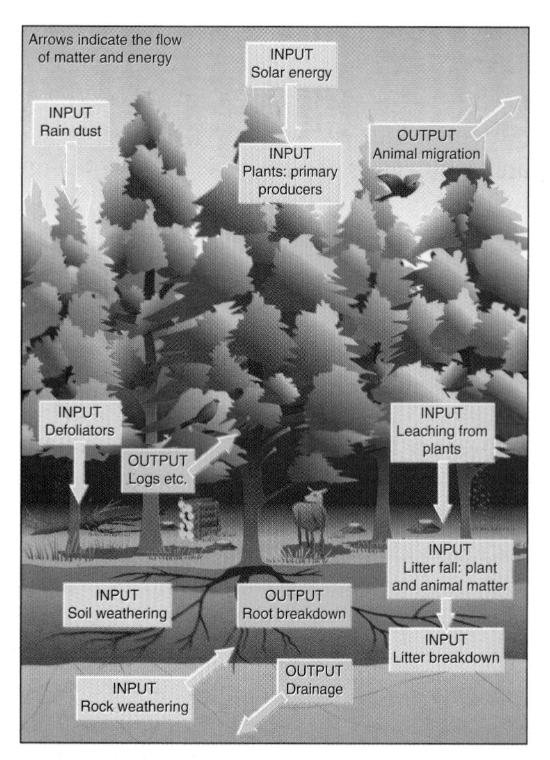

▲ **Figure 8.36** The forest ecosystem.

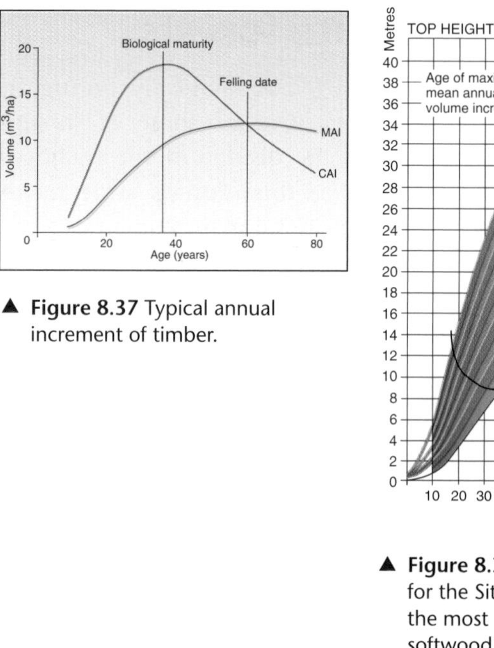

▲ **Figure 8.37** Typical annual increment of timber.

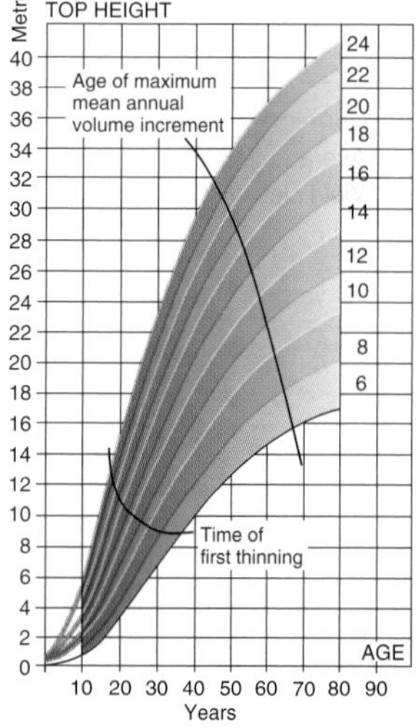

▲ **Figure 8.38** Yield class table for the Sitka spruce, one of the most common softwood species grown on British plantations.

Look at Figure 8.38. If 1ha of trees 25 years old has an average height of 5m, into which band will they fall? The band number – in this case 6 – means that the hectare of wood should increase in volume at a rate of 6m^3/year. Compare this with a height of 14m after 20 years. This would fall into yield class 24, showing that it would grow with an annual increase of 24m^3. The yield class system helps to determine potential productivity. Yield class 6 will be felled after 65 years with a total volume 65 x 6m^3 = 390m^3.

The forester manages this ecosystem by producing timber for sale. Best-quality timber, used for making furniture, may fetch £50/m^3, while the lowest grades (£15/m^3) are used for wood pulp. Trees grow at a predictable rate, which depends on conditions within the ecosystem. If these change, so does the rate of growth. Assuming that conditions remain stable, the current annual increment (CAI), or the annual addition of extra timber, can be plotted. A graph for calculating forest returns is shown in Figure 8.37. The highest point of the curve shows biological maturity, but the tree continues to grow after this, so it is not the time to harvest the crop.

The felling date is calculated by working out the average growth rate of the tree since it was planted, the mean annual increment (MAI). The point on Figure 8.37 where the curves cross is the felling date. From then on, the land would yield more wood if the tree was felled and the area replanted than if it was left standing. If ecosystem conditions are stable, the predictable growth rate of any species can be calculated using a yield class table – see Figure 8.38. The age of 'maximum mean annual volume increment' is the best time to fell. These tables exist for all commonly grown timbers.

Fieldwork can usefully compare two sites with different environmental conditions. Two sites in mid-Wales were studied, one site planted at 300m above sea-level, and the other at 700m. The site at 300m produced more than double the volume of timber in two-thirds of the time. At the time, the lower site would have grossed £51 500/ha compared with £6090/ha for the higher site. The results are shown in Figure 8.39.

Productivity	Stand 1	Stand 2
Current age (year)	16	26
Average height (m^{-1})	11.71	7.91
Yield class (m^3 ha^{-1} yr^{-1})	24	8
Current standing crop (m^3 ha^{-1})[a]	310	102
Age at foresters maturity (year)	46	64
Remaining growth period (year)	30	38
Projected yield (m^3 ha^{-1})	1030	406
Projected value	£51 500	£6090
Environmental factors		
Altitude (m)	300	700
Aspect	ESE	ESE
Geology	Old red sandstone	Old red sandstone
Soil type	Brown earth	Peaty gley
Soil pH	5.1	4.2
Drainage	Good	Moderate
Air temperature (°C)	9.6	7.0
Soil temperature (°C)	5.5	0.2
Wind speed (average km/h min^{-1})	Calm	15
Humidity %	76	79
Number of trees (ha^{-1})[b]	2125	2500

Notes:

[a] The current standing crop is calculated by fieldwork. The method is to establish the volume of an average tree using the formula:

$$\text{Vol } m^3 = \left| \frac{\text{girth}^2}{4\pi} \times \text{height} \right| \times \text{Form factor which for conifers is 0.5}$$

This value is then multiplied by the number of trees per hectare.

[b] The total number of trees per hectare is established by calculating the average number for a sample area.

▲ **Figure 8.39** Fieldwork results.

Why are conifers so common in upland Britain?

The expansion of conifers into upland regions as an option for farmers is discussed in Chapter 10. Figure 8.40 gives some indication of why coniferous, or evergreen, species are increasing in areas under cultivation. Three plant species – pine, beech, and oak – are compared in terms of their nutrient uptake. Pine is the only conifer of the three. In all respects, it is less demanding than beech or oak, and the difference in uptake of calcium and potassium is particularly marked. If conifers are less demanding of soils, it follows that they are more suited to upland areas than are other species.

Conifers are therefore able to survive on a low-nutrient soil. This has an impact on the rest of the nutrient cycle. The leaf biomass produced in conifers is low in nutrients, and when it falls as litter it decomposes only slowly. It is of lower food value to decomposers than other litter. Its decomposition therefore releases nutrients back to the soil slowly. Upland regions of Britain are cooler and wetter than surrounding lowlands, and soil leaching is greater. Conifers therefore maintain their own low-grade nutrient cycle.

Species	Nitrogen	Phosphorus	Potassium	Calcium
Pine	45	5	7	29
Beech	50	13	15	96
Oak	87	7	79	95

◀ **Figure 8.40** Nutrient uptake of different tree species (kg/ha).

1 Study Figure 8.39. Explain the differences in yield between the two sites. Which specific environmental influences would influence the ecosystem to account for this difference in biomass?

2 Refer to the nutrient cycle in Chapter 7. Construct a nutrient cycle diagram to show the relative proportions of material held in the biomass, litter, and soil in a coniferous forest, and the relative flow levels between them.

3 What seem to be the arguments in favour of upland forests and plantations in Britain? What seem to be the factors against them?

Ideas for further study

1 Investigate a managed woodland near you. Consider how the mix of economic activities affects the forestry ecosystem. Evaluate the way it is managed. Try to do some fieldwork to see if the management plan and priorities are working to the benefit of the ecosystem.

2 Keep a resource file of news cuttings which illustrate development projects in woodland regions in different parts of the world. Evaluate these against the criteria for assessing those you have studied in this chapter. What are the issues? Who seems to be controlling the projects? In whose interests do the projects seem to be working?

3 Monitor the progress of the proposed National Forest in central England. What seem to be the issues? Who stands to gain? Who stands to lose?

Summary

- Woodland ecosystems depend on a number of interrelated physical and human processes. The interaction between these gives a woodland its character.
- Rainforests are under threat in most areas where they are traditionally found.
- There are disagreements about the rate of deforestation in rainforests at present. Different groups perceive the issue in different ways, and use different evidence. Other people rely upon hearsay and myth.
- Forests have social and economic value, as well as an unknown value in the field of medicine.
- Human pressures act on forests and threaten the forest resource base.
- Forest projects that acknowledge the needs of the people who live closest to them are more likely to be successful in maintaining an ecological balance.

References and further reading

L. Brown (ed.), *State of the World Yearbooks*, Earthscan for the Worldwatch Institute.

N. Calder, *Spaceship Earth*, Channel 4 Books, 1991.

UNEP, *Caring for the Earth: A Strategy for Sustainable Living*, Earthscan, 1991.

A. Cowell, *The Decade of Destruction*, Channel Four Books (plus videos), 1990.

S. Hecht and A. Cockburn, *The Fate of the Forest*, Verso, 1989.

The Korup Project, *Plan for Developing the Korup Project National Park*, ODNRI & WWF, 1989.

G. Lean and D. Hinrichsen, *Atlas of the Environment*, WWF, 1992.

G. Monbiot, *Amazon Watershed, the New Environmental Investigation*, Michael Joseph, 1991.

9 Changing grasslands

A world overview of grasslands

Figure 9.1 shows the area of the world covered by grasslands. It is estimated that grasslands covered about 40 per cent of the Earth's land surface prior to the impact of people and their domestic animals. Today, the proportion is much lower – about 27 per cent at best. The term 'grassland' is rather ambiguous. The word probably encourages images of featureless grass plains. There are such areas, as Figure 9.2 shows, but the typical grassland landscape is more varied. Figure 9.2 was taken in the area known as the Prairies, in the mid-west of the USA. Today, most of the mid-west grasslands have been ploughed and are cultivated for grain crops such as wheat. The landscape is anything but natural. However, it is only recently that it has become like that. The passage in Figure 9.3 is from a book about the westward migration of people across the USA in the 19th century.

The landscape of the Prairies shown in Figure 9.2 is very much affected by human activity. There are few areas left in the world where natural grasses grow to 2m in height. Most have been ploughed or cultivated in some way. In fact, very few areas known as 'grassland' are purely grass. Most are mixed woodland and grass, and as tree densities increase there is a fine distinction between wooded grassland and forest.

▲ **Figure 9.2** The Illinois Prairies. Most former grasslands in the USA and Canada are cultivated for grain production.

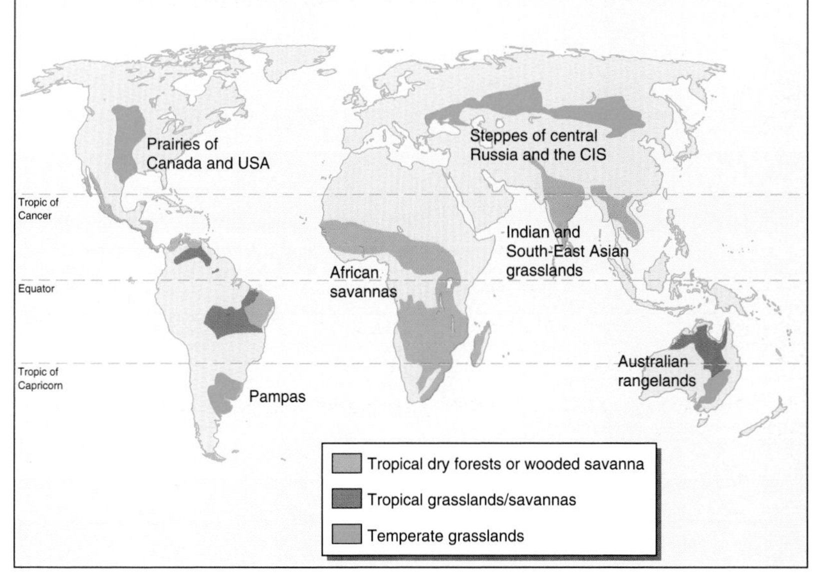

▲ **Figure 9.1** The distribution of grassland biomes (after Udvardy, 1975).

Particularly impressive was the Grand Prairie of central Illinois. No traveller failed to be impressed by the magnitude of that vast plain, stretching away as far as eye could see, and covered with a six-foot growth of grass that billowed gently in the wind or, in the spring, dazzled onlookers with its color-splashed carpet of wild flowers. Yet the prairie country was shunned by the first (white) settlers, whose frontier technique was adjusted to a wooded country.

▲ **Figure 9.3** Extract from R. A. Billington, *Westward Expansion.*

Types of grassland

Figure 9.1 shows two broad types of grassland.

- *Temperate grasslands* The Prairies, together with the steppe grasslands and the Australian grasslands, or rangelands, are found in areas of temperate climate (described in Chapter 4).
- *Tropical grasslands* These are different and are of two types: wooded savannas and dry savannas.

Tropical grassland landscapes are all wooded to some extent. Wooded savannas occur where rainfall totals are higher and permit a range of trees to survive. Dry savanna grasslands have fewer trees and though woodland is still part of the landscape, it is not enough to qualify as 'forest'. An example is shown in Figure 9.4, which is a view of the Masai Mara grasslands in Kenya. Notice the presence of trees in the landscape. Where the climate becomes drier still, trees become very isolated and grassland gives way to semi-desert. Tropical grassland climates are described in Chapter 5.

Grasslands have in many cases been colonized and cultivated, altering the natural landscape and vegetation. These grasslands are being affected more and more by human activity. The grasslands of the Masai Mara are criss-crossed by tracks followed each year by migrating wildebeest. Population pressure means that such migrations are being confined to certain routes and areas by boundaries, and the grassland ecosystem to which wildebeest are confined is suffering under the pressure. This is a worldwide pattern. Estimates vary but the percentage of land area believed by different authorities to be covered by savanna and temperate grassland is shown in Figure 9.5.

▲ **Figure 9.4** The grasslands of the Masai Mara National Park in Kenya.

	Whittaker and Likens (1975)	**Atlay, Ketner, and Duvigneaud** (1979)	**Olson, Watts, and Allison** (1983)
Savanna	15.0	22.5	24.6
Temperate grassland	9.0	12.5	6.7
Total grassland (million km^2)	24.0	35.0	31.3
Grassland as % of world land area	16.1	23.7	20.7
Grassland as % of world land area excluding Antarctica	17.9	26.5	23.1

▲ **Figure 9.5** Estimates of the area of the world's grasslands.

Measurement of the exact amount and proportion of grassland in the world is difficult. Land use statistics provided by the Food and Agricultural Organization (FAO) on a country basis include 'pasture', but this is not clearly defined, and does not distinguish between natural pastures, and those that are newly created from woodland or arable land and managed in rotation with crops. Two trends are clear:

- there used to be considerably more natural grassland in the world than there is now
- the area is continuing to diminish.

1 a) Use your atlas to find the world's savanna areas, on maps showing continental or global patterns of natural vegetation. On a world map outline, shade these and name them.

b) Describe the location of the savanna grasslands of the world.

c) Compare these with climatic zones. What are the characteristics of the climates where savanna grasslands are found?

2 Now repeat this exercise for temperate grasslands of the world.

3 In your atlas find maps which show population distribution, and look for some up-to-date annual population growth data. Try to predict which areas of grassland in the world are most under threat from population increases. Annotate your maps to show these areas.

What are 'grasslands'?

Grassland can be described as a type of vegetation that is subjected to periodic drought, dominated by grass and grass-like species, and growing where there are fewer than 10–15 trees per hectare. The definition is vague, and words used for grasslands are different depending on which part of the world you are in. They are referred to as *steppes* in Eurasia, *prairies* in North America, *llanos*, *cerrados* or *pampas* in South America, *savannas* in Africa, and *rangelands* or even 'bush' in Australia. Whatever their name, adaptation to drought conditions is a common feature of all grassland communities. Although almost all are affected by human activity, they seem to have developed in two areas.

- Areas where tree growth is prevented by edaphic, or climatic, factors. These are limited to areas with soils that are low in nutrients and/or rainfall. In such areas, grasslands are natural and are probably climatic climax communities.
- Areas where, over a long period, browsing (or eating) by wild herbivores has prevented further tree establishment and growth, and where wild species are still present and outnumber domestic livestock. In such areas, animals act as arresting factors in preventing woodland expansion, and grasslands are not climatic climax plant communities.

The best examples of natural grasslands formed where soil and climate favoured the growth and survival of grass and herbaceous species, rather than trees, are in northern South America and South Africa. The impact of large herbivores was greater in other areas of the world, such as the savannas of Africa, the steppes of Asia and eastern Europe, and the prairies of North America. In Africa, the large herbivore community was dominated by antelopes and zebras, in Eurasia by gazelles, goats, bison, and wild horses, and in North America by deer and bison. The effect of these animals was to prevent further development of forest by grazing. It is likely, therefore, that without animals present, many areas which we now know as grassland would be forest.

In both cases, ecological factors have lasted long enough for plants and animals to adapt and establish a balance. Two features of natural grasslands are that:

- vegetation is unsown by people, and
- the balance between plant species has not been sufficiently interrupted by human activity.

Many grasslands are very diverse in plant species, and in some areas their diversity is nearly as great as that of tropical rainforests. The variety of species in the grasslands of South America alone is shown in Figure 9.6. Animal species diversity is generally low, although only vertebrates are well recorded. For example, worldwide there are about 477 species of

Continued on page 204

Continued from page 203

birds regarded as primarily adapted to grasslands – less than 5 per cent of the global total. Grassland birds tend to be dispersed, to enable them to take advantage of sparsely distributed food resources, in an environment where the climate is unpredictable. Many species migrate over large areas, especially for water. Therefore it is difficult to conserve grassland birds through the management and protection of wildlife areas, in contrast to tropical rainforests where virtually no species migrate. Many grassland species would by now be extinct were it not for their ability to adapt to a mixed farmland environment.

Similarly, a total of 245 of the world's mammal species – about 6 per cent of the world's total – are considered to be adapted mainly to grassland environments. The majority of large mammals such as elephants, giraffe, and zebra are grazers and/or browsers, whilst small mammals and birds are mostly seed-eaters or omnivores. Carnivores such as cheetah, lion, and hyena are few in number.

There are also vast numbers of smaller mammals and insects. In the African savanna and Australian rangelands, termites are important, and may consume up to one-third of the total annual production of dead wood, leaves, and grass. Their biomass may be as high as 22g/m^3 – more than twice that of the greatest densities of vertebrates on Earth, found for example in the migrating herds on the Serengeti plains in Tanzania.

Formation	Area (km^2)	No. of trees and shrubs	No. of subshrubs, half-shrubs, herbs, vines, etc.	No. of grass species	Total no. of species
Cerrado in north-western São Paulo	50	45	175	17	237
Cerrado in western Minas Gerais	15 000	*c.*200	*c.*330	73	*c.*600
Whole cerrado region	2 million	429 (774)[1]	181	108	718 (1063)[1]
Rio Branco savannas	40 000	40	87	9	136
Rupununi savannas	12 000	*c.*50	291	90	431
Northern Surinam savannas	*c.*3000	15	213	44	272 (445)[2]
Central Venezuelan llanos	3	69 (16)[3]	175	44	288
Venezuelan llanos	250 000	43	312	200	555
Colombian llanos	150 000	44	174	88	306

Notes [1] Total flora including other plant formations [2] Total flora including bushes [3] Number of savanna trees excluding groves

▲ **Figure 9.6** Variety of plant species in the grasslands of South America.

1 Using information in Chapter 7 to help you, construct a savanna food web.

2 Based on the food web you have drawn, construct a series of trophic levels for the savanna.

3 Show how both the food web and trophic levels might vary as:
   a) rainfall increases
   b) rainfall decreases.

The grassland ecosystem

Chapter 7 describes how the productivity of ecosystems can be measured, through net primary productivity (NPP) and biomass. Figure 9.7 shows NPP for different grasslands compared with a selection of other world biomes. It clearly shows that all grasslands fall way behind tropical rainforests in their NPP and biomass. However, grasslands are very significant. Chapter 8 shows how biomass accounts for most retention of energy within the rainforest ecosystem. Little energy is retained within soils, and plant litter is rapidly consumed by forest fauna.

	Area (million km²)	Average NPP (g/m²/yr)	Total NPP (billion tonnes/yr)	Average biomass (kg/m²)	Total biomass (billion tonnes)
Savanna	15.0	900	13.5	4.0	60
Woodland and shrub	8.5	700	6.0	6.0	50
Temperate grassland	9.0	600	5.4	1.6	14
Tropical rainforest	17.0	2200	37.4	4.5	765
Desert and semi-desert	18.0	90	1.6	0.7	13

▲ **Figure 9.7** A comparison of net primary productivity (NPP) of savanna and temperate grasslands and other major biomes.

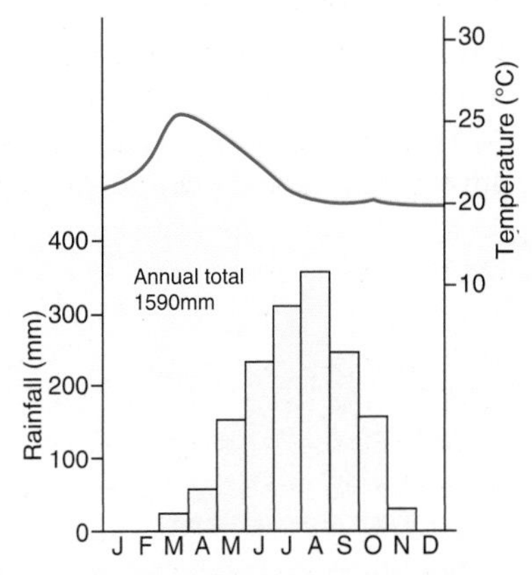

▶ **Figure 9.8** Climate graph for the West African wooded savanna. Here the dry season lasts for approximately three to four months and the total rainfall is over 1600mm. The dry savannas of the Sahel may have dry seasons of eight months and totals of 250mm.

Savanna and other grasslands are different. Many trees survive the drought of a long dry season by losing their leaves, thus reducing transpiration. Litter accumulates on the ground at the start of the dry season, and the lack of rain ensures that it is not decomposed as rapidly.

The seasonal nature of rainfall in the world's grasslands (Figure 9.8) means that litter accumulates between one year and the next, and is only recycled as an energy source back into the soil very gradually. Lower rainfall totals also reduce the loss of soil nutrients through the washing out of nutrients beyond the reach of plants, a process called **leaching**. The single period of plant growth during the rainy season uses only part of the store of soil nutrients. As a result, there is a fairly even balance between the nutrient balance held by soil, biomass, and litter in savanna grasslands (Figure 9.9).

True temperate grasslands are different again. Temperatures vary much more through the year than in tropical areas. A lack of sufficient rainfall during the spring growing season prevents tree survival, which in turn reduces total biomass. Long grasses – called **steppe** or **prairie grasses** – are therefore the natural vegetation. There is much production of litter each autumn as grasses die back, but decomposition in this cooler climate is slow. As a result, a smaller proportion of nutrients is retained in the biomass than in the savanna. Nutrients accumulate in the soil in such a dry climate. The single period of growth each spring is only sufficient to take up some of these nutrients, and the soil therefore remains a major store of nutrients within the system.

Temperate grasslands are none the less as varied in their form as tropical grasslands. True temperate grasslands such as the prairies and steppes are virtually treeless; most temperate grasslands are wooded to some extent. The Australian temperate grasslands that feature in the case study on pages 216–22 are wooded, though human activity is reducing the extent of woodland here as it is in tropical savanna regions.

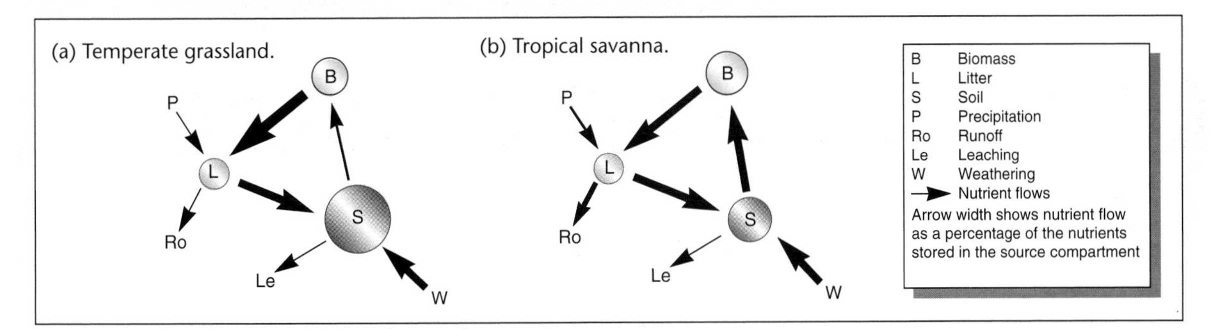

▲ **Figure 9.9** Nutrient circulation in savanna and temperate grasslands.

Types of savanna in Africa

In Africa, everything that is not forest, desert, or montane (mountain) vegetation is regarded as savanna. Not all of this is grassland, however. Much is cultivated and even more is probably savanna woodland. A high proportion of the continent supports dry, semi-natural vegetation in vast, unbroken stretches, and Africa has by far the largest proportion of the world's savanna or wooded grassland.

Because of the sheer size of the continent, and the difference in climate patterns between the areas north and south of the Equator, it is not accurate to speak of a single type of savanna in Africa. Figure 9.10 shows four categories of savanna. Examples of three of these are shown in Figure 9.11. Study the map, and read the section in Chapter 5 on climate in Africa. This shows how increasing seasonality and decreasing amounts of rain further from the coast create more demanding conditions in which plants have to adapt to survive. Plant communities of the dry savanna margins of the Sahara Desert, for instance, have to survive six to eight months of drought, and rainfall totals as low as 250mm annually, whereas the wooded savanna communities on the fringes of the West African rainforests may only suffer a deficiency of water for one to three months, and receive rainfall totals of over 1200mm. Savanna categories are decided on the basis of proportions of grassland and trees or forest.

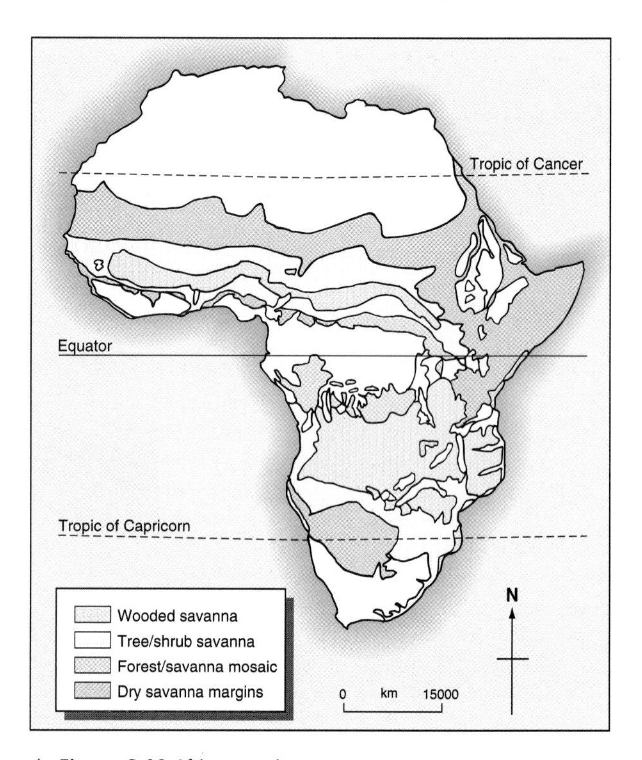

▲ **Figure 9.10** Africa – main savanna types.

Wooded savanna

Trees are most dense, growing in a network of grasses. These areas coincide with areas of higher rainfall. Often the forest is sufficiently dense to be known as 'dry forest', and may blend with rainforest. However, in areas where drought is more likely, more of the trees are deciduous, and they are interspersed by thickets of tall 'elephant grasses' which may be over 2m tall.

(a) Wooded savanna.

▶ **Figure 9.11** Some of the main savanna types.

Tree and shrub savanna

Trees and shrubs are more intermittent, but the landscape still appears wooded from a distance. The trees are not so tall, and may be replaced by smaller species or shrubs, spaced intermittently among grasses.

(b) Tree and shrub savanna.

Dry savanna margins

Expanses of grassland are broader, and tree survival is more dependent on local factors, such as a river course. Most grasses are drought-resistant, while most trees and shrubs have the ability to withstand long periods without rain. There are more deciduous species of acacia, which lose their leaves to withstand drought, and most bushes are thorny.

(c) Dry savanna margins.

Plant species in the different grasslands vary too. Menaut (1983) produced maps which show diversity of species in areas of 10 000km^2. He estimated the average number of species in each area, and called it 'areal richness'. The higher the figure, the more diverse and rich the area is in number of species – see Figure 9.12. The presence of large mammals is important in the ecology of African grasslands. Fire has also been a major influence in the evolution of the savanna. Natural fires, caused by lightning, limit the accumulation of dry organic matter, and favour the survival of some species at the expense of others. People have also been burning the grassland for cultivation for at least 50 000 years.

Study Figures 9.10, 9.11, and 9.12.

1 Is there a relationship between savanna types in Africa, and areal richness zones?

2 The average areal richness in African savanna is 1750 species – not far below that of the rainforests with 2020 species. How can this be explained? Consider variations in climate across savanna regions.

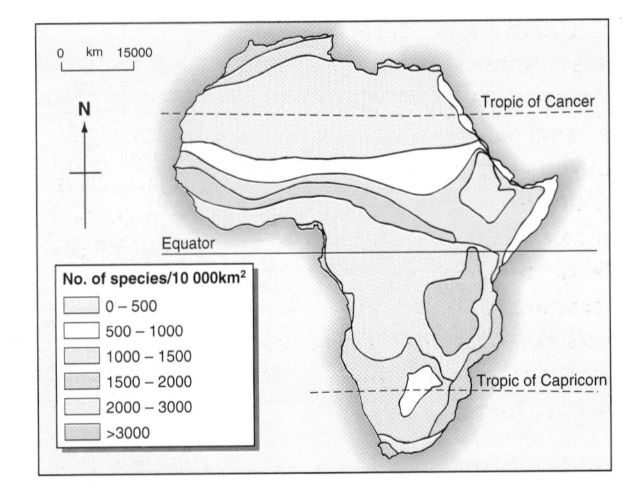

▲ **Figure 9.12** Africa – zones of areal richness (after Menaut, 1983). Areal richness is measured by the estimated number of plant species per 10 000km^2.

The savanna in West Africa

In Africa, people have burnt the savanna for thousands of years to improve stock-grazing and make hunting of wild game easier. Burning has increased as population has increased. Towards the end of the 19th century, when population began to grow more rapidly, the savanna grew through forest clearance, burning, and massive increases in the number of cattle. Much of this was due to settlers who imported European cattle, which were susceptible to tropical diseases such as rinderpest. Around 1900, this disease spread in seven years from north Africa to the far south, killing 90–95 per cent of domestic cattle as well as many wild species. The result was widespread human starvation and abandonment of vast areas of grazing land.

Subsequent regeneration of scrub and woodland allowed the spread of the tsetse fly. Colonial governments made repeated attempts to eradicate the tsetse fly. One method was removal of the woodland and cover that the fly needs to survive. This allowed indigenous herbivores and their predators to extend their range. These enlarged distributions have been maintained, assisted recently by the control of poaching, and the establishment of national parks and protected areas. Grassland in these areas may appear natural, but is recent and strongly influenced by human activity.

Increases in cotton cultivation in many semi-arid areas of Africa have involved large-scale use of insecticides such as dieldrin, with little regard for species that are not a threat to the crop. Cotton crops give little food for birds or mammals, and water control schemes for irrigation intercept water that previously flooded river valleys and thus provided feeding areas for many birds. The loss of grassland is a serious problem for those European breeding birds that spend the winter in West Africa, because their habitat is confined on its southern edge by forest, farmland, or the Atlantic Ocean.

The Fulani people of West Africa

Cattle-herding is a traditional occupation of the Fulani people in countries of West Africa. They migrate with their cattle across the borders of Mali, Niger, and Burkina Faso. Until the 1970s this movement rarely took the Fulani south of latitude 12°N, because of the presence of cattle diseases there. However, since then they have moved to the northern areas of Nigeria, Ghana, and Ivory Coast – see Figure 9.14. Concentrations of Fulani cattle are now highest in an area within 150km radius of the city of Katiali.

▼ **Figure 9.13** Fulani with their Zebu cattle.

▼ **Figure 9.14** The movement of Fulani people.

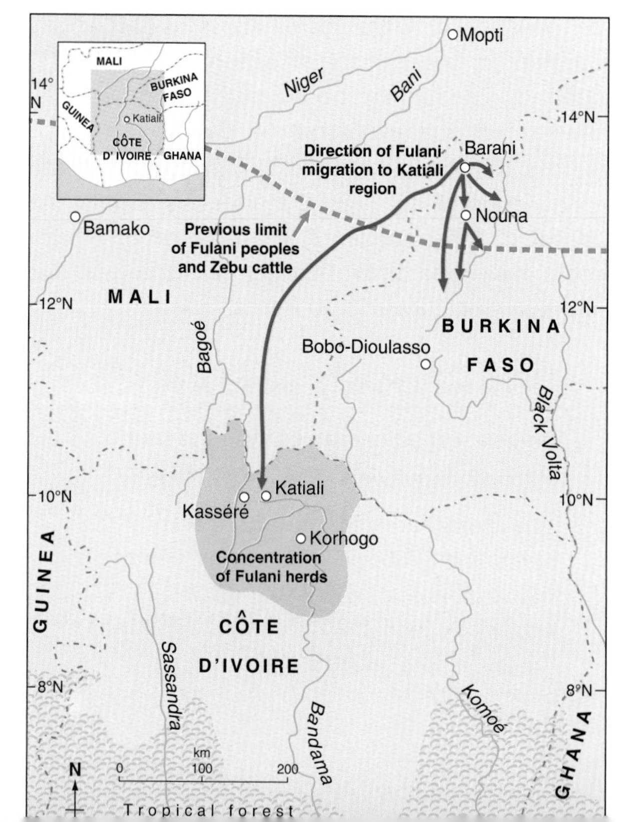

▲ **Figure 9.15** The effects of intensive grazing in a dry area. The cattle here appear to have eaten everything at ground level. Without protection from grass roots, soils soon erode either from the effects of wind or rainfall runoff.

Migration southwards took place first during the initial sequence of droughts within the Sahel during the late 1960s through to the mid-1970s, and again during a second bout in the mid-1980s. Drought creates a problem for cattle-herders not only because there is a shortage of water, but also because there is a lack of grazing. Grassland areas are especially prone to overgrazing (Figure 9.15).

Drought is not the only reason why the herders migrated into Ivory Coast. In that country, cattle were not taxed, market prices for animals were higher, and veterinary services were free. Between 1966 and 1984, the number of Fulani cattle in Ivory Coast increased from 38 000 to 255 000. They now account for one-third of all livestock in the country, and over half of domestic beef production. This movement poses a number of issues and questions.

- How do the Fulani manage their cattle? This region is traditionally hostile to their breed of cattle.
- What is the impact on the savanna ecosystem of the increasing numbers of cattle?
- What is the attitude of the people and government of Ivory Coast to the arrival of large numbers of immigrants?
- Are there better ways of managing the savanna ecosystems, which would fulfil the needs of all sectors of the Ivory Coast population?

Fulani management of cattle

The influence of rainfall

Figure 9.16 shows the rainfall for the northern part of Ivory Coast. The dry season between mid-October and early April is almost always rainless. At some time during April, in most years, the rains arrive, rising sharply to a peak in August, after which they decrease and eventually end in the first few days of October. To understand the rainfall pattern and what causes it, you should refer to Chapter 5.

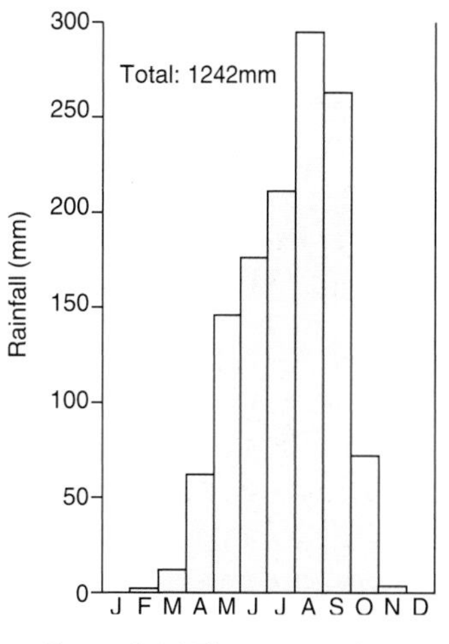

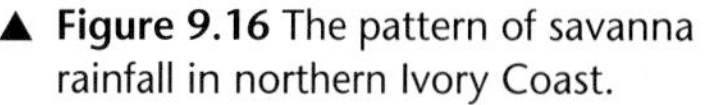

▲ **Figure 9.16** The pattern of savanna rainfall in northern Ivory Coast.

1 Make a sketch of the graph in Figure 9.16. Annotate it with labels to show:

a) when grass growth is likely to be rapid

b) when the newest and best-tasting grasses are available for cattle to feed on

c) period of drought

d) period when grasses are likely to be tough and woody and difficult for cattle to eat.

2 a) Why is migration to find grass and water in such a climate a preferable option to that of remaining in one place?

b) Which is a better way of managing a grassland ecosystem: to allow migration, or to force cattle herders to stay in one place? Consider the arguments on both sides.

3 Chapter 5 describes how the arrival of the rainy season is unpredictable – in some years rains may arrive early and continue late, in other years the reverse is true. What pressures must this place on herders such as the Fulani?

The tsetse fly

The Fulani herders migrate seasonally with their cattle. There are complex reasons behind this migration. If movement were just a matter of obtaining water and grass, cattle-herders would move to any place where they could find these two things. However, one of the greatest dangers to cattle and herders is trypanosomiasis, a disease more commonly known as 'sleeping sickness', which affects both cattle and people. Cattle continue to die from this disease, as they do from brucellosis and other tick-borne parasitic diseases.

Trypanosomiasis is carried by the tsetse fly. The habitat of the tsetse fly is largely confined to forests along the course of rivers. It breeds during the rainy season, and its numbers and extent vary greatly according to the rain, rising sharply with the arrival of the rains and falling as the dry season approaches. Their extent varies too with cattle distribution and cultivation of land. Tsetse flies are common where forests remain untouched, close to streams and pools which last the dry season, and where abundant wildlife survives. Changes to the landscape through burning for cultivation, deforestation, hunting, and cultivation itself reduce the range of the fly.

Conflicts of interest

The Fulani try to avoid the tsetse fly by moving their herds to places close to or actually within areas that have been cleared for cultivation. Here tsetse fly densities are much lower. However, this takes herds of cattle close to cultivated land. There can be bitter conflict over crop damage, and both the short-term and long-term movements of cattle bring herders into conflict with cultivators. There is increasing tension between the cattle-herders and the indigenous Senufo and Malinke peoples in northern Ivory Coast, who are traditionally cultivators. The government of Ivory Coast wants to reduce this conflict, but it also welcomes the increased supplies of domestic beef that the Fulani have brought – these have helped to stabilize supplies and prices.

1 Annotate your sketch of Figure 9.16 further with labels describing the times of particular danger from the tsetse fly.

2 What possible strategies exist to keep herders and cultivators apart? How should these be implemented? Who should do it?

3 Would the strategies you suggest help to reduce the tension between Fulani migrants and the indigenous Senufo and Malinke peoples?

What issues arise from cattle migrations?

So far, this study has looked at the general migration of Fulani herders southwards into Ivory Coast. The Fulani herders migrate with their cattle through the year depending on supplies of grass and water. Here we look at three families and their migration pattern, in order to understand why migration takes place, and its implications. Each family was studied for one year during the 1980s, from the June of one year until June of the following year.

Family 1 – Moumouni Sangare

Moumouni Sangare was born in Mali, and he entered Ivory Coast with his family in 1969. During the year of this study, he managed a herd of 159 animals, belonging to his family and those of his two brothers. The 159 cattle consisted of 52 milking animals and a bush-grazing herd of 107 animals. For eleven of the twelve months, Sangare's brothers and their children guarded the milk cattle near their camp. For the remaining month, a hired herder was paid US $30 per month to guard both herds. The herds were grazed during the day and brought into corrals at night.

Figure 9.17 shows how the herds migrated during the year. Site 1 is where the family were based in June. During June and July, the first two months of the rainy season, the bush herd grazed about 6km south of site 1, while the milk herd stayed close to site 1. From August until January, the bush herd were grazed on fallow fields of cultivators south of N'Ganon (sites 2–6). The savanna grasses become woody at this time and less palatable to cattle, and so the herders moved their cattle to places where grasses were likely to be more tender. Cattle were watered from tributaries of the Bandama river rather than from the river itself, because of the threat of parasitic disease in the moist lowlands. At night, corrals of brushwood were established on higher ground away from the tsetse fly. As the dry season began, tributaries dried up, and the threat of tsetse fly diminished, so cattle drank from the river itself.

In October, early fires were set to encourage regrowth of the savanna and to obtain game. While this was of benefit to cultivators, it reduced the pasture available to the milk herd. However, regrowth was established by November at site 2, and the milk cattle returned. During the early part of the dry season, in November and December, water was scarce and cattle had to be moved frequently. This period was marked by frequent incidents of damage to local crops of maize, groundnuts, and rice, and consequent compensation claims from cultivators. Moumouni Sangare had to pay a total of US $60 to two farmers whose fields had been damaged by cattle.

By January, dry streambeds and poor grazing meant that the cattle faced long treks. Moumouni decided to move the camp north to be close to the Badenou river. Both the camp and the night corrals for the milk herd were moved north, to site 5, in early January. Between then and early June, the bush-grazing herd moved to eighteen different locations. Movement uses up cattle energy, and exposes them to infection, so the family needed to reduce movement where possible, especially of the milk herd. On three occasions, at sites 12, 15, and 16, cattle had to be moved because there was competition with other herders. When there are greater numbers of cattle in an area, the threat from ticks and flies is increased, and overgrazing becomes a problem. In March the bush herd was moved closer to the milk herd near Katiali. As the rainy season approached, in late March, the camp returned to its old location south of Katiali, at site 1. The bush herd moved further away to the north, to an uninhabited area (site 20).

▼ **Figure 9.17** The movement of cattle belonging to the family of Moumouni Sangare during one year.

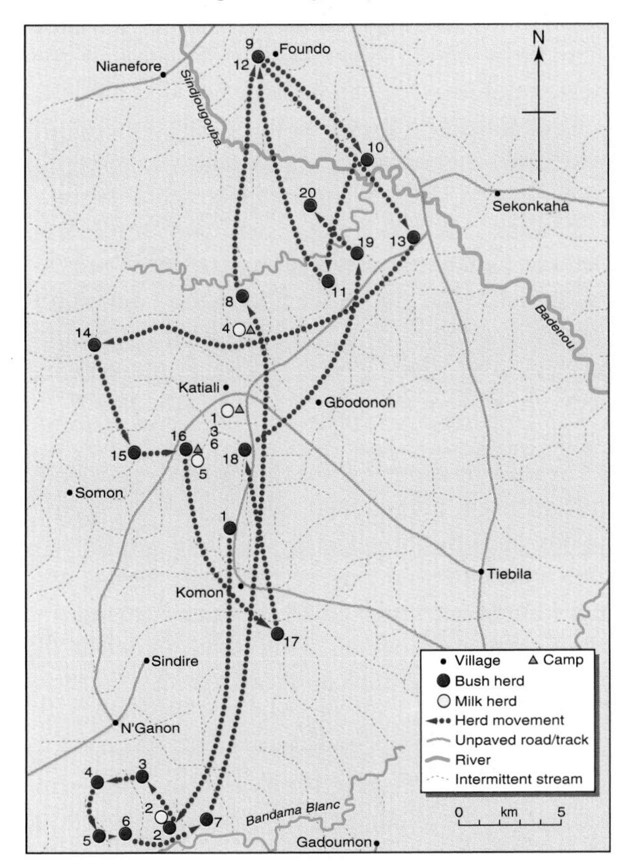

1. How significant is water as a factor in locating campsites? Explain your answer, referring to evidence in the text and Figure 9.17.
2. Why should milk cattle travel shorter distances than bush cattle?
3. Which group of cattle:
   a) places the greater demand on the savanna ecosystem
   b) is more likely to cause overgrazing
   c) is more likely to involve conflict with other land users?

   Explain your answers.

Family 2 – Samboya Sidibe

This family was also born in Mali. Samboya Sidibe migrated to Kasséré in Ivory Coast in 1969. The family remained there until 1981 when conflict with Senufo forced them to move to Pitiegomon, 18km east of Katiali. Having been attacked there too, they moved to the point shown as site 1 in Figure 9.18, which was uninhabited. The camp then moved to a new location – site 2 – north of Katiali, close to where milk could be sold and grain purchased. Samboya Sidibe and his brother owned 362 cattle, one of the largest herds in the region. Like Moumouni Sangare's herds, they consisted of milk and bush cattle. Three migrant labourers were employed to guard the animals. At this time, several animals died from fevers.

To avoid infection, the bush cattle were moved to site 3, only to return to site 2 when a labourer left. More animals died of fever and diarrhoea. Between July and November, 22 cattle died. A move east and then south-west to site 7 by early November was caused by a search for uninfected river water, though the family often found themselves near cultivated land. Eventually the milk herd found untenanted land and stayed there, between sites D and F. Between November and March, the family moved the bush herd from sites 7 to 12, almost 30km from the base camp at site 1. Rain in the following two months encouraged better grazing, and the bush herd returned to site 14, joining the milk herd in late May.

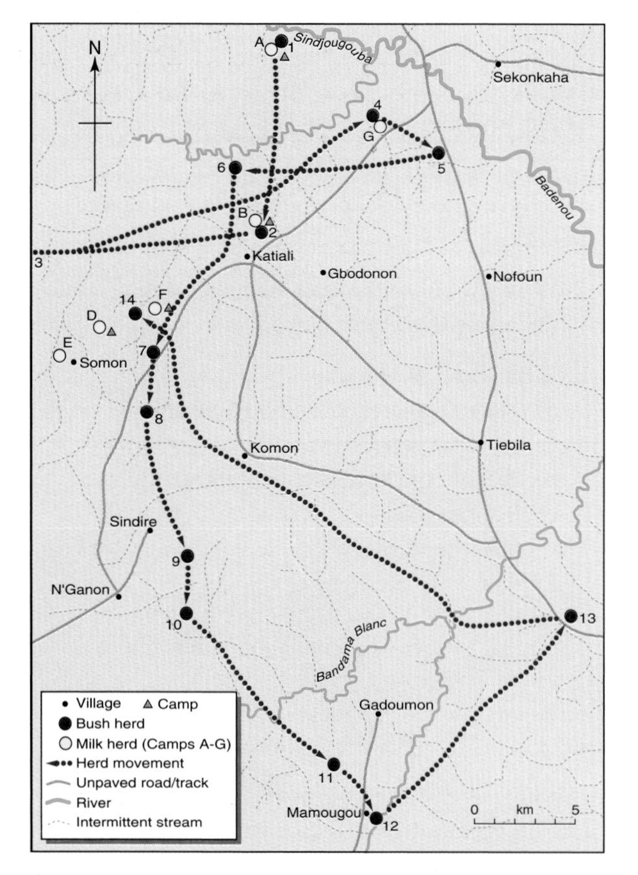

▲ **Figure 9.18** The movement of cattle belonging to the family of Samboya Sidibe during one year.

Family 3 – Demba Sangare

With an uncle, a brother, and their families, Demba Sangare moved from Mali to Ivory Coast in 1972. He and his brother combined their herds and managed them with four hired herders. Between them they had 427 cattle – 109 in the milk herd (herd 4 on Figure 9.19), and 318 in the bush herd. The bush herd was subdivided into three herds of approximately 100 cattle each (herds 1–3 on Figure 9.19). Most responsibility for daily movement and grazing fell on the hired herders. Demba's family visited the herds to monitor disease and ticks, and to ensure that there was no movement onto cultivated land. Why should the bush herd be subdivided?

Figure 9.19 shows the grazing areas covered by Demba's herds. Movements are more complex than those of the other families because of the greater numbers of cattle involved. In June, the family campsite is shown as site 1. The presence of fields and other herds forced the family to move all herds south to site 2. The search for good pasture and water took them to sites 3–6 until late October, when they were moved to site 7 to be vaccinated. In mid-November, the migration becomes complex. The three bush herds were re-divided into two groups – their movements between November and late May are shown in Figure 9.19.

The milk herd during this time grazed on stubble in harvested fields. Near site 2, fields became muddy and the cattle fell ill. The herd was moved southwards to join one of the bush herds, and moved between sites 3 and 8. With each change of camp came a change of night corral. Other herds moved too, eating fruits and leaves from trees such as the baobab and mango. Herders also removed branches from trees with edible leaves during this time. One animal strayed into a cotton field and became caught in a trap; it died and Demba had to pay $270 in compensation to the owner.

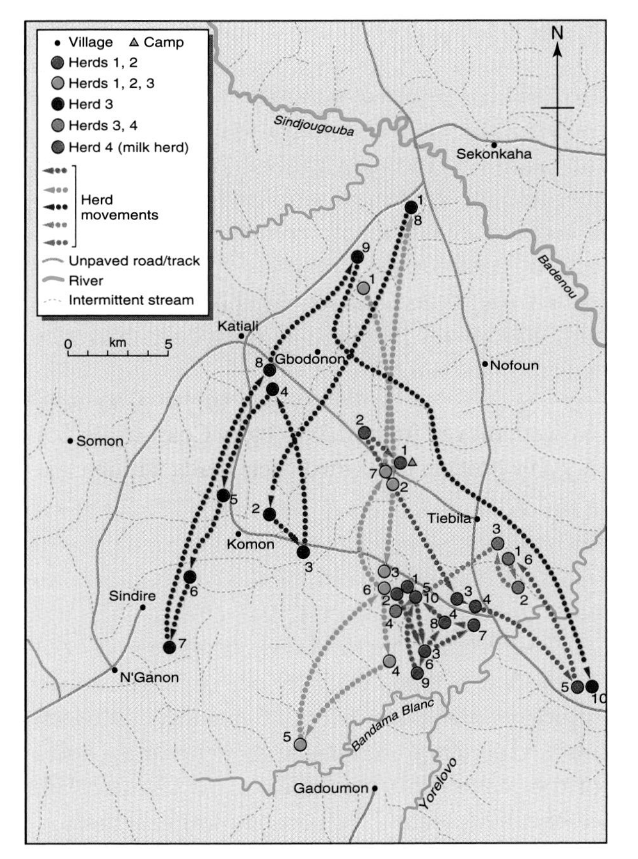

▲ **Figure 9.19** The movement of cattle belonging to the family of Demba Sangare during one year.

1 Measure the distances covered by each of Demba's herds on Figure 9.19. Follow these steps.

a) Trace the routes taken by herd 1. Measure the total distance travelled.

b) Repeat this for herds 2, 3, and 4 (the milk herd).

c) Which herd travels the least distance? Why? Given that movement occurred on 180 days, what was the average distance travelled each day? Is this significant?

d) Which herd travels the furthest? What effect would this have?

2 Assess the impact on the savanna ecosystem of each of the herds.

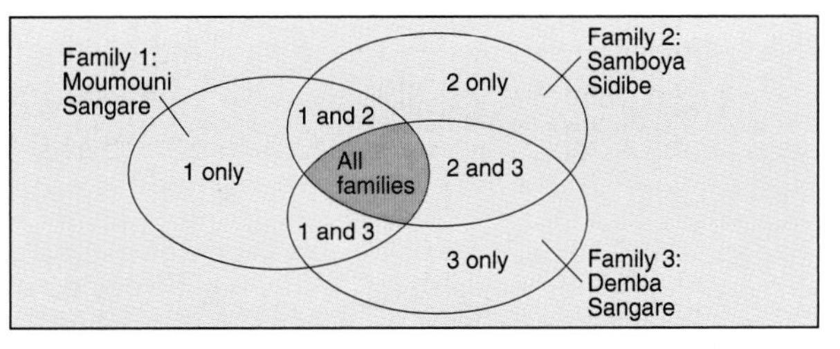

▲ **Figure 9.20** Venn diagram to illustrate the factors that cause Fulani families to move.

What patterns emerge?

1 Work in pairs. Draw a large copy of Figure 9.20. Each oval represents one family.

a) Read each case study and list the factors that cause each family to move.

b) Identify factors common to all three families and place these in the shaded area of your Venn diagram.

c) Place other factors either in the individual family spaces (if they apply only to one family), or in the areas of overlap (if they are shared by two/three families).

2 You now have a means of identifying the most and least important factors that may force a family to move. Which problems need to be addressed most often/most urgently? Can these be resolved without intervention? If not, who should intervene, and how?

3 a) Which of the problems facing families, if left unresolved, threaten the savanna ecosystem?

b) In what ways do they threaten it?

4 What may be the possible effects of doing nothing about these problems?

5 In pairs, consider the possible social, economic, and ecological impacts which would be posed by the following measures:

a) widespread spraying to rid the region of tsetse fly

b) government grants to fence off cultivated land

c) drilling boreholes to provide year-round water supplies in areas most affected by drought

d) reducing the number of cattle to a maximum of 50 per family.

6 Should any of these measures be adopted? If not, suggest other possible strategies and justify these.

The Australian drought

Study Figure 9.21. Australia is a huge country with a relatively small population of 18 million. The volume and variety of cheap fresh foods there are unrivalled in most countries. In recent years, however, food supplies and incomes from farming have been under threat from drought. Figure 9.22 is a report from the *Financial Times* concerning the effects of a prolonged drought which by October 1994 had lasted for nearly four years. Chapter 5 explains the possible causes of this drought, and shows how it may have been part of a change in climate known as 'El Niño'.

Droughts in Australia tend to be cyclic. In the 1980s there were droughts in southern and eastern Africa, northern India, north-east Brazil, the USA, and Australia. Australia's wheat production in 1982 was down by 37 per cent on the annual average between 1977 and 1981. In areas where rainfall averages between 250mm and 635mm of rain per year, wheat accounts for over 60 per cent of the cropped land, so this represents a major reduction. However, the effects were not evenly spread. Wheat production in Victoria was only 16 per cent of the average, and in New South Wales only 29 per cent. These two states contain two-thirds of the population of Australia. Between 1864 and 1984, Australia experienced ten major droughts and a series of minor ones. Six of the major droughts have occurred since 1950. The droughts of the early 1990s seem to be part of another series.

▲ **Figure 9.21** Melbourne fruit and vegetable market.

However, drought is part of a much larger problem. In the last 30–40 years, land across the eastern half of Australia has slowly degraded and become poorer in quality. This study focuses on the south-eastern part of Victoria and New South Wales, in an area of temperate grassland which occupies Australia's largest river basin, the Murray–Darling (see Figure 9.23). The two major rivers – the Murray and the Darling – and their tributaries drain an extensive area, and flow from the Great Dividing Range westwards to the Murray estuary south-east of Adelaide. This area includes Australia's four most populous states and the Capital Territory, and the rivers flow over large areas of grassland.

Official forecasts for Australian wheat production were lowered again yesterday because of the continuing drought, which is crippling key growing areas on the east coast. The country's overall winter crop production is now expected to be the smallest for 12 years.

According to the Australian Bureau of Agriculture and Resource Economics, the government forecasting and research agency, total winter crop production is likely to more than halve from last year's 28.2m tonnes, to just 13.08m tonnes. Wheat production, meanwhile, is forecast to be a mere 8.3m tonnes.

If the latter prediction proves correct, this will be the smallest harvest since 1972–73 and a reduction of more than 50 per cent on the 1993–94 figure. Abare had already cut its forecast for winter crop production several times, as key planting seasons were missed.

▲ **Figure 9.22** From the *Financial Times*, 26 October 1994. This article by Nikki Tait in Sydney was typical of many reports in late 1994 which highlighted the problems of the prolonged drought in eastern Australia.

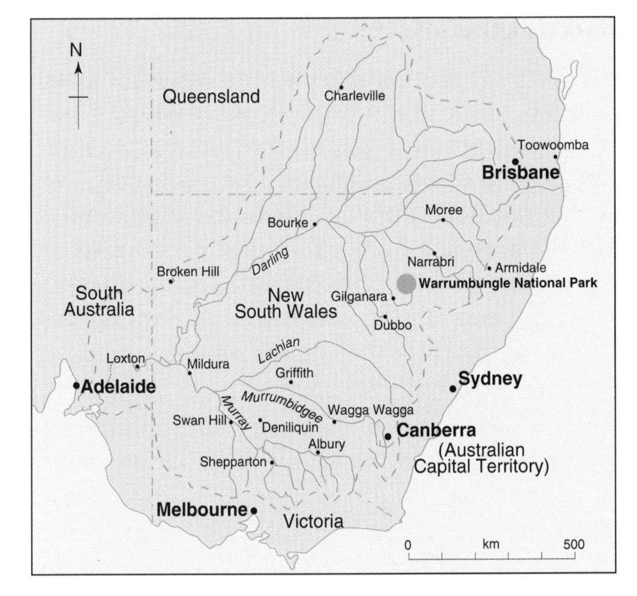

▲ **Figure 9.23** The Murray–Darling river basin.

▲ **Figure 9.24** Aerial view of orchards near Griffith, northern Victoria.

When this area was first settled by Europeans in the 19th century, the lower parts of the basin consisted of temperate grassland – more than 21 000 km^2 in Victoria alone. Now, much of the basin is used either for cattle and sheep grazing on the upper slopes, or as irrigated cropland in the places close to the river – see Figure 9.25. As a result of multi-million-dollar investment in damming tributaries to provide water for irrigation, this is now the most intensively cultivated region of Australia. Figure 9.24 shows orchards near Griffith. Intensive fruit growing has replaced the natural grasslands as a result of irrigation, and northern Victoria and southern South Australia are among Australia's most productive wine-producing regions.

Such changes require investigation. The droughts of the 1990s have not yet seriously threatened supplies of water, but they have occurred at a time when irrigation is having adverse effects on land quality. While cattle and sheep have died in their millions as a result of the drought, there are deeper issues which will persist beyond the arrival of any rains which break the drought.

▼ **Figure 9.25** The irrigated areas of the Murray valley.

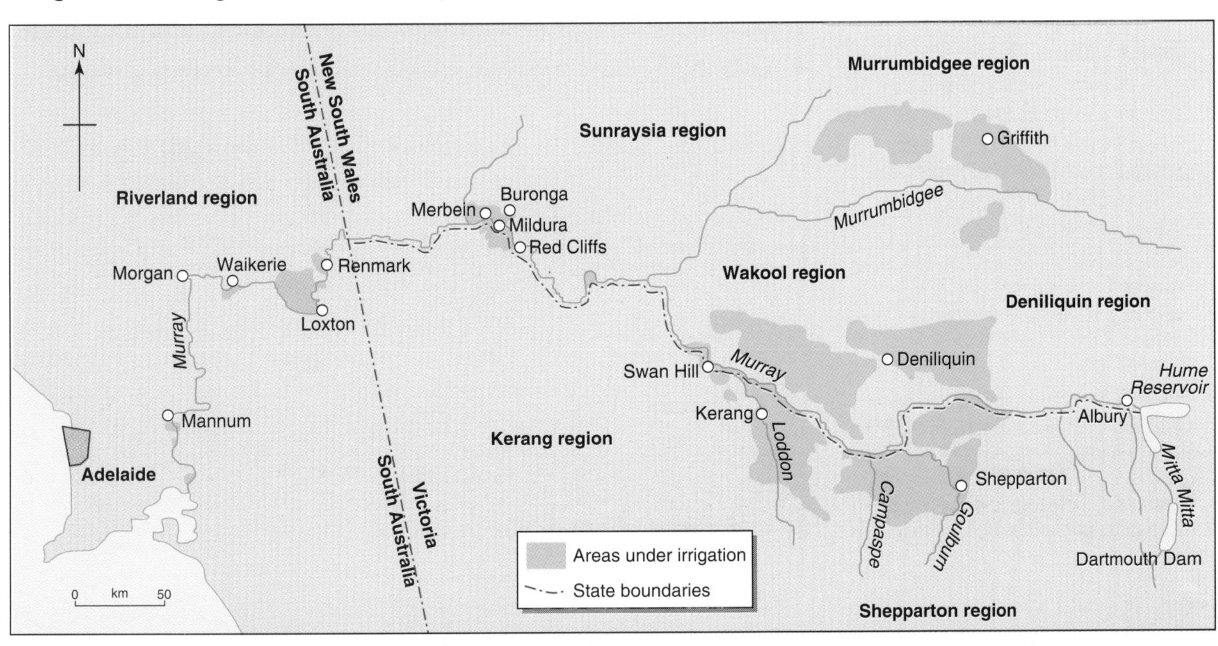

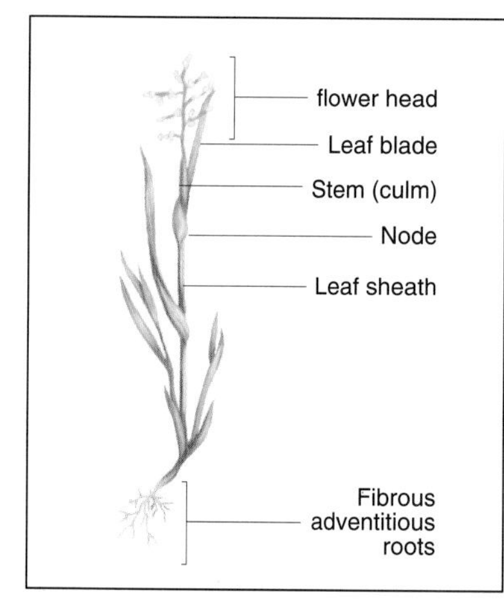

(a) Annual grass developing from a single stem.

(b) Tussock grass forming a cluster of roots around the base from which the grass develops.

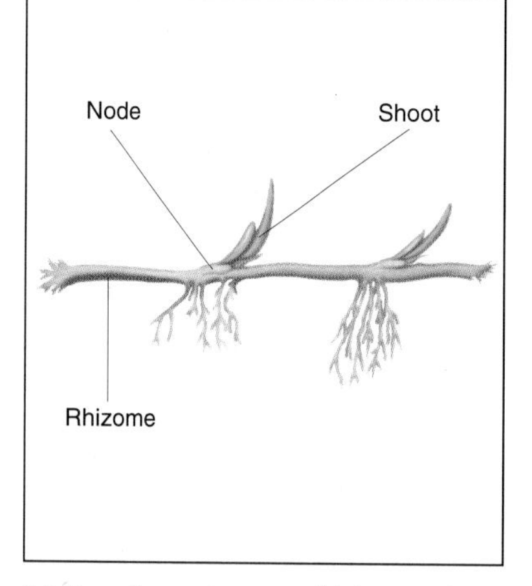

(c) Creeping mat grass which spreads through underground stems.

▲ **Figure 9.27** The shape and form of three grasses.

Temperate grasslands in south-eastern Australia

The Murray–Darling river basin consists of two broad natural vegetation belts: the upper slopes of the Great Dividing Range which are mostly covered in eucalypt forest, and temperate grasslands. Figures 9.26 and 9.28 show contrasting views of temperate grassland. Figure 9.26 is a view of the Warrumbungle National Park in New South Wales. Find this on Figure 9.23. The landscape consists of broad areas of grass, interspersed with isolated trees. The dead tree in the foreground of the picture is symptomatic of a part-forested grassland in decay. Grazing by cattle has prevented the growth of saplings, so in time the woodland will disappear.

▲ **Figure 9.26** Grassland in the Warrumbungle National Park in New South Wales.

Notice the grasses in the foreground of Figure 9.26. Figure 9.27 shows different grass forms – a comparison with Figure 9.26 shows that the Warrumbungle National Park grassland consists of tussock grasses. Tussock grasses may grow up to 2m tall, and have deep root systems which are designed to obtain water from greater depth. Compare this with the grazing areas of eastern New South Wales shown in Figure 9.28. These are mat grasses typical of species encouraged by farmers and grazed extensively by cattle and sheep. The grasses are much smaller and may be no more than 50cm tall. In both cases, the woodland is 'relic', or decaying, and in both cases the roots are a significant part of the ecosystem. In dying back in winter, the stems form a mat in which seeds lie dormant until spring.

◀ **Figure 9.28** Grassland area of eastern New South Wales near Dubbo. Notice how the grass forms a continuous sward or cover, unlike the tussock species where individual plants are visible.

The temperate grassland ecosystem

Australia's geographical isolation has produced unique ecosystems unlike those found anywhere else. Until the introduction of species brought by European settlers 200 years ago, there were few carnivores among the indigenous mammals of grassland areas except the dingo, a member of the dog family which probably arrived with the earliest Aboriginal settlers 40 000 years ago. The fauna – animal and bird population – consists of grazing animals such as kangaroo, of which there are 45 species, browsers such as koala, omnivores such as possums, burrowing species such as wombats, and a wide variety of bird life including parrots, cockatoos, and kookaburras.

The source of food for most of these species was the natural vegetation. Kangaroos are extensive grazers, and both surface plants and animals make Australia's ecology of worldwide importance and interest. Studies of different grasslands worldwide have shown that although grasses make up 90 per cent of ground cover, they account for only one-quarter of grassland species. The structure of temperate grassland is simpler than that of tropical savanna. Emergent trees and shrubs form one layer, and grasses another.

A third layer lies below the surface, and one that is vital to the ecosystem. At periods of maximum growth, plant biomass on the surface may only be 20 per cent of the total. The root system contributes a large amount of organic matter in the soil which in turn supports large numbers of soil animals, especially decomposers. The energy value of decomposers alone can equal that of surface vegetation. Between one-quarter and two-thirds of net primary productivity (NPP) is consumed by grazing animals. These include grazers above the ground, and rodent burrowers below ground.

1 Construct a diagram showing a food web for indigenous species of animals and plants in Australia's grasslands. Include surface and subsurface animals. Use the diagram of a food web in Chapter 7 (Figure 7.2) to help you.

2 Add to this diagram the following species, which were introduced by European settlers:
- farming stock such as cows and sheep
- introduced species such as foxes, cats, pigs, goats, horses, starlings, and rabbits.

3 Assess the impact of such species on the food web.

Climate and the human impact on the grassland ecosystem

The factor that dominates the grassland ecosystem is climate. Figure 9.29 is a climate graph for Adelaide. Remember that this is compiled from averaged data, and that both annual and daily variations may be as important as the trends revealed by the graph. For instance, the daily summer maximum temperature may be 7° above that shown, and often reaches temperatures greater than 40 °C. Similarly, winter minimum temperatures include frequent frosts, especially inland. Rainfall amounts vary across the river basin. While Adelaide has over 500mm per year, Mildura has approximately 350mm, and the north-eastern part of the basin about 250mm.

This climate poses problems for farmers. Lowest rainfall totals are in the summer when evaporation rates are highest, and rivers soon reach low levels (Figure 9.30). Yet grasslands offer grazing opportunities for cattle and sheep which few early settlers could resist. The provision of water was, and still is, partly solved by windpumps, which draw water from underground – see Figure 9.31. The water is pumped from water-bearing porous rocks – aquifers – deep underground. This source of water is known as artesian water. It is sustainable as long as the volume of water pumped out by farmers is balanced by the amount of rainfall which finds its way underground. It is not sustainable for large volumes of water, and is already under serious threat as a sustainable store. User demand exceeds rainfall supply by several times. In addition, the maintenance of cattle and sheep around water points often increases pressure on grazing land, so that overgrazing is a problem.

▼ **Figure 9.29** Climate graph for Adelaide. This climate pattern is typical of Australia's temperate grasslands, although annual rainfall totals vary. Rainfall generally decreases inland, rising again towards the Snowy Mountains.

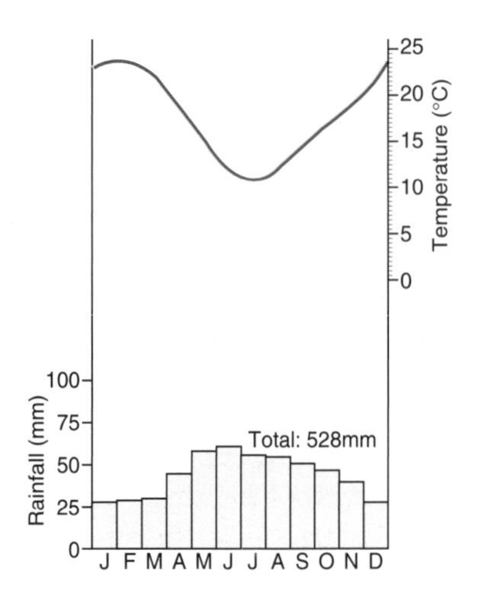

▼ **Figure 9.30** The Castlereagh river, at Gilgandra in New South Wales, at the beginning of the drought in winter 1992. The normal size of the river channel is evident from the extent of the sediment in the river bed. When the river is in full spate, these sediments and most of the grass beyond the river are covered.

▼ **Figure 9.31** Windpump in northern Victoria. The water is drawn from underground using wind power, and is collected in the corrugated metal storage tank.

The impact of irrigation on the grassland ecosystem

Figure 9.25 shows the extent of irrigation in the Murray river valley. Irrigation of parts of the Murray–Darling basin began during the late 19th century. Like most of the world's major river basins, the majority of the basin is much drier than its source. The highland catchment of the Murray river is only 2 per cent of the total area of the whole basin, but contributes one-quarter of its water runoff. The water travels across and is used within the lower parts of the basin. How does the grassland ecosystem measure up to the intensity of land use that irrigation brings?

Soil salinity in the Murray basin

Irrigation has created a problem known as salinity in the Murray river basin. Salinity is a measure of the amount of mineral salts in water. Figure 9.32 is a graph of salinity along the Murray. Reading from left to right on the graph, salinity levels change between Hume reservoir in the upper parts of the basin (shown on Figure 9.25) and the mouth of the river. Notice the three lines on the graph: the blue one for the average flow between 1969 and 1982, the red one for low levels of flow in 1980/81, and the green one for 1974/75, a year with high levels of flow.

1. Explain the fact that when river levels are low, salinity is high, and vice versa.
2. What does Figure 9.32 suggest about the tolerance of plants as opposed to that of people towards salinity? What might be the effects of increased salinity?
3. Draw a sketch map of the irrigated areas of the Murray valley shown in Figure 9.25. Decide on a colour scale for different levels of salinity, and show salinity along the river between Hume reservoir and the Murray estuary. Which areas appear to be most under threat?

▼ **Figure 9.32** Salinity in the Murray river.

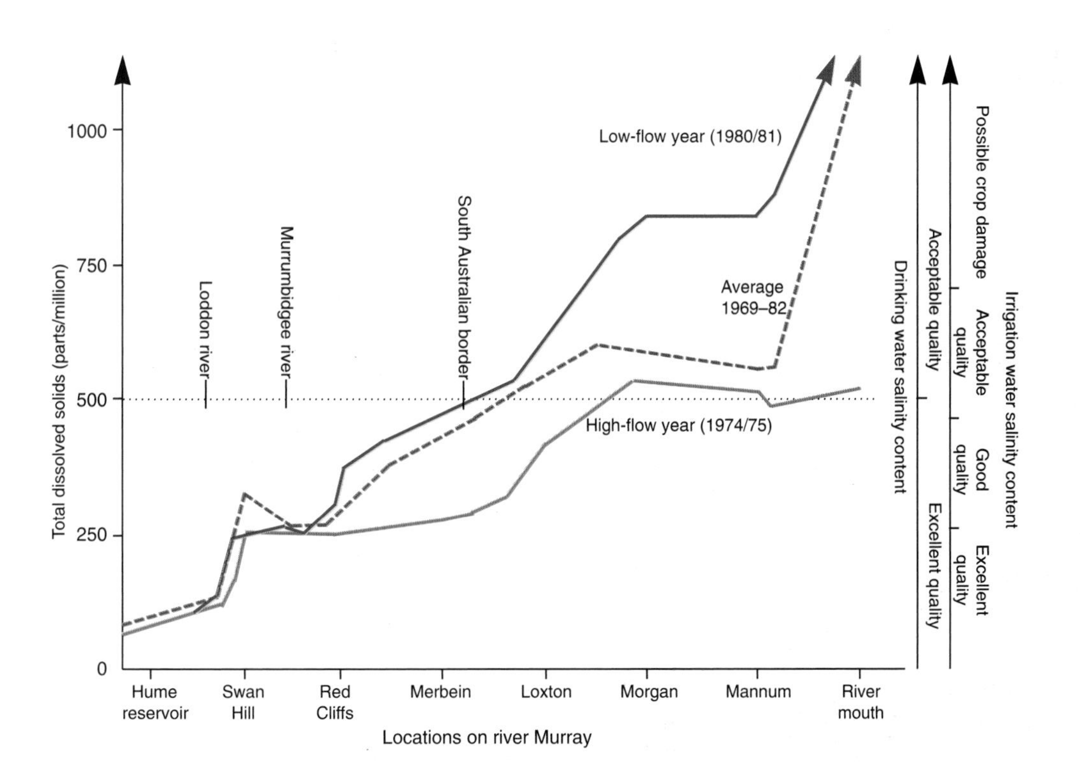

Salinity – 'The salt of the Earth'

Salinity is a particular problem in major irrigation schemes. All water contains mineral salts. The most common are sulphates and chlorides of sodium, calcium, and magnesium. Evaporation of water molecules increases salt concentration, especially in hot climates. Irrigation increases the problem, when water is channelled out of a river and applied to farmland. The application of water dissolves mineral salts within the soil. As water drains through the soil, it 'leaches' or washes out mineral salts, and carries them into the irrigation channel and back to the river. As the river flows further downstream, more and more of its water is used for irrigation, and salinity increases rapidly.

Increased salinity is most problematic where irrigation is so intensive that the water table – the area of saturated rock or soil – lies close to the ground surface. Concentrations of sulphates and chlorides precipitate, or crystallize, from the water. These can be 'flushed out' by reducing the water level, but in doing so clay and organic particles are lost and the soil structure is destroyed. Also, plants are unable to tolerate high levels of salt, and unless irrigation is carefully managed soils may become so alkaline that they become unusable.

Soil irrigation and salinity in Shepparton, Victoria

Study Figure 9.33, which is an article from *The Shepparton News*, a local newspaper in a small town of 28 000 people in northern Victoria. The *Lonely Planet* guide to Australia (1989) describes Shepparton's location as being 'within a prosperous irrigated fruit and vegetable growing area'. This news article has a different view.

Increased rates, deteriorating roads, a higher incidence of divorce – even difficulties in burying the dead. They are all effects of the growing salinity problem in the Goulburn Valley.

The recently released draft management plan to deal with the problem is being explained at a number of meetings.

The plan, developed by the Salinity Pilot Program Advisory Council (SPPAC), makes a number of daunting predictions, apart from the obvious effect salinity will have on the environment.

It is expected there will be a 4.3 per cent rise in Shepparton City rates if the plan goes ahead. The plan is estimated to cost $880 million.

Money aside, SPPAC Chairman John Dainton brought home the reality of the problem with some chilling accounts.

'In a number of cemetery locations in the region, it is already difficult to bury the dead with high water tables making it necessary to pump graves prior to the coffins being lowered to their final resting place,' he said.

'We have all experienced what salinity, high water tables and poor drainage do to our road network.

'Rodney Shire estimates it is spending $1.2 million annually on road edging and foundations.'

Mr Dainton said that if the problem was not addressed roads would continue to self-destruct at an unprecedented rate.

The snowballing effects of these losses on processors, suppliers and the service sector in the region would be devastating.

SPPAC estimated this reduction in wages and salaries paid would manifest itself in job losses – 3500 in this city by the year 2020.

Perhaps the most frightening effect would be on the entire social, environmental, and economic fabric of the region.

'The breakdown of family structures which includes child abuse, divorce and suicide can be expected when a community becomes stressed,' Mr Dainton said.

'We can also expect illness rates to increase, a migration of our talented youth from the area, and an increase in the need for counselling at all levels.

'If nothing is done to arrest the alarming rise in water tables and the ensuing salinisation processes, the problems now faced by many landholders will multiply rapidly as the true extent of the problems unfolds,' Mr Dainton said.

The SPPAC report also envisages that river salinities will rise dramatically as a result of the dryland saltload doubling over 100 years from 196 000 tonnes to 460 000 tonnes a year.

▲ **Figure 9.33** From *The Shepparton News*, 4 September 1989.

The maps in Figure 9.34 show the extent of waterlogging of soils in the Shepparton region in 1982 and 1990. By 1990, over 50 000 hectares of the irrigated area – one-quarter of all irrigated land in this region – had a water table less than 2m below the surface. Unless management practice is changed, this will increase to 90 per cent by the year 2020. Long-term productivity is estimated to fall by 30 per cent as salinity increases. In a region where orchards and irrigated pasture are the most significant part of the economic base, this is serious.

▼ **Figure 9.34** Water table contours for the Shepparton region.

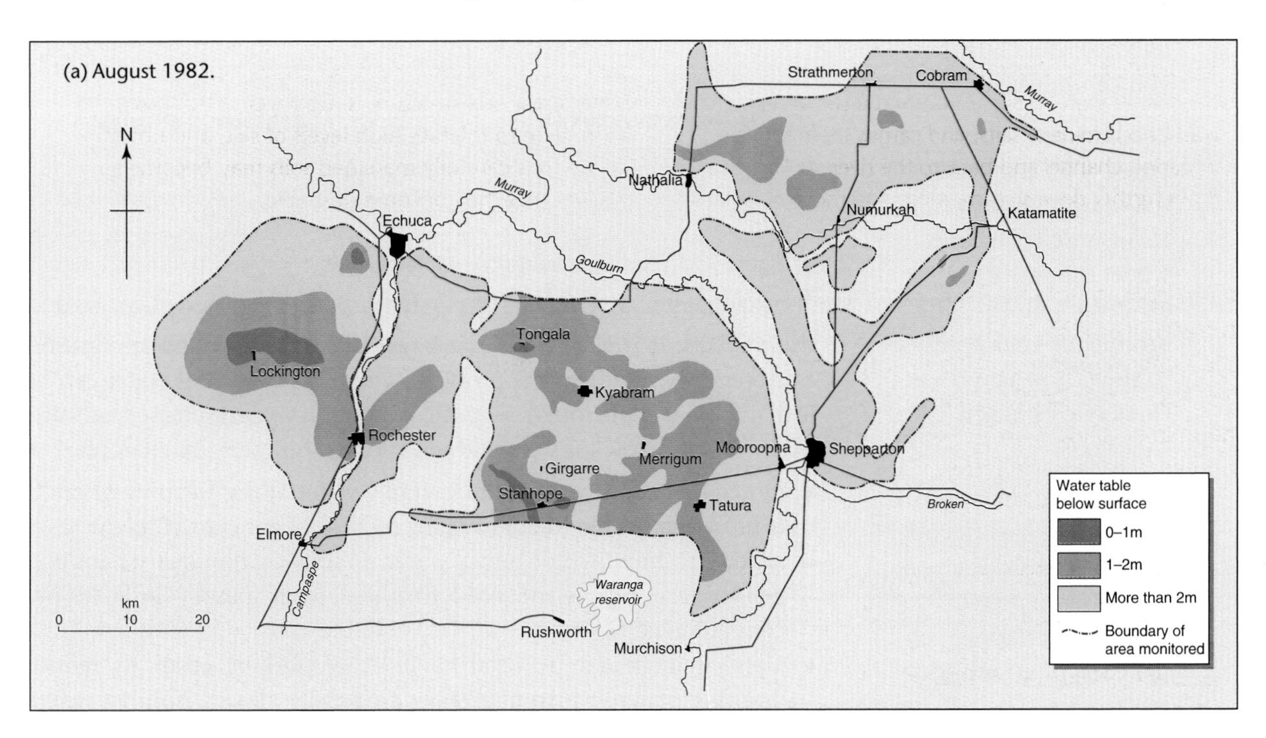

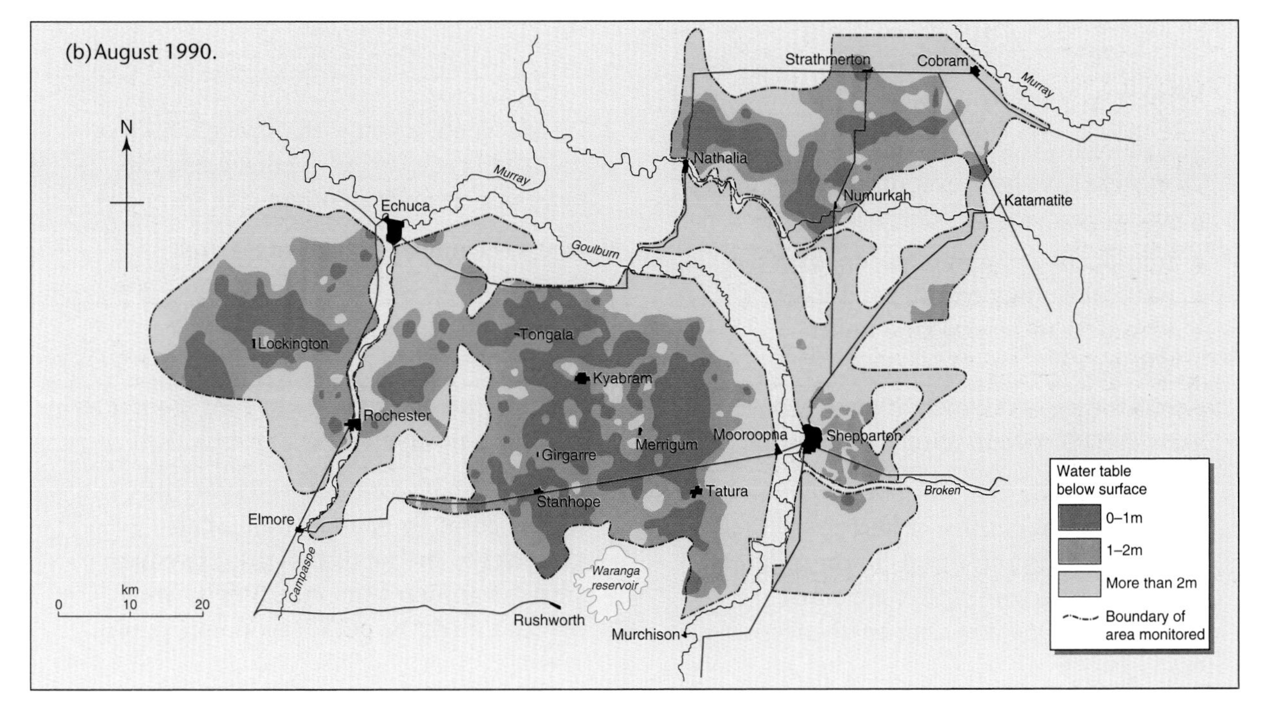

1 Make a copy of Figure 9.35 in the middle of a large sheet of paper. Draw pointers on your 'futures wheel' to show the effects of salinity on 'People', 'Services', the 'Economy', 'Ecology', and any other headings you can think of.

2 Read the article in Figure 9.33 and study the maps in Figure 9.34. Show on your futures wheel the predicted effects of salinity. Link up the different points to show how you think these effects might happen.

3 Add predictions of your own on the impacts of salinity. In this way you are trying to predict what might happen, on the basis of information available.

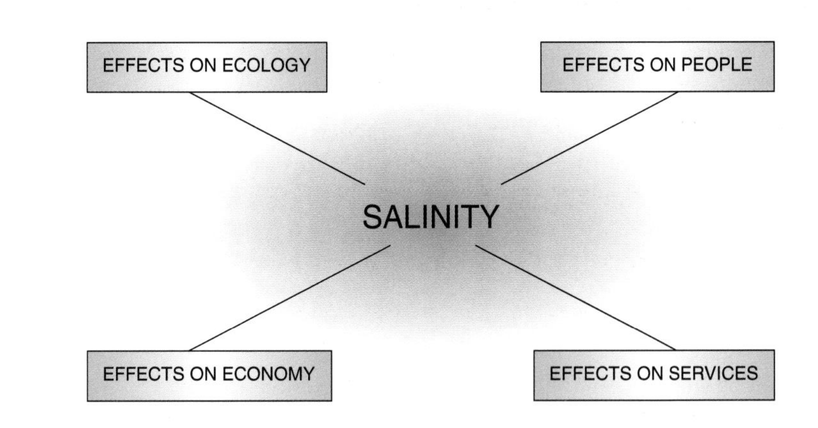

▲ **Figure 9.35** The effects of salinity – a futures wheel.

Strategies for managing saline soils are varied both in number and in their effects. It is clear now that large-scale irrigation and application of water only increases the problem. Over 90 per cent of water can be lost in this way, through seepage, evaporation, or increased salinity. Many farmers now look to new irrigation techniques, for example:

- *furrow irrigation*, which concentrates water along furrows where it is more easily able to get to plant roots – 50 per cent efficient
- *trickle irrigation*, which pipes water directly to the area where it is needed, around the roots of a fruit tree, for instance – up to 90 per cent efficient
- *lasers*, which can be used to level ground in order to reduce drainage of water into natural or artificially created hollows where waterlogging might become a problem.

Which way now?

The following threats to the grassland ecosystem have been examined in this study:

- recent droughts
- introduction of non-indigenous species
- pastoral farming of cattle and sheep
- irrigation, increasing land under cultivation.

Assess the impact of each of these on the grassland ecosystem, using the following procedure.

1 Form groups of three or four people. Compare the four threats to the grassland ecosystem according to their social, economic, and environmental effects.

2 Decide whether you wish to weight each of the criteria equally, or whether some are more important than others.

3 Assess the impact of each of the four threats. Try to rank the threats from least destructive to most destructive of the ecosystem.

4 Discuss with other people in your group the following statement:

'In a period of economic recession, the threats posed by natural and human processes to the survival of the ecosystem are less significant than the uses to which the land might be put for economic return.'

Ideas for further study

1 In groups, select other grassland areas of the world for study. Examples include the dry savanna and scrublands of southern Africa, or the Masai Mara in Kenya and Tanzania. In your study, focus on:
- the ecosystem, its components and flows
- the extent of human activity and whether it poses a threat to the ecosystem
- how such human activity might be or is being managed.

2 In the 1990s, debates have arisen from proposals by governments of African countries to limit numbers of animals such as elephant, whose numbers have increased sharply following protection measures. These have met with opposition. To what extent should overseas governments interfere with national policies on ecological management?

Summary

- The term 'grasslands' is used for a variety of different biomes, few of which are simply grass-only. There is no such thing as a typical grassland, only a spectrum between wooded and unwooded areas.
- Grasslands are ecologically the second most productive biome in the world, and among the richest in terms of plant and animal species.
- Grasslands are under threat. Both tropical and temperate grasslands are subject to drought, to modification by human activity, and to intensifying human demand.
- Attempts to exploit grassland areas are likely to create substantial ecological problems, unless they are carefully managed in the future.

References and further reading

BBC *Wildlife* magazine contains articles about major ecological issues, including those of the large grasslands of the world.

A. Grainger, *The Threatening Desert – Controlling Desertification*, Earthscan Publications, 1990.

G. O'Hare, *Soils, Vegetation and Ecosystems*, Oliver and Boyd, 1988, reprinted 1994.

I. Simmonds, 'A global first: primary productivity', *Geography Review*, January 1995.

10 Farm landscapes and ecosystems

1 Compare the patterns shown in Figures 10.1a and 10.1b. Use atlas maps to describe relationships between types of agriculture, and geology, relief, climate, soils, and natural vegetation in the UK.

2 Describe the characteristics of areas where livestock and cereals agriculture are practised.

Agriculture is one of the most important human activities. It has probably had the biggest impact on natural ecosystems and landscapes. Many factors determine the types of agriculture practised and their effects on ecosystems: for example, the nature of the land and climate, the demand for products, the technology available, and local culture. Different types of agriculture have different effects.

How farming changes ecosystems

When land is used for agriculture, the natural ecosystem is changed. In a natural ecosystem, biomass remains, decomposes, and contributes to cyclic processes within the system. In farming, biomass is harvested and removed from the ecosystem. One of the most important changes is the way in which natural components, such as nutrients, are replaced with artificial materials such as fertilizers. As a result, natural ecosystems are much altered. Figure 10.1 shows the distribution of natural vegetation and agriculture in Britain. Natural vegetation is what would be present without the influence of people or human activity. Compare these maps with atlas maps of relief, climate, and geology, to see how natural vegetation is closely linked to other physical factors.

▼ **Figure 10.1** The distribution of natural vegetation and agriculture.

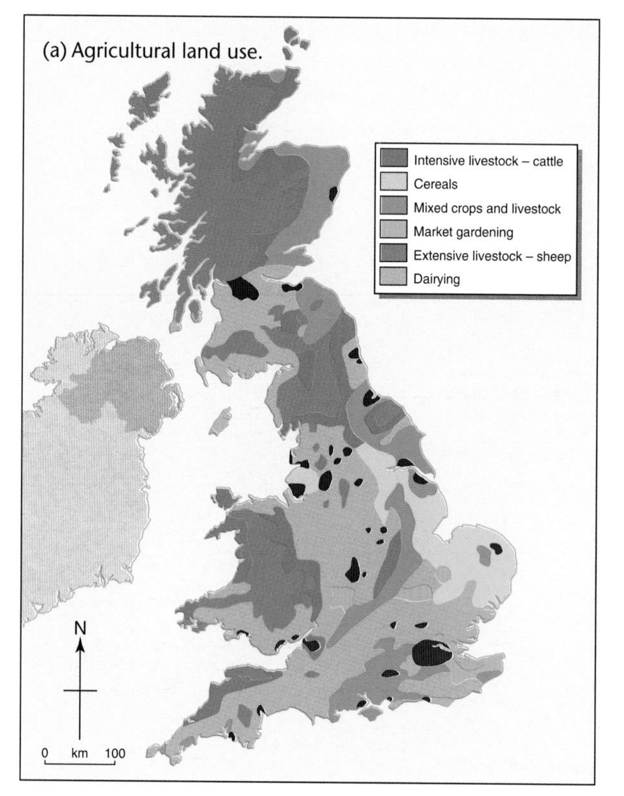

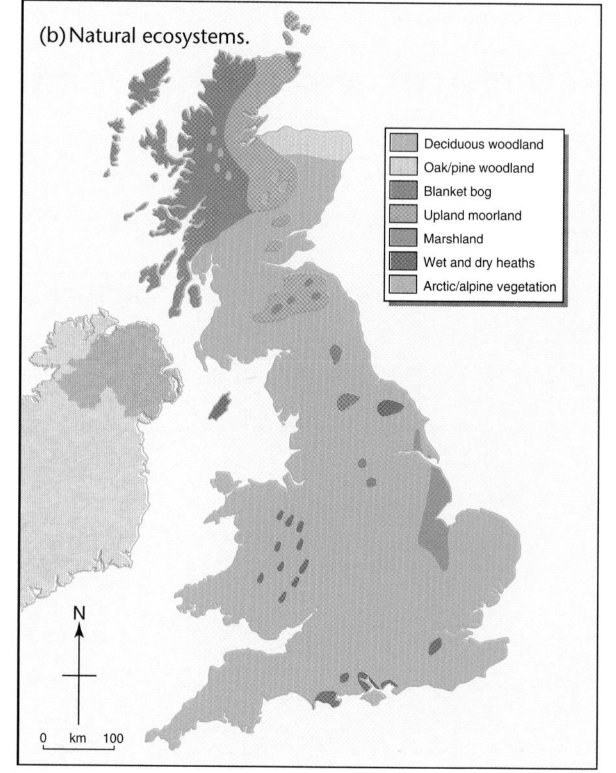

Some agricultural systems are very different from natural ecosystems. Factory farming of pigs or chickens, for example, takes place in tightly controlled indoor environments. Temperature, light, and food are regulated to maximize levels of production. Most farm systems are influenced by the natural characteristics of an area, whilst at the same time having some impact on the local environment.

1 Select another country or region, and research its agricultural patterns in upland areas. How is the distribution of agriculture linked to the physical characteristics of the area? How has the environment been adapted to allow for farming?

Differences between natural and agricultural ecosystems

1 Form groups of three or four people. Copy the Venn diagram in Figure 10.2. Study the statements listed. Some of these describe natural ecosystems, some describe agricultural ecosystems, and some apply to both.

   a) Categorize the statements as either natural, agricultural, or both.

   b) Arrange the statements on your diagram.

   c) Put any statements that you find difficult to fit in the diagram outside it.

   d) Present your diagram to the rest of the group for discussion. Justify your categorization.

2 **a)** Think of an example of an agricultural weed. Is this species a weed when it occurs in a natural ecosystem? What are 'weeds' and 'pests'?

   b) What is meant by a 'self-regulating system'?

   c) What sorts of hazards may alter ecosystems?

   d) What is the difference between 'productivity' and 'biomass'? How is productivity increased in agricultural systems?

   e) How does agriculture alter nutrient cycles?

   f) How far do abiotic factors affect productivity in agricultural systems?

   g) Why do many farm systems involve a limited food chain, and have plants at the same stage of development?

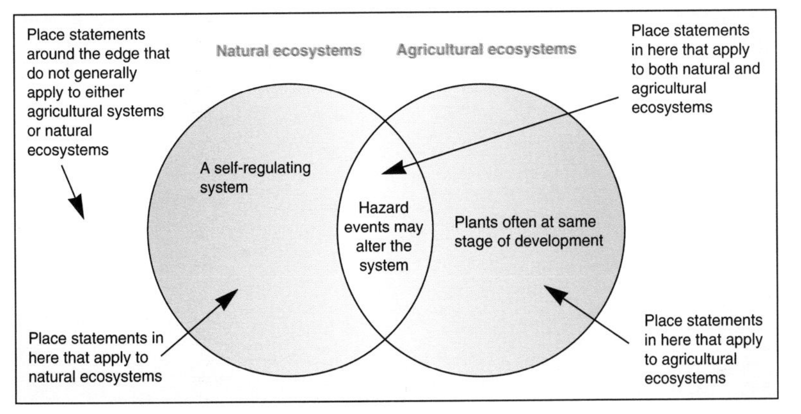

- Weeds and pests present
- A self-regulating system
- Hazard events may alter the system
- Diversity of species
- Governed by the aims of people
- Biomass production may be increased
- Nutrient cycles intact
- Limited amounts of nutrients available
- Permanent vegetation
- One or a few dominant species
- Abiotic factors determine productivity
- Plants at different stages of development
- Nutrient cycles broken
- Variety of animals at the highest trophic level
- Nutrient input can be increased
- Often vegetation cover is intermittent over time
- People are the top consumers in the food chain
- Plants often at same stage of development

▲ **Figure 10.2** Venn diagram for comparing natural and agricultural ecosystems.

Farming the rainforests of Kilimanjaro

1 a) Using an atlas, draw an annotated sketch map of this part of Tanzania. On your map include the location of Kilimanjaro, and the natural vegetation found in this part of Africa.
b) Why does the large belt of tropical rainforest not extend to the east coast?
c) Why do you think the slopes of Kilimanjaro support montane rainforest, while the surrounding lowlands are not forested?
d) Compare the population density of the Kilimanjaro region with that in the rest of Tanzania.

2 Use the text and Figures 10.3 and 10.4 to draw a nutrient flow diagram (as shown in Figure 7.8) of the agroforestry system practised by the Chagga people.

Agroforestry in Tanzania

Some agricultural systems do not differ much from the natural ecosystems within which they take place. The Chagga people practise a form of agroforestry on the southern slopes of Kilimanjaro in Tanzania. This imitates the stratification of natural forest by using several layers of trees, bushes, and vegetable plants (Figure 10.3). Each cultivated species is selected according to available light conditions. The crowns of some original rainforest trees help to protect the soil, ground surface, and lower vegetation from tropical storms, while their roots help to hold the soil in place. Livestock are part of the farm system, grazing on the ground plants and consuming waste material. Their dung is vital for manure. Even irrigation channels are put to a double use, providing a habitat for fish which add another source of protein to the diet of the farmers and their families. This system is so successful that it supports the highest rural population per square kilometre in Tanzania.

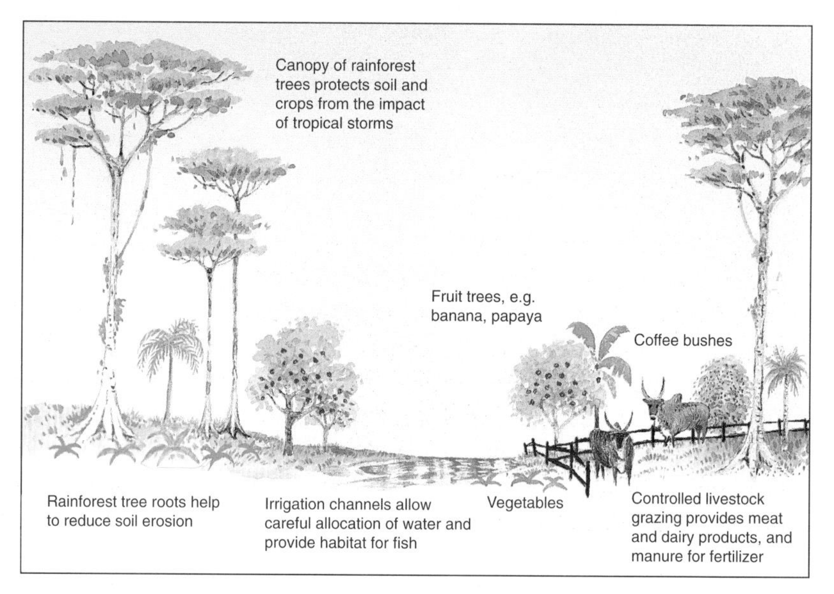

▲ **Figure 10.3** The system of agroforestry practised by the Chagga people of Tanzania.

▶ **Figure 10.4** Agroforestry practices of the Chagga people.

What other factors affect agriculture?

Technology

Natural characteristics of an area influence agriculture, but the development of farm technology has reduced the dominance of this. Now genetically engineered crop plants are grown in areas that were previously unsuitable. Irrigation systems have allowed deserts to bloom in the USA and the Middle East, for example.

Cultural values

Cultural values – reflected in consumer opinion, for example – have always been important in farming. A recent trend in the UK has been the increased availability of organic produce and 'conservation-grade' meats. This has created opportunities for farmers to capitalize on the higher prices that consumers are prepared to pay for this type of meat. Another trend has been customer desire to have year-round supplies of fruits and vegetables, which has led to increased imports of fresh food from other countries.

Links with industry

Agriculture is now linked closely with industry, for example by controlling the production, marketing, and distribution of agrochemicals. Many farm products are now sent direct to processing factories. These links can be highly developed. In East Anglia and Lincolnshire, for example, much farming is carried out under direct contract with firms such as Bird's Eye and Findus.

Farming and politics

Political decisions, at a variety of levels, are often the cause of changes in agricultural practice. In the UK the post-war drive for self-sufficiency led to great intensification of farming. The EU Common Agricultural Policy (CAP) has played a major role in farm planning. In the 1980s, the introduction of milk quotas placed limits on the volume of milk that farmers could produce in the EU. Upland farmers were especially hard-hit by this, as milk forms a substantial part of their income. Grants were made available for farmers to receive sums in excess of the market price to slaughter their dairy cattle. In future, the international General Agreement on Tariffs and Trade (GATT) is also likely to affect farmers.

Changing upland environments

In order to look at the complex relationships between ecosystems and agriculture, the next part of this chapter looks at upland areas of Britain. It focuses on the Pennines, and looks at the issues involving ecosystems, farming, and landscapes.

Agriculture and landscape in the Yorkshire Dales

The Yorkshire Dales National Park forms part of the central Pennine chain of uplands in England. Much of the national park is farmed (Figure 10.5). Sheep farming is dominant in the north of the area, and cattle in the south where the land is less rugged.

◀ **Figure 10.5** Typical Yorkshire Dales landscape.

▲ **Figure 10.6** Rough moorland grasses in the Yorkshire Dales.

Open moorland covers 56 per cent of the national park. 'Moorland' is used to describe the rough upland grasses, either on alkaline limestone soils or acid soils which overlay the millstone grit, which is a form of hard sandstone. Limestone areas are well drained and support a variety of lime-loving, short turf grasses that remain green throughout the year. This type of vegetation is preferable for grazing. Species of grass on acidic soils are less palatable and include rushes which turn brown and die off over the winter. These rough grasses can be seen in Figure 10.6.

Where acidic soils are well drained on the plateaus in the east and the north of the Dales, the vegetation is mainly from the heather family. Many of the peaty soils are found on the Yoredale sequence of rocks, which consists of alternate layers of limestones, shales, and sandstones. This variation produces alternate waterlogged and well-drained areas. On permeable limestone, streams sink underground, re-emerging where the limestone meets shales and sandstones. The influence of geology is clearly seen in Figures 10.7 and 10.8.

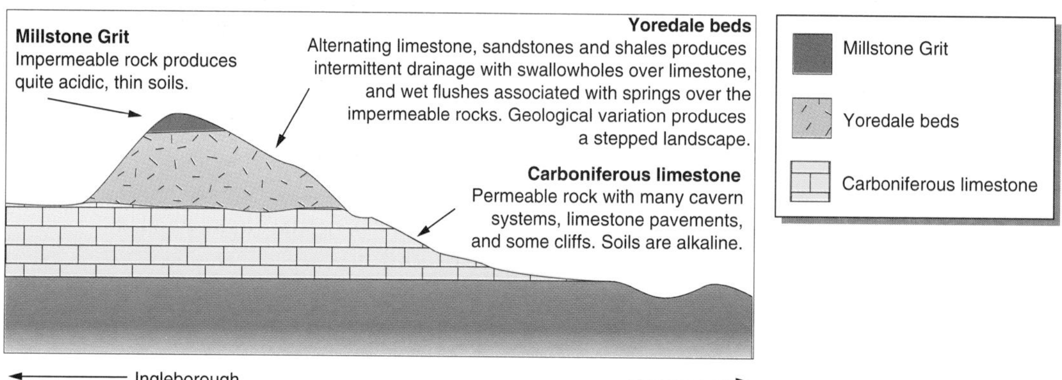

▲ **Figure 10.7** Key geological characteristics of the Yorkshire Dales illustrated by a section of the Ingleborough area. Ingleborough is one of the highest peaks in the Pennines.

▶ **Figure 10.8** Links between vegetation and geology in the Yorkshire Dales. Two distinct areas are shown on this photograph. In the distance, shorter and greener turf grasses are visible. These correspond with limestone. The brown moor grasses can be seen below these, with small exposures of black peat. These are the acidic soils derived from sandstone.

▼ **Figure 10.9** Flow diagrams to show the impact of deforestation on soil development in upland England.

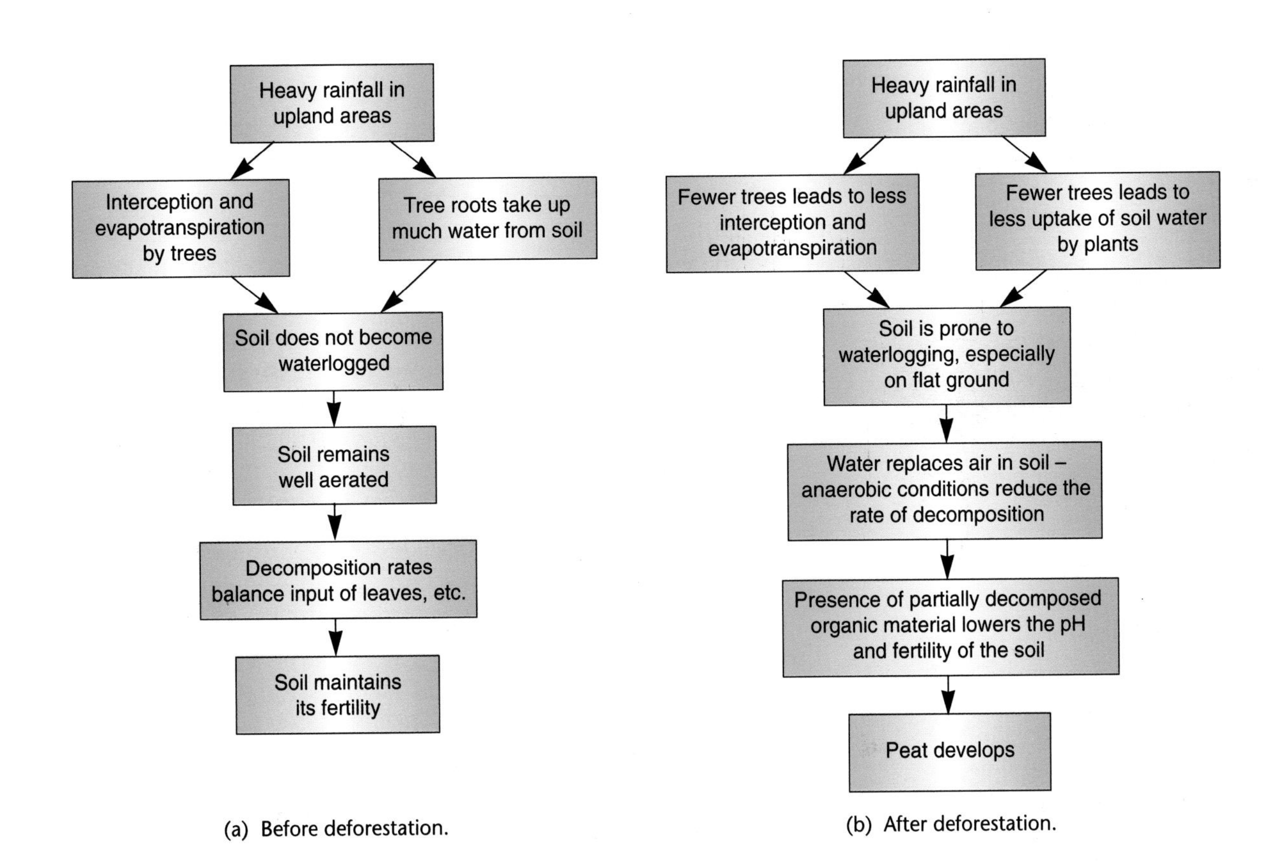

(a) Before deforestation.

(b) After deforestation.

A large proportion of the land now supporting acid-tolerant plants was originally covered by deciduous woodland. Since the Neolithic period this has gradually been cleared for timber and grazing land. The long-term consequence of this has been a reduction in the fertility and pH of the soil, resulting in less potential for native tree re-establishment. The flow diagrams in Figure 10.9 explain these ecosystem changes. Here there is extensive sheep grazing, sometimes at a level of only 2.5 sheep per hectare. Where animal stocking levels are too high, the heather is replaced by grasses. Heather has been managed by a process of regular, controlled burning which destroys the unpalatable woody stems of the heather and allows re-growth of young shoots which are more valuable as a food source for livestock.

As this form of agriculture is relatively unprofitable, other uses are often found for this land. One of the most common is the rearing of grouse for game shooting, which can take place alongside sheep grazing. Recently, more land has been converted to commercial forestry. However, only conifers will survive on these thin acidic soils. Many of these trees are non-native to this country, such as the Sitka spruce. Even indigenous species find conditions here difficult, and the moors have to be drained before planting.

Of the remaining land, 42 per cent has been enclosed for agriculture. This includes pastures on the valley sides which are grazed throughout the year. On the lower land near the farm buildings there are meadows. Some of these are of great conservation interest, supporting wildflowers such as the wood cranesbill and the buttercup, which flower in the spring. This ecosystem is maintained with little fertilizer, and a strict management regime. After flowering and setting seed, the tall grass is cut during the summer for winter fodder, either as hay or silage, and livestock are then allowed to graze. Where the system has been intensified with applications of fertilizer or re-seeding, many wildflowers have disappeared. This has been necessary in some areas to increase productivity of the meadows. During cold winters with frequent snow cover, winter feed prices are high, and many farmers now get two crops of hay within one summer in order to reduce feed costs.

Factors influencing the landscape

1 Form groups of three or four people. Study Figures 10.5, 10.10, and 10.11. Consider the factors that may have influenced this landscape. Place your ideas into the following categories:
- natural factors
- social factors
- economic factors
- political factors
- historical factors.

Use an atlas to research the location, relief, climate, and other characteristics of the Yorkshire Dales.

2 Decide which of the factors are the most important. Put them into a rank order. Reach an agreed rank order with the rest of the group.

3 Annotate a copy of Figure 10.11 to show how the factors you identified have influenced the landscape.

Farming and landscape change in the national park

National park authorities in England and Wales have two main aims:

- to conserve and enhance the special quality of the landscape, and to care for the habitats, historical features, and agriculture systems within the national parks
- to provide access for people, and facilities for them to visit and enjoy the countryside.

These aims clearly affect farming, and create conflict within the Yorkshire Dales. However, just because the area is a national park does not mean it is immune from changes to farming. Many economic and political pressures influence activities here. These have knock-on effects for the landscape and the way of life of the 18 600 people who live in the Yorkshire Dales. Changes to the landscape and ecosystems may affect the number of visitors to the national park – currently about 10 million people a year.

The development of the Yorkshire Dales landscape

The Yorkshire Dales is characterized by its dry-stone walls, traditional farm buildings, and field barns. These are not all of the same age. Some of the dry-stone walls may have their origin in prehistoric and Roman times. The Cistercian monks improved and extended farmland in the 12th and 13th centuries, further enclosing the landscape for relatively intensive livestock rearing, mainly to produce wool. Sheep-folds were built on the moors to provide shelter from the hard weather conditions.

At the same time, the remaining area was farmed under a feudal system of open fields and common land. As the urban population grew, rural communities had to support them, which meant increased opportunity for income on the farms. The feudal system broke down, and enclosure saw land divided into small farms. The small rectangular fields of the present-day landscape began to emerge. Moorlands were further enclosed during the 18th century.

◀ **Figure 10.10** Some of the effects of enclosure on the Yorkshire Dales landscape. The boundary between improved pasture and rough moorland is clearly visible as a wall running from left to right across the centre of this photograph. The moorland remains unenclosed and is unaffected by the lime that is applied each summer to the pastures in the valley bottom.

The barns were built at this time to store fodder and provide shelter for cattle which grazed far from the main farm buildings. The use of field barns has declined as cattle farming has become restricted to the valley floors. Many of the barns have fallen into disrepair, as have many stone walls. High maintenance costs have led to the replacement of stone walls with wire fences. Where repairs are made, it is often with the help of grants from the national park authority or the Countryside Commission.

▼ **Figure 10.11** Outline landscape sketch.

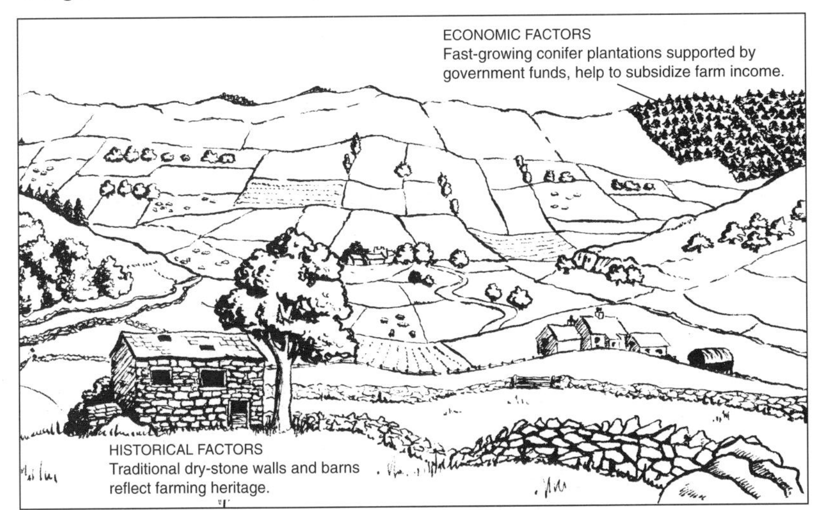

Positive features	Bi-polar score					Negative features
	+2	+1	0	−1	−2	
varied						monotonous
colourful						drab
etc.						etc.

◀ **Figure 10.12** Grid for bi-polar analysis of rural landscapes.

Evaluating rural landscapes

As agriculture changes so does its impact on landscapes and ecosystems.

1 In pairs, decide what features make a rural landscape attractive, and list these. Then find words or phrases that mean the opposite. For example:
   - attractive/unattractive
   - varied/monotonous
   - colourful/drab
   - streams or rivers/no streams or rivers.

   When you have agreed on 8–10 pairs of statements, make a grid for bi-polar analysis as in Figure 10.12.

2 Study Figures 10.5 and 10.14a. List the variety of habitats here.

3 How much is each habitat affected by the agricultural system?

What will the landscapes of the Yorkshire Dales look like in the future?

This section considers the effects of agricultural changes in the Yorkshire Dales on the landscape. Figure 10.14a combines some features of the present landscape. Figures 10.14b, c, and d show what the area could look like, depending on what happens to the agriculture of the area in the future.

The figures are designed to show the links between economic process and landscape change – what we see around us usually has economic reasons.

When you read the scenarios, look at Figures 10.14b–d showing landscapes of the future. Decide whether future changes are generally positive or negative, using Figure 10.13 as a basis.

▼ **Figure 10.13** The impacts of agricultural changes on the environment.

Landscape	**Impacts of change on the:**					
	Natural environment		**Social environment**		**Economic environment**	
	+VE	–VE	+VE	–VE	+VE	–VE
(a)						
(b)						
(c)						
(d)						

Note +VE means positive or beneficial impacts –VE means negative or detrimental impacts

▼ **Figure 10.14** Present and possible future landscapes of the Yorkshire Dales.

Situation 1 – the present day

Today the income of rural communities comes from agriculture and tourism. Agriculture is supported by subsidies, though this is not enough to prevent some dereliction. Meadows are cut for hay or silage. To get a reasonable income there is some over-stocking of animals, which has led to overgrazing of moors and woodlands. However, the landscape is important for recreation and tourism.
Annual cost of maintenance: £9.0 million

(a) The present day.

Situation 2 – a 'wild' landscape

The area has been sold to a conservation organization. There is no agriculture and the vegetation has been allowed to diversify. Visitors have access to all areas except those of particular wildlife interest. The owners have a management strategy to maintain the ecological balance of the area. There is opportunity for employment in countryside management and recreation.
Annual cost of maintenance: £8.1 million

(b) A 'wild' landscape.

(c) An 'intensive' landscape.

Situation 3 – an 'intensive' landscape

Government farm subsidies are withdrawn, so that farmers have to decide how to maximize profits. The system is geared towards food production. Wealthier farmers buy out small farms, and have access to the latest technology and breeding techniques, allowing the creation of large livestock ranches. Meadows are replaced by large fields which are used for silage. Traditional buildings and field boundaries are removed. Livestock have damaged deciduous woodlands, which are replaced by conifer plantations.

Annual cost of maintenance: £10.4 million

(d) A 'semi-abandoned' landscape.

Situation 4 – a 'semi-abandoned' landscape

Government farm subsidies are withdrawn, and upland farmers cannot compete with farmers on more favourable land. Many farmers have sold up and been forced to leave. Some buildings are converted for uses such as tourist accommodation, while others become derelict. Accessible farmland is intensified. To maintain their income, some farmers turn to farm-based tourism or forestry.

Annual cost of maintenance: £2.8 million

1 Study Figures 10.14b–d. These show three possible future landscapes in the Yorkshire Dales. Each has been drawn according to what might happen in future. Present your impressions of each landscape, using your bi-polar grid (Figure 10.12).

2 Now complete another chart, for today's landscape (Figure 10.14a).

3 **a)** Where does the present-day landscape come in each group's rank order ?

b) Which landscapes have been ranked highest? Why is this?

c) Which landscapes have been ranked lowest? Why is this?

d) How has each group defined attractive rural landscapes?

Which is the preferred landscape?

1 Write a report about agriculture and landscape change in the Yorkshire Dales. Consider each future scenario, decide on the best one for the national park, and justify your decision. Include the annotated sketches and tables that you have already completed.

2 Consider the survey carried out in the Yorkshire Dales in 1990 (the results are presented in Figure 10.15). This survey attempted to find out views of residents and visitors on landscape quality and possible changes to it. The sketches that you have studied were used in the survey. When asked which landscape they preferred, 47.2 per cent of visitors and 50.2 per cent of residents chose the present landscape. Of the others, the 'wild' scenario was the most popular, and the intensive farming system was the least popular. How does this compare with your views?

3 The survey was used to find out which features people would like to see more of, less of, or about the same as the present.
   a) Comment on the results of the survey.
   b) How should planning authorities within the national park use these results?

4 How much should the question of cost influence any planning for landscapes of the future?

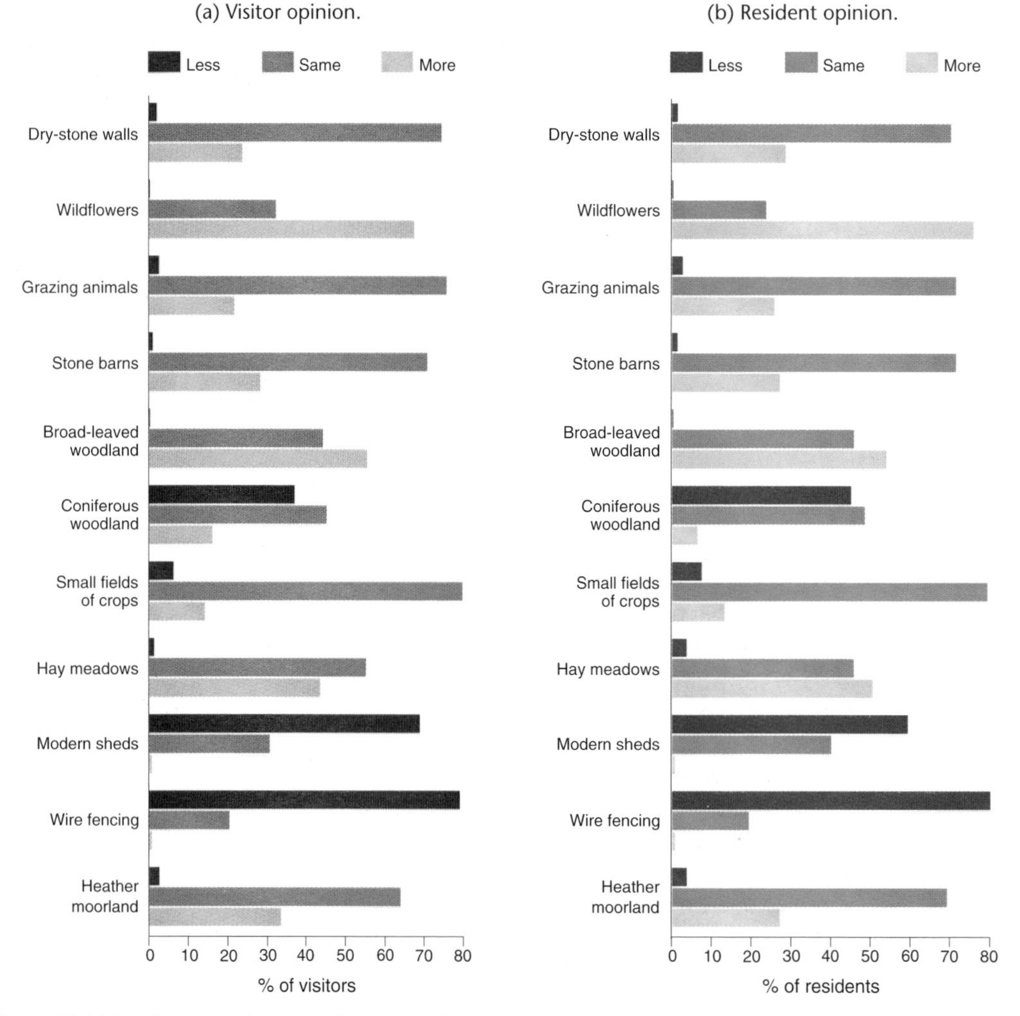

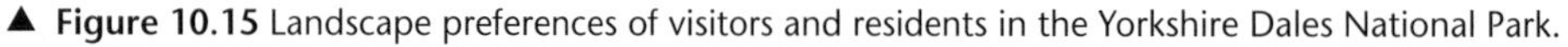

▲ **Figure 10.15** Landscape preferences of visitors and residents in the Yorkshire Dales National Park.

Government responses to conservation issues in the UK

The Countryside Stewardship Scheme

The Countryside Commission offers management agreements to enhance and conserve important English landscapes, their wildlife habitats, and history. Payments are made for changes to farming and land management practice that produce conservation benefits or improved access and enjoyment of the countryside. Eligible landscapes include traditional orchards, uplands, waterside areas, historic landscapes, and the coast. Agreements run for ten years. Payments are made when farmers agree to follow environmentally sensitive management plans, and enhance access to the land.

Set-aside

Set-aside schemes originated in the USA and were brought into the EU Common Agricultural Policy in 1988. They are designed to balance intensification of farming and to reduce food surpluses. The main cause of EU food surpluses – of grain, butter, and beef – has been the use of subsidies to guarantee prices to farmers for certain types of produce. To persuade farmers to reduce output, compensation was needed. This comes in the form of set-aside grants. The initial scheme applied only to land that had been previously used for growing arable crops, and involved each farmer in a commitment to reduce the area of arable land by 20 per cent for five years.

The scheme has its critics. Some farmers have taken their most marginal land out of production and increased intensity of farming on the rest, without decreasing productivity or profits. There have been complaints that set-aside land is allowed to become derelict, supporting weeds such as nettles and thistles – recreational users object to this.

Some issues were addressed in the CAP reforms of the early 1990s, when rotational set-aside was introduced. This requires arable farmers to take some land out of production before claiming payment, but they now have to rotate set-aside land every six years. Land must therefore be maintained in a condition suitable for cropping. The EU has passed legislation to give each government responsibility for providing funds for environmental management of set-aside land. In the UK, this is provided by the Countryside Premium which gives further incentives to farmers to manage their land. Examples of management include hedgerows and habitats attractive to ground-nesting birds.

Environmentally Sensitive Areas (ESAs)

The MAFF has designated 43 areas of the UK as ESAs. These areas are important for their wildlife, historic features, or landscape. Farmers within ESAs can opt into the scheme and claim payment for reducing the intensity of agriculture to promote conservation. Advice is given by English Nature, and the Countryside Commission. Current ESAs include the Lake District, the Pennine Dales, and the Norfolk Broads. The distribution of ESAs in England and Wales is shown in Figure 10.16.

Nitrate Sensitive Areas (NSAs)

Nitrate pollution of surface and groundwaters occurs where nitrate fertilizers are applied to crops or grass and are not all taken up by plants. After the 1991 EU Nitrate Directive, the UK government set up ten NSAs covering 11 000ha of land where water supplies might become polluted by nitrates. Payments are made to farmers to reduce nitrate leaching, by reducing the use of fertilizers and manure. Autumn-sown crops are encouraged so that crop cover over winter allows plant uptake of nitrate, when high rainfall could cause leaching.

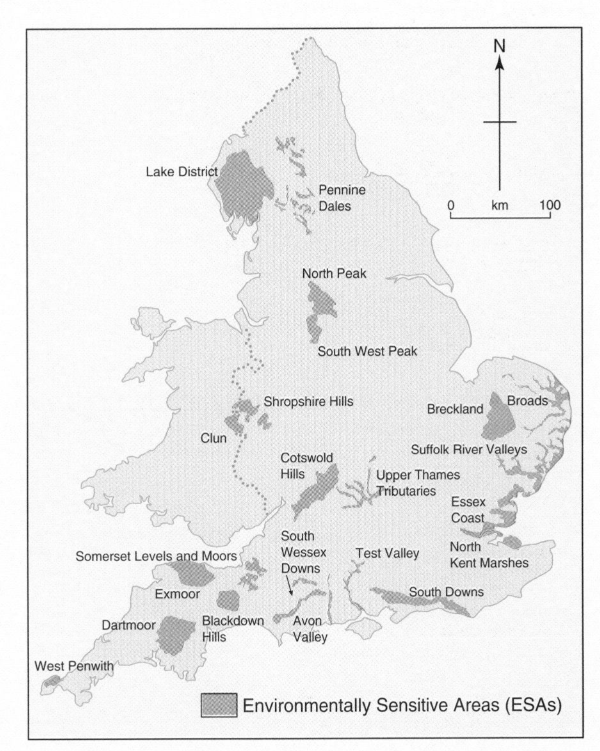

▲ **Figure 10.16** Environmentally Sensitive Areas in England and Wales.

Case study: Bradley Farm, Northumberland National Park

Bradley Farm is a 200ha hill farm in the south of the Northumberland National Park. It is owned by the National Trust, and farm-buildings and land have been rented by Julian and Lesley Acton since 1987. section of Hadrian's Wall runs along the norther boundary of the farm, and there are many visitors t the area. The location of Bradley Farm is shown i Figure 10.17.

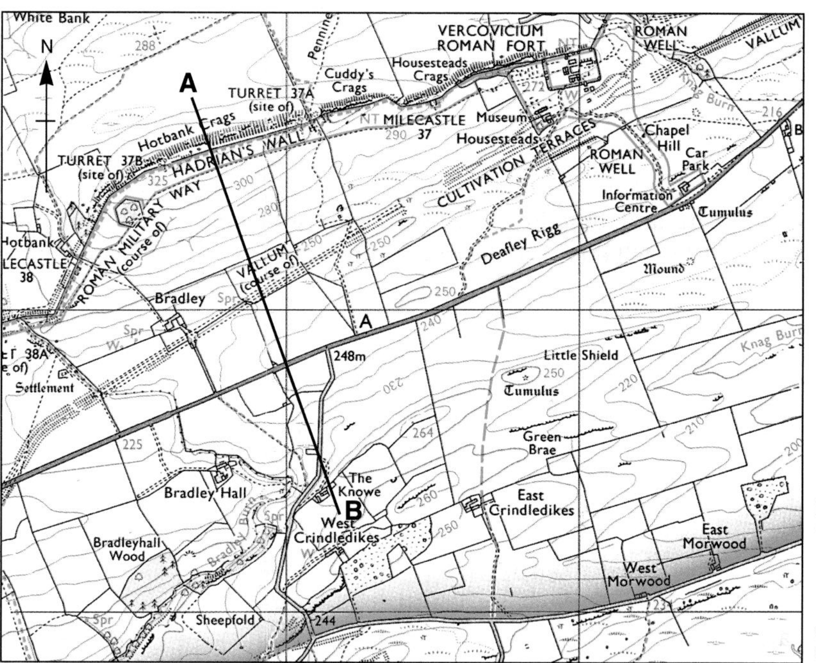

▶ **Figure 10.17** The position of Bradley Farm on a 1:25 000 OS map extract.

▼ **Figure 10.18** Aerial view of part of the land of Bradley Farm, Northumberland.

Physical characteristics and habitats

Bradley Farm is south-facing, and lies between 180 and 300 metres above sea-level. The soils, vegetation, and landscape are directly affected by the geology. This consists of limestone and sandstone into which dolerite has been intruded, forming the Whin Sill. The Whin Sill forms the ridge on which Hadrian's Wall was built, and the smaller ridges are formed by the sandstone. The limestone has been eroded to form scarp faces and valleys – see Figure 10.18.

Three types of land use are found here. On sandstone and dolerite, soils are acidic and are covered by bilberry and heather. Limestone soils are alkaline and support better grazing, and herbs including wild thyme, harebell, and clover. Wetland areas of varying sizes occur where springs soak the soil, or where boulder clay or peat are found. Here the main plants are rushes and sedges.

Most of the vegetation has been deliberately altered to increase farm productivity. Over large areas, heath has converted to a grass-based ecosystem, and now contains species such as sheep's fescue and mat grass. Limestone grasslands are well grazed and are mostly closely cropped turf. There are several areas of improved pasture where soils are deeper and where fertilizers have helped to increase productivity. About one-fifth of the farm is suitable for ploughing, being relatively flat and stone-free, and is used for hayfields. Species have been selected to provide grass for hay and silage. In addition to agricultural land there are mature and young plantations containing deciduous and coniferous trees, and rich ground flora. These act as shelter belts and are maintained as wildlife habitats.

The farm system

Bradley Farm is entirely pastoral, supporting about 40 cattle, 800 ewes, 50 rams, and up to 1200 lambs and 40 calves in spring and summer. The farm system and year are illustrated in Figure 10.19. Lambs and calves are born in the spring, so that additional grazing requirements can be supplied during maximum grass growth in late spring and summer. The weather is also warmer then, which helps to reduce stock losses when the young animals are most vulnerable. During the winter, grass grows at a reduced rate, and silage and hay are fed to the animals at this time. Sheep remain outdoors all year, but cattle are housed in sheds from late autumn. They are given commercial feed, mainly grain. Although some lambs are sold for meat, many are sold to lowland farmers who 'finish' them on quality pasture.

INPUTS

Livestock
- Rams (added to flock of 50) from other farms to improve breeding
- Average of 4 cows (from dairy herd) for breeding. Aberdeen Angus bull (hired)

Other capital costs
- Cattle feed
- Vet and medicines
- NPK fertilizer
- Machinery purchase, repair and fuel
- Insurance
- Labour

Financial inputs
- Countryside Stewardship Scheme
- Hill subsidy and premiums
- National park grants

PROCESSES THROUGHOUT THE YEAR

Spring	Summer	Autumn	Winter
Pasture management Unimproved grassland grazed by cattle →	→		
Unimproved grassland grazed by sheep →	→	→	→
NPK fertilizer applied to inbye (improved pasture and hayfields) →	Improved pasture grazed by ewes with twin lambs; Grass grows until cut in July (grass production exceeds demand) →	Hay and silage stored (silage is fermented grass kept in airtight plastic-covered bales) →	Hay and silage used as winter feed (grazing demand exceeds grass production)
Hayfields grazed by sheep until late May →			
Sheep management Lambs born, suckle ewes and eat grass from two weeks old →	Continue to suckle until August. Ewes with twins moved to improved pasture →	Lambs gradually sold for slaughter, or to lowland farmers →	200 ewe lambs retained to replace old breeding stock
	Mature sheep sheared →	200 oldest ewes sold; Remaining ewes with rams in November	
Cattle management Calves born, suckle cows (mostly outdoors) →	→	Calves weaned, housed indoors. Fed silage and grain →	Calves gradually sold for 'finishing' to lowland famers; Cows brought inside, fed silage

OUTPUTS

Livestock
- About 1000 lambs for slaughter or fattening
- 200 old ewes for slaughter
- 40 calves sold on for fattening
- Wool

Conservation benefits
- Landscape features improved, added and maintained, e.g. woodland, walls
- Ecosystems become more diverse, e.g. unimproved grassland, wetland, hedgerows

▲ **Figure 10.19** Bradley Farm system and year.

The influence of outside agencies

Figure 10.20 illustrates how outside agencies affect the farm economy, by dictating practice and by subsidizing livestock, landscape conservation, and habitats. Many other organizations have an interest in the farm, and influence decision-making. For example, the geology means that the land is part of a Site of Special Scientific Interest (SSSI). Permission from English Nature has to be sought before carrying out many farm operations, including woodland management, construction or removal of tracks and walls, and spreading fertilizer. Half of the farm is part of Housesteads Roman Fort scheduled ancient monument. Key archaeological features include the fort, wall, remains of turrets, and a milecastle along the wall, and cultivation terraces and settlement remains outside the fort. This means that any land disturbance has to have consent from English Heritage, including digging holes for trees and gate posts, or repairing drains. If consent is given, a National Trust archaeologist must be present to observe the digging. Before any work can begin on sheep pens or barn alterations, the farmer has to gain permission from the National Trust and the National Park Planning Authority. The National Rivers Authority (NRA) has responsibility for checking that farm activities do not pollute water courses, or extract too much water.

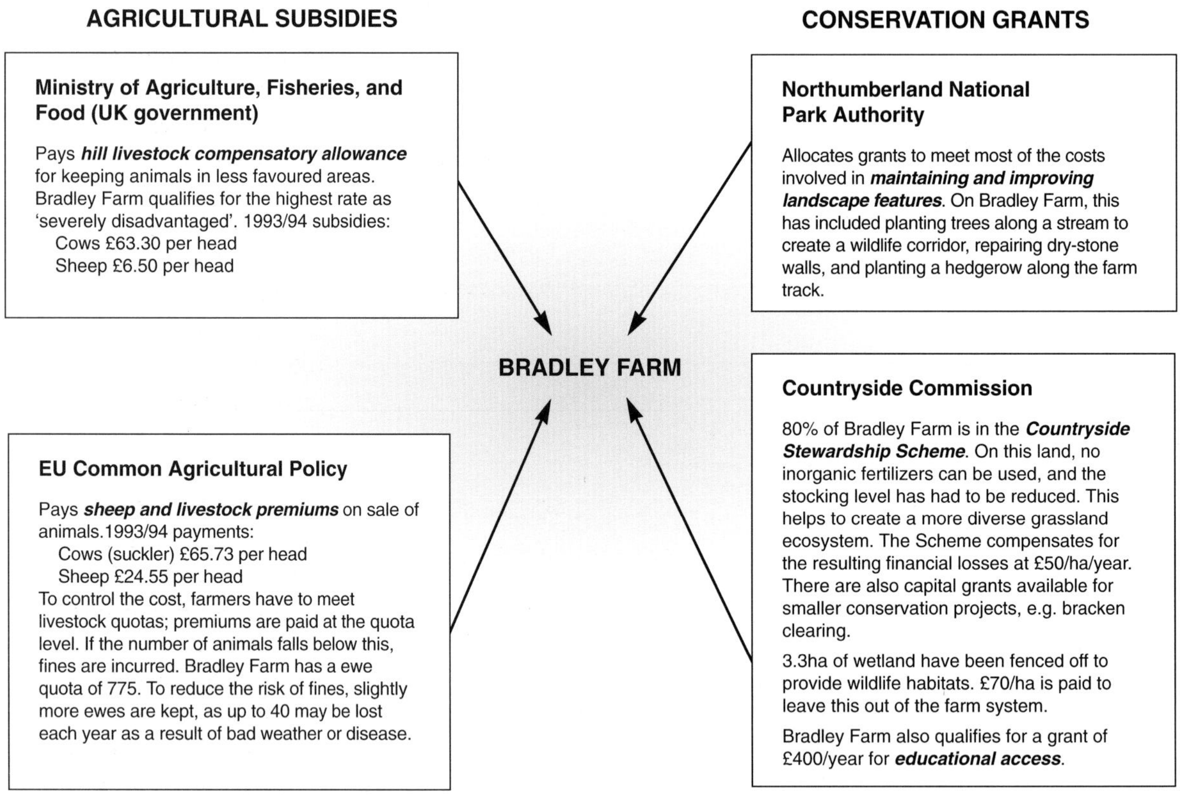

AGRICULTURAL SUBSIDIES

Ministry of Agriculture, Fisheries, and Food (UK government)

Pays ***hill livestock compensatory allowance*** for keeping animals in less favoured areas. Bradley Farm qualifies for the highest rate as 'severely disadvantaged'. 1993/94 subsidies:
Cows £63.30 per head
Sheep £6.50 per head

EU Common Agricultural Policy

Pays ***sheep and livestock premiums*** on sale of animals.1993/94 payments:
Cows (suckler) £65.73 per head
Sheep £24.55 per head
To control the cost, farmers have to meet livestock quotas; premiums are paid at the quota level. If the number of animals falls below this, fines are incurred. Bradley Farm has a ewe quota of 775. To reduce the risk of fines, slightly more ewes are kept, as up to 40 may be lost each year as a result of bad weather or disease.

BRADLEY FARM

CONSERVATION GRANTS

Northumberland National Park Authority

Allocates grants to meet most of the costs involved in ***maintaining and improving landscape features***. On Bradley Farm, this has included planting trees along a stream to create a wildlife corridor, repairing dry-stone walls, and planting a hedgerow along the farm track.

Countryside Commission

80% of Bradley Farm is in the ***Countryside Stewardship Scheme***. On this land, no inorganic fertilizers can be used, and the stocking level has had to be reduced. This helps to create a more diverse grassland ecosystem. The Scheme compensates for the resulting financial losses at £50/ha/year. There are also capital grants available for smaller conservation projects, e.g. bracken clearing.

3.3ha of wetland have been fenced off to provide wildlife habitats. £70/ha is paid to leave this out of the farm system.

Bradley Farm also qualifies for a grant of £400/year for ***educational access***.

▲ **Figure 10.20** Outside agencies involved in the economy of Bradley Farm.

▼ **Figure 10.21** Geographical features of Bradley Farm.

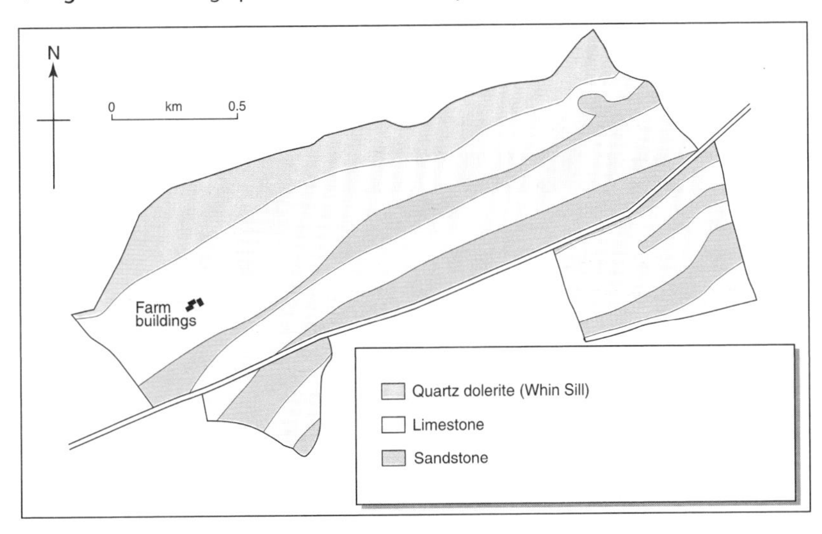

(a) Geology.

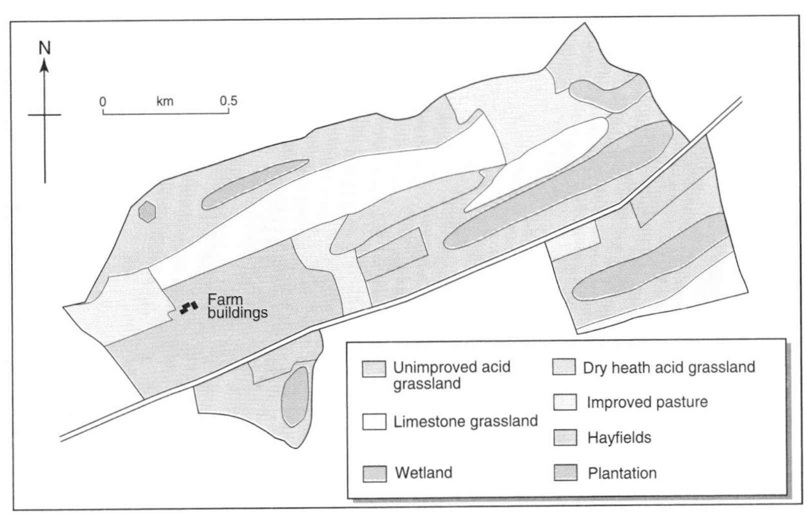

(b) Land use.

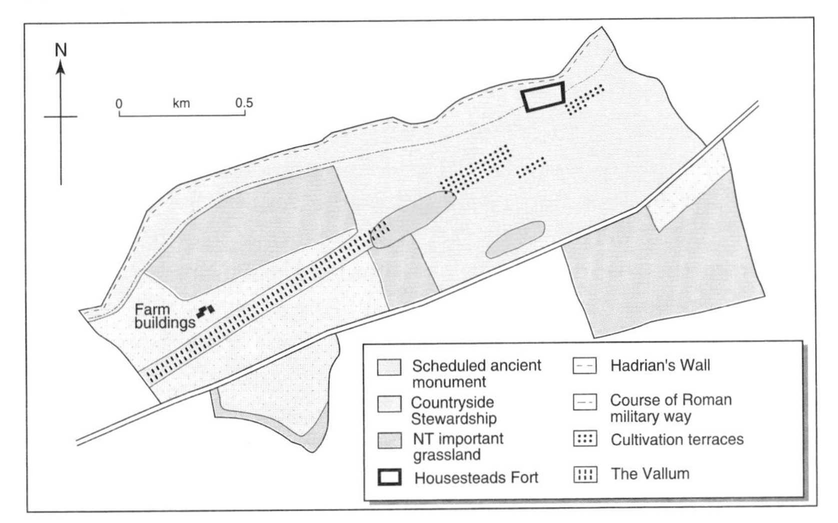

(c) Archaeological and conservation features.

Characteristics of Bradley Farm

1 Using Figures 10.21a–e, draw annotated overlay maps to show links between geology and types of land use.

2 Use Figure 10.17 to draw a cross-section along line A–B. Add symbols and labels to show relationships between relief, geology, drainage, and habitats.

3 a) How closely do habitats on Bradley Farm follow the rock types?
  b) How has grassland management influenced habitat distribution?
  c) How do conservation measures on Bradley Farm influence grassland management?

4 Consider the *sheep farming* element of Bradley Farm. Draw a systems diagram for this, to show *inputs, processes, outputs*, and *feedback*.

5 Using Figure 10.20 and the text, calculate the value of grants and subsidies received by Bradley Farm in a year, excluding capital grants for projects. Include:
- the Countryside Stewardship Scheme on 80 per cent of the land, and wetland conservation

- the educational access grant
- the hill livestock compensation for all *adult* animals
- sheep and livestock premiums on the sale of lambs and calves.

6 Farm management, grants, and subsidies have allowed Bradley Farm to improve the wildlife value of its ecosystems. Show the factors that have allowed this to happen, on a star diagram.

7 Would Bradley Farm make more profit if it were free of interference from outside agencies? Are there arguments for allowing farmers this freedom?

8 From what you have learned, do farmers gain or lose from:
  a) Britain's membership of the EU
  b) organizations such as English Nature and the National Trust?

Read through all the information on upland areas. Write an essay on: 'There is no justification for spending millions of pounds each year on managing landscapes. The landscape and all those who live in it should be left to the forces of economics.' Discuss.

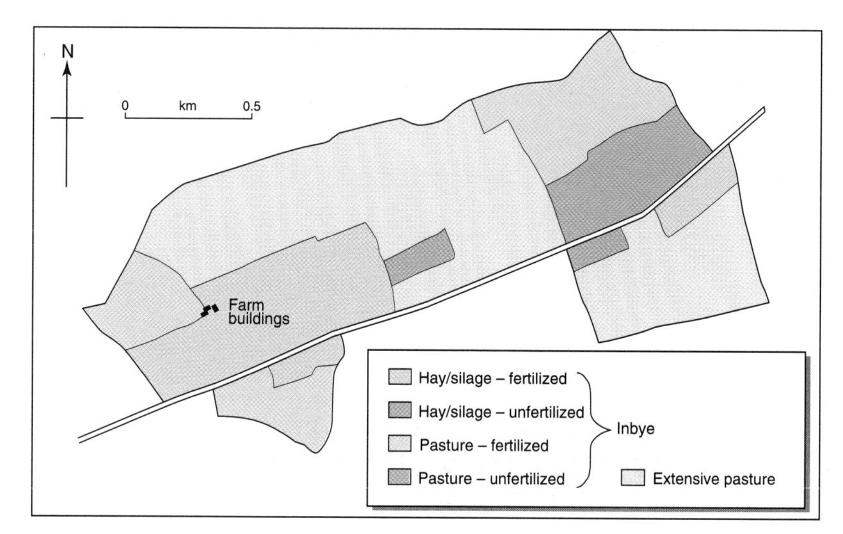

(d) Grassland management.

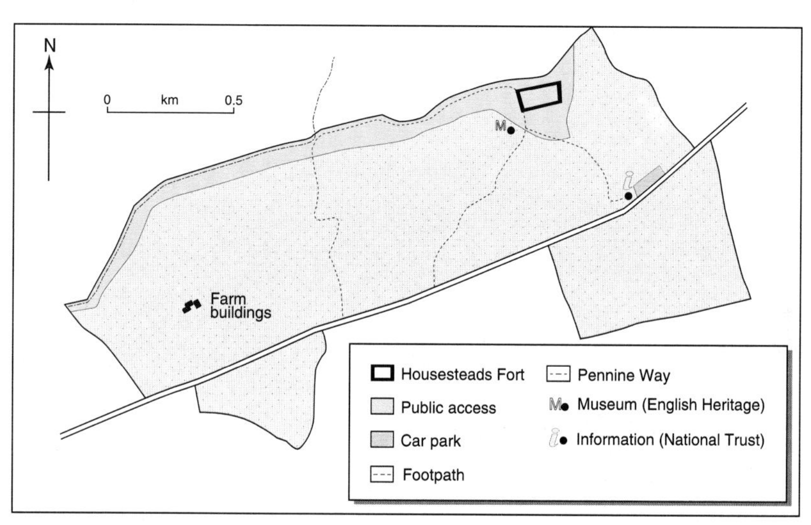

(e) Recreation.

Conflicts and opportunities created by tourism

The other main impact on the farm is tourism. Although Northumberland National Park is the least visited, Hadrian's Wall acts as a 'honey-pot site'. Housesteads Fort attracts 130 000 visitors each year, with 60 000 walks being taken along a 1.6km stretch of Hadrian's Wall. This footpath is part of the Pennine Way, which is used by 10 000 walkers each year. Two smaller footpaths cross the farm from the road to the Wall. Visitors present problems such as footpath erosion, open gates, litter, unleashed dogs, and walkers missing footpaths. The Actons spend three hours a week on tourist problems in the summer. There are some benefits from tourism for the farm economy such as bed and breakfast, cutting the grass around the fort, and clearing access roads by snow plough.

Lowland Britain: the impact of farm intensification on the Norfolk Broads

Changes in the Norfolk Broads

The Norfolk Broads is a rural area facing pressure from human activities, such as changes to agricultural systems, growing population, and increasing recreational use. Conflicts arise between different land and water users, and there is evidence that the landscape and wildlife value of the area is declining.

The Broads are artificial shallow lakes, resulting from peat extraction during the medieval period. The marshes have been traditionally managed for grazing. The land has been drained for centuries to reduce flood risk and to allow greater agricultural use. The wetlands are enhanced by drainage ditches crossing the fields and linking with streams and rivers. This traditional landscape is shown in Figure 10.22.

▲ **Figure 10.22** The traditional landscape of the Norfolk Broads.

What changes are affecting the Norfolk Broads?

Farming and ecology

Agricultural change in the Norfolk Broads is difficult to separate from other changes. EU subsidies have encouraged farm modernization, particularly the use of electric pumps to create drier soils for cereal growing, in turn leading to inputs of agrochemicals. The ecology of both grazing marshes and open water has changed as a result of using fertilizers. On land this has reduced the diversity of wildflowers and marsh vegetation, with losses of insects, birds, and small mammals at higher trophic levels. The most serious impact on the water ecosystem has been eutrophication – see Figure 10.23.

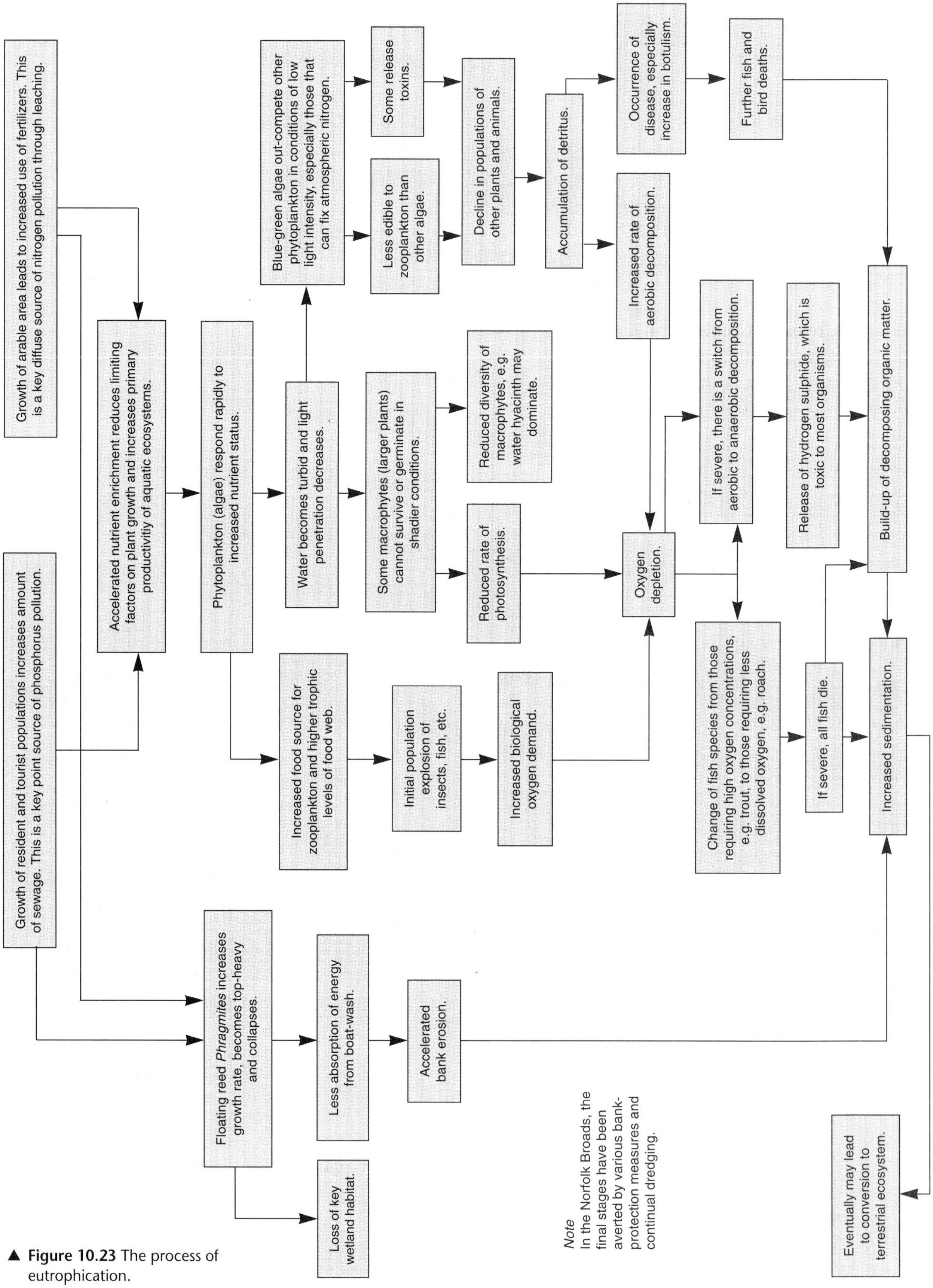

Growth of resident and tourist populations increases amount of sewage. This is a key point source of phosphorus pollution.

Growth of arable area leads to increased use of fertilizers. This is a key diffuse source of nitrogen pollution through leaching.

Accelerated nutrient enrichment reduces limiting factors on plant growth and increases primary productivitiy of aquatic ecosystems.

Phytoplankton (algae) respond rapidly to increased nutrient status.

Floating reed *Phragmites* increases growth rate, becomes top-heavy and collapses.

Loss of key wetland habitat.

Less absorption of energy from boat-wash.

Accelerated bank erosion.

Increased food source for zooplankton and higher trophic levels of food web.

Initial population explosion of insects, fish, etc.

Increased biological oxygen demand.

Water becomes turbid and light penetration decreases.

Some macrophytes (larger plants) cannot survive or germinate in shadier conditions.

Reduced rate of photosynthesis.

Reduced diversity of macrophytes, e.g. water hyacinth may dominate.

Blue-green algae out-compete other phytoplankton in conditions of low light intensity, especially those that can fix atmospheric nitrogen.

Less edible to zooplankton than other algae.

Some release toxins.

Decline in populations of other plants and animals.

Accumulation of detritus.

Increased rate of aerobic decomposition.

Occurrence of disease, especially increase in botulism.

Further fish and bird deaths.

Oxygen depletion.

Change of fish species from those requiring high oxygen concentrations, e.g. trout, to those requiring less dissolved oxygen, e.g. roach.

If severe, there is a switch from aerobic to anaerobic decomposition.

Release of hydrogen sulphide, which is toxic to most organisms.

If severe, all fish die.

Build-up of decomposing organic matter.

Increased sedimentation.

Eventually may lead to conversion to terrestrial ecosystem.

Note
In the Norfolk Broads, the final stages have been averted by various bank-protection measures and continual dredging.

▲ **Figure 10.23** The process of eutrophication.

Tourism

Tourism has resulted in problems of bank erosion, increased by the wash from power boats on the rivers and broads. Sediment levels in the water have risen as a result of bank erosion, and because soil has been washed from arable land, which is more vulnerable to soil erosion than damper marshes. One example of habitat destruction is the loss of many of the reedbeds fringing the broads.

Can the issues be resolved?

Several solutions have been attempted. Some solutions do not involve farming or tourism, for example dredging waterways and protecting river banks. One solution which may deal with several issues is the re-establishment and maintenance of reedbeds, especially close to sewage outfall pipes. These slow down the water, trap sediment, and, over a period of time, the reeds take up excess nutrients from the river. They also absorb wave energy from the boats, thus reducing bank erosion, and provide a wildlife habitat.

The Norfolk Broads was one of the first Environmentally Sensitive Areas, and farmers have benefited from compensation schemes to reduce intensity of farm systems, especially fertilizer input. This has resulted in the conversion of some areas of arable land back to grazing marshes.

Investigating water pollution

Although thorough chemical analysis of water to detect the presence of pollutants is not feasible without expensive high-tech equipment, it is possible to study water quality based on a range of criteria. The following are particularly relevant to investigations of eutrophication.

- Using commercially available water-testing kits to measure levels of phosphates and nitrates. Ask your science teacher or technicians about suppliers and types.
- Determining the biological oxygen demand (BOD) to indicate the degree of micro-organism activity during decomposition of accumulated organic matter. As decomposition rates increase, so does uptake of oxygen from water. This limits the amount available to other organisms. There are laboratory methods to analyse BOD involving the incubation of water samples. Ask your science teacher. In the field, you can use the presence of particular species as biological indicators. For example, where water is clean with a low BOD, trout and mayfly may be present. As BOD increases, these cannot survive, but chubb and dace can. At higher BOD levels, coarse fish such as roach can do quite well. There are various published biological indicator indices using easily sampled small organisms such as worms, shrimps, and larvae. One such system is the Trent Biotic Index.
- Comparing the volume of suspended material in the water – known as its **turbidity** – by filtering water or making a subjective visual comparison. *Filtering* involves weighing filter paper before and after the water is filtered, to determine the weight of residue left. Weighing should be done once the paper has dried. *Visual comparison* involves placing a glass or plastic measuring cylinder onto a piece of paper with a cross drawn on it. Shake the water sample, and then slowly pour it into the cylinder until you can no longer see the cross through the bottom of the cylinder. At this point, measure the depth of water in the cylinder. The more turbid the water, the smaller the measurement will be.

Other changes in lowland farming in the UK

Since the 1940s there have been many changes to UK arable farming as a result of the desire for self-sufficiency and changing technology. Land-use changes have resulted in increased farm size and larger fields. Crops too have changed. Oats have decreased in area, while sugar beet, wheat, and oilseeds have increased. These changes are the result of:

- the introduction of genetically altered strains of crop that are more productive, or are adapted to particular growing conditions
- the development and adoption of inorganic fertilizers and pesticides, which have increased yields by improving the nutrient status of soils, and eliminated pests – there has been a massive increase in the use of nitrogen fertilizer
- specialized farm machinery and greater availability of fuel has increased farm efficiency.

The cost of these inputs has caused amalgamation of many small farms into fewer, larger units to maintain financial returns. The desire to increase efficiency has often led to a monoculture system. Monoculture is the production of a single crop or product. This has resulted in a reduction in wildlife diversity. Thousands of kilometres of hedgerow have been removed to increase field size and to allow access by larger machinery, and the use of pesticides and fertilizers has altered food webs.

The fertilizer debate

Farm systems remove nutrients from the soil as produce is harvested. Farmers replace these nutrients to ensure continued yield. Their options include organic fertilizers, such as farmyard manure, or inorganic fertilizers. Crop rotation also contributes to an improvement in soil nutrient content. There is much debate about the best methods – see Figure 10.24.

'Good old-fashioned muck-spreading is more natural than all these chemicals,' said Mr Tim Lowe, washing down his manure-splattered tractor cab. 'We're not keen on nitrates and just use what's necessary.' From his farmyard in Ratcliffe Culey in south Leicestershire, he has seen the debate over nitrates – which leach into water supplies – polarize the farming community, as it has in other areas.

'The land can only take up so much nitrate – there's bound to be leakage into the waterways,' said Mr Lowe. 'We're not like these farmers who use a tremendous amount of fertilizer. They're producing three or four tonnes of corn an acre, whereas 20 years ago if you got two tonnes it was marvellous.

'Then there are those farmers who haven't got natural manure. They've got to rely on nitrates. The poor old farmer gets it in the neck whatever he does.'

But for many farmers the nitrate issue is simply one of supply and demand, with the Government adopting a hypocritical stance. Mr Peter Garland, who has farmed near Ratcliffe Culey for 30 years, has little sympathy with the Ministry of Agriculture's plight. He said: 'The ministry have been advising us to put extra nitrogen on to get bigger crops and now they're saying there's a problem and we've got to reduce. In fact, we are reducing because of the cost of nitrates . . . I've been using nitrates for 30 years and I still use farmyard muck as well. We don't have a problem with nitrates round here. It's the light, sandy soil where the nitrates go straight through into the water courses that's the problem. It would have a big impact on our profits if we had nitrate restrictions. Yields would be depressed.

'There's a lot of talk about organic farming, but if we all went organic there wouldn't be enough food produced to feed the country. I don't think the general public understand. With organic food, they've got to put up with caterpillars and insects crawling over their lettuce leaves.'

▲ **Figure 10.24** From *The Guardian*, 21 September 1989.

The issues are clouded by misconceptions. Inorganic fertilizers are agrochemicals, like pesticides. However, pesticides introduce chemicals that are new and harmful to the ecosystem, while fertilizers consist of essential chemicals already present in some form. Inorganic fertilizers are easy to handle and store, and allow precise distribution. Organic fertilizers may be available on mixed farms; these improve certain properties of the soil, such as water retention and cohesiveness.

Inappropriate use of inorganic fertilizers can cause pollution through leaching, but this is also true of organic fertilizers. Leaching occurs when soluble nutrients dissolve in soil water and are washed below the root zone. It reduces soil fertility and causes nutrient enrichment in groundwater and surface water. The amount of leaching depends on the quantity and solubility of nutrients in the soil. Unlike nitrogen, phosphorus is almost insoluble in soil. Other factors include soil characteristics such as pore structure and permeability. Soils consisting of coarse sand-sized grains, which have large pore spaces, are most susceptible to leaching.

Nitrate losses can be reduced by good farm practices, such as careful timing of the application of fertilizer, and sowing winter crops early in order to avoid leaving soil bare in winter.

The impact of agricultural improvement on UK grasslands

Pasture is an important landscape feature and habitat in the UK. It supports a wide variety of herb and grass species, which cope well with regular grazing and mowing, and relatively low soil fertility. Uses and areas of grassland have changed dramatically since the 1940s as farming has intensified. Draining, ploughing, re-seeding, and fertilizer application have converted pasture to arable land, or have increased grass yields to increase animal stock levels. These changes are shown in Figure 10.25.

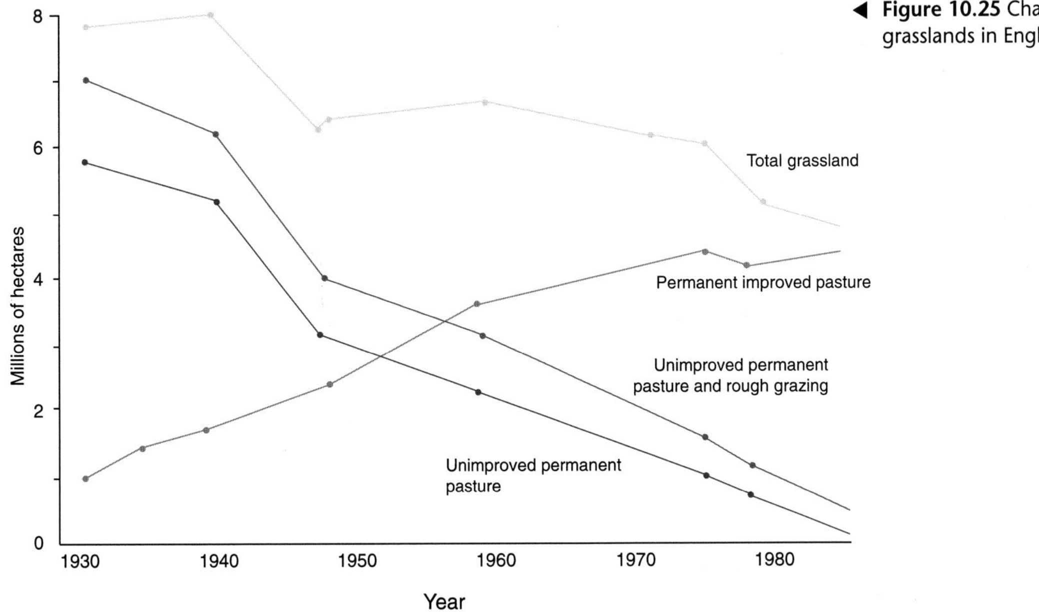

Figure 10.25 Changes to lowland grasslands in England and Wales.

Meadow pasture has altered during the 20th century. While farmers improve grassland to give good grazing, urban dwellers may prefer to see meadows full of mixed grasses and flowers, such as cowslips. Figure 10.26 shows how meadows change when farming seeks to improve grassland. Figure 10.26 categorizes plants into three groups: competitors, stress-tolerators, and ruderals. Plants are put into each category according to the conditions in which they thrive, and to their growth and reproductive patterns. The environments in which plants grow are shown as either those which show degrees of stress for plants – such as extreme cold, low fertility, or drought – or disturbance caused, for example by grazing, trampling, or drought. Notice that drought both stresses plants and disturbs them. Few plants survive high levels of stress and disturbance, but combinations of the two provide conditions that favour particular types of plants. This relationship is illustrated in Figure 10.26.

Many plants that thrive on unimproved grassland are ruderals. Here grazing or mowing patterns provide some disturbance. To cope with this, meadow grass has to develop rapid reproductive mechanisms between periods of disturbance. The plants themselves are therefore generally small. Traditional meadows are characterized by short-lived wildflowers and grasses. If the grassland is on infertile soil, only those grasses that tolerate stress can exist. The balance between disturbance and stress produces an ecosystem with species diversity where no single plant type dominates.

One result of improving pasture for production by adding fertilizer is to remove stress. Where there is no grazing before the hay is cut, disturbance levels are reduced. In this situation, competitors monopolize the environment by growing quickly and producing vast quantities of biomass to spread over the ground. This requires rapid nutrient uptake, and results in shading out of other plants. Ruderals and stress-tolerators cannot compete under such conditions, and competitors, for example rye-grass, soon dominate. As seeding is less urgent in such conditions, and disturbance is low, flowers are less important, and the character of the meadow is altered. Improved grasslands tend to be less visually appealing, and support fewer consumers, such as insects and birds. The effect of changing meadows has been to make these areas species poor.

Summary

Compare the issues affecting upland and lowland Britain. To what extent are they similar? Can the issues be resolved by similar actions?

STRESS INCREASES e.g. cold, low fertility, low precipitation →

DISTURBANCE INCREASES e.g. grazing, trampling, drought ↓

COMPETITORS	STRESS-TOLERATORS
• Growth is not restricted by stress or disturbance, so productivity is high. • Plants are often large, with an extensive lateral spread. • A high proportion of nutrients are rapidly incorporated into the vegetative structure. • There is a high turnover of leaves and roots. • Flowers and seeds are generally formed quite late in the growing season.	• High levels of stress limit growth rates. • Leaves are often small. • Many plants are evergreen. • A high proportion of nutrients are found in storage structures in the leaves, stems, and/or roots. • Flowering is intermittent over a long life-span.
RUDERALS • Regular disturbance means that no plants have an extended opportunity for growth. • Leaf production is explosive over a short period. • Plants tend to be small, with a limited lateral spread. • Flower production is important, and seeds are set early. • Seeds often contain a high proportion of nutrients, and can remain dormant for long periods.	**NO VEGETATION**

▶ **Figure 10.26** Grime's theory of plant strategies.

Impact of irrigation in Pakistan

Pakistan has a rapidly growing population, but much of the environment is difficult for agriculture. Only a third of Pakistan's total area is cultivated, and 80 per cent of this is possible only because of extensive irrigation systems, which use water from the river Indus and its tributaries. But agriculture is vital. It employs over half the workforce, and its products account for 25 per cent of its GDP. However, even though the Green Revolution of 30 years ago brought chemical fertilizers, irrigation techniques, and hybrid seeds, food production has not met the needs of the population. The staple diet is wheat, and 1.5 million tonnes are imported each year (20 per cent of total domestic wheat production). Demand for food is now exerting such pressure on agricultural systems that in some areas natural, social, and economic environments are unable to cope.

Large-scale irrigation began in the 1850s, when Pakistan was a British colony, with the construction of barrages and dams across rivers and valleys. The trapped water was transported by canals to farmland. Before independence in 1947 there was evidence of environmental problems. However, the split from India at independence destroyed much of the infrastructure on which Pakistan depended, and economic development became vital. As a result, irrigation systems were extended and farming became more intensive. Although new water transfer schemes are still being constructed, they are usually associated with multi-purpose dams and the increase in irrigated land has slowed down since the late 1970s. This is partly the result of rising costs, but also because many suitable locations have already been irrigated, and many schemes have brought as many problems as benefits.

Irrigation in Pakistan was first achieved by re-directing surface water away from rivers during the wet season. Canals were dug at right-angles from natural water channels, and, when river levels rose, water spilled into artificial channels onto the fields. Canal gradients were gentle so that water flowed slowly, and sediment was not carried onto the land. The canal network is now 65 000km long. Water was also tapped from alluvial deposits using gently sloping shafts that allowed water to flow out of the ground without using pumps – this is known as the *karez* system. The main problem is that water is most available when it is least needed, during the wet season. Since the 1960s there has been an increase in the use of tube-wells (Figure 10.28), which extract groundwater. These are powered by electricity from integrated HEP/irrigation systems from dams such as those at Tarbela Reservoir on the river Indus and Mangla Reservoir on the river Jhelum.

▼ **Figure 10.27** Pakistan.

▼ **Figure 10.28** A tube-well in use in Pakistan.

Irrigation: problems and solutions

Irrigation has been introduced along with other changes in technology, to increase the production of wheat, rice, sugarcane, and cotton. However, there are problems. For example, in Achakzai, a village in north-west Baluchistan, a government drive to install tube-wells using loans and subsidies to farmers led to the neglect of the traditional *karez* system, because farming families no longer shared in its maintenance. Although the tube-wells supplied more water for a short time and allowed the local economy to boom, constant groundwater extraction led to a lowering of the water table. The wells dried up, and farming suffered as the *karez* system fell into disrepair. Land around the village no longer supported the population, and the settlement has declined.

By contrast, high water tables may also be a problem. Salinization and waterlogging are the causes of greatest concern in Pakistan. Both result from poor management of traditional irrigation systems. Many channels carrying water are unlined and much water has seeped into the ground over the years. This has raised the water table in many areas by 0.5m a year. In the past, the water table was 25m below the ground surface. Now, 8 million of the 13.5 million hectares that are irrigated by canals have a water table within 3m of the surface. Soils become waterlogged, plant roots rot, and crops fail.

As the water rises towards the surface, it carries with it dissolved salts of chlorides, carbonates, sulphates, and nitrates, which are found in deposits in semi-arid regions, and accumulate where there is waterlogging. These collect in the upper horizons of soil as water is evaporated under high temperatures. This is called salinization. It limits the range of crops that can be grown, and reduces the yield of those that can. The replacement of wheat by rice in some areas is the result of such soil changes, as rice is more tolerant of salt and waterlogging. Figure 10.29 summarizes some of the other environmental impacts of irrigation.

The Pakistan government recognizes problems associated with irrigation, and has devised schemes to reduce the damage. For example, where there is waterlogging and salinization, tube-wells have been drilled and water pumped out of the ground for agricultural use. There are at least 7000 such wells in operation. Other engineering solutions include the installation of drainage systems, and lining canals. This reduces waterlogging, and creates silt traps to reduce sedimentation of water channels. Altering farming techniques is another option, and includes leaving land fallow to allow the rains to flush salts away, or rotating crops with those that tolerate higher salt levels or have the ability to transpire water from deep root zones.

Investigating the irrigation issue

1 a) Using an atlas, geographical data resource books, and/or CD-ROM, create a datafile about Pakistan. Include details on landscape, climate, natural vegetation, and river basins. Refer to population statistics such as density, birth rates, and distribution, and socio-economic data such as calorie intake, health care, literacy rates, income or GDP, and trade.

b) Present the data using maps, tables, and graphs to summarize information. Find corresponding data for an economically developed country such as the UK or the USA.

2 Use information in the text and in Figure 10.29 to analyse the effects of irrigation in Pakistan.

Whatever solutions are found, the need for food cannot be denied. Water resources have to be more efficiently managed in future, so that wastage and environmental damage are reduced. Perhaps the answer lies in small-scale irrigation schemes, with individual farmers taking responsibility for water. Alternatively, genetic engineering may allow the introduction of seed that is more suited to the alternate monsoon and dry seasons.

Impact on:	Possible changes	Causes	Results
A Soil	**1** Formation of salt crust at or just beneath the surface	Salts deposited from rising groundwater, or poorly drained irrigation water. Encrustation likely if salts dry out.	Impermeable salt pan through which water and plant roots cannot penetrate. Salt can be blown a great distance, and contaminate other areas.
	2 Erosion of topsoil	Soil may be removed as ground is levelled in preparation for irrigation. Excess water on land can wash fertile soil down slopes.	Reduced depth of soil for root zone. Sedimentation of streams and water transfer channels, clogging up the irrigation system.
	3 Increased rate of humus breakdown	Intermittent wetting and drying of soils, especially after drainage to combat waterlogging.	Reduced soil fertility. Soils dry out more easily without organic matter to hold moisture. This may enhance erosion.
B Water	**1** Siltation of streams	See A2.	See A2.
	2 Pollution of surface water and groundwater	Water used in agriculture may contain high levels of agrochemicals, and salts from contaminated land. Eventually much of this returns to rivers or aquifers.	Impact on plants and animals in surface water. Problems with safety of water withdrawn for human consumption or for further agriculture.
	3 Collapse of aquifer	Where water is extracted from the ground by pumping, the aquifer may collapse owing to pressure from rocks above.	Subsidence of ground. Aquifer cannot be recharged to supply more water.
	4 Intrusion of seawater into aquifer	Where water is extracted near the coast, the voids left in the aquifer may be invaded by seawater.	Further use of groundwater is limited by the salt content.
C Living organisms	**1** Increased overall plant growth, and extended period of plant cover	Availability of water allows higher levels of plant productivity, unless other environmental changes limit increase.	Improved agricultural productivity. More vegetation cover, especially during the dry season, helps to protect the soil from wind erosion. Greater input of organic matter to soil increases fertility.
	2 Plants and animals unable to tolerate contaminated soil and water	See B2.	Reduced biodiversity.
	3 Increase in agricultural pests and water-related diseases	Standing water provides an ideal habitat for insects and other pests. Water may be contaminated by pathogenic organisms.	Damage to crops by pests. Increased risk of people contracting diseases such as typhoid, dysentery, and malaria.
	4 Change in spatial distribution of habitats, e.g. reservoirs, water channels, and river banks	Physical alteration of landscape to allow irrigation.	Dispersal of organisms will change. This may bring problems (see C3), or aid conservation by new features acting as wildlife corridors for migration of species.

▲ **Figure 10.29** Environmental impacts of irrigation in Pakistan.

Ideas for further study

1 Select a local farm or farming area for study. Use the following sequence of enquiry questions.
  a) What kinds of farming are going on? Can the farms be represented as systems?
  b) What human and physical factors affect the types of farming?
  c) What impact is this kind of farming having on the ecosystem?
  d) How has this type of farming altered the landscape?
  e) What is the balance of positive and negative effects of this kind of farming on the ecosystem? Can negative effects be reduced?
2 Study an upland area in a contrasting environment, for example the Alps (tourist pressure), or a wilderness area such as the mountains of South Island in New Zealand.
3 Find out about the ecological impacts of a large irrigation scheme. Investigate its purpose, methods, levels of success, and the balance of positive and negative environmental impacts, for example:
  - Snowy Mountains, Australia
  - Aswan Dam and the Nile, Egypt
  - Syr Darya and Amu Darya river basins in Kazakhstan and Uzbekistan (both of which enter the Aral Sea).

Summary

- Agriculture almost always changes how natural ecosystems operate.
- Agriculture depends upon physical processes, but is greatly affected by human factors. These include social, economic, and political preferences for the ways in which land is organized and used.
- The use of land for agriculture may result in conflict with ecosystems. Land management strategies such as hedgerow removal, fertilizer application, and irrigation may have an impact.
- Agriculture makes a major impact on the landscape. Changes in agriculture are usually reflected in changes to the landscape.

References and further reading

T. Burt, 'Update: The nitrate issue', *Geography Review*, 1993 Vol.6 No.5

N. Coles and L. White, 'Running winds and shifting sands. Farm life in the Rajasthan Desert', *Geographical Review*, 1990 Vol.4 No.2

Countryside Commission, *The Countryside Premium for Set-aside Land*

L. Heathwaite, 'Eutrophication', *Geographical Review*, 1994 Vol.7 No.4

S. Trudgill, 'Organic farming and the soil', *Geographical Review*, 1991 Vol.4 No.4

K. Willis and G. Garrod, *Landscape Values: A Contingent Approach and Case Study of the Yorkshire Dales National Park*, University of Newcastle upon Tyne, 1991.

Ecosystems and human activity: Summary

Key ideas	Explanation	Examples
1 Nature of ecosystems	Ecosystems consist of structured webs or systems at a range of scales, which includes living organisms and their material environments of soil, air, and water. These components are linked by movements of energy and nutrients.	• Local small-scale examples (sand dunes, moorland in Chapter 7) or biome examples (rainforest or savanna in Chapters 8 and 9)
	Functional analysis.	• Food chains and webs, population pyramids and trophic levels
	Structural analysis.	• Nutrient recycling and energy flow through trophic levels
	Distributional analysis.	• Population and limiting factors, e.g. how far rainforests can support large numbers of people
	Productivity and global patterns.	• Net primary productivity (NPP) and biomass at the small scale • Either woodland or grassland ecosystem
2 People as a component of ecosystems	People are an increasingly important component in ecosystems at all scales and in all locations, possessing potential for dominance due to increasing numbers, adaptability and technological expertise. Human impact on ecosystems: • short-term • long-term.	• Small-scale, short-term change, e.g. heather burning and dune management in Studland Heath • Larger-scale and longer-term change, e.g. rainforest deforestation, irrigation and salinity in the Murray–Darling basin in Australia
3 Ecosystems are dynamic	Ecosystems may exist in a relatively stable state or may be subject to change through natural processes or the influence of human activities such as agriculture.	• Small-scale change, e.g. pressures on Studland Heath, oil spillage in Alaska, in leisure pursuits on Kinder Scout • Large-scale change, e.g. deforestation or grassland management. Irrigation and salinization. • Pressures brought by increasing numbers of cattle in savanna regions
	Plant colonization and ecological succession.	• Plant succession theory
4 People attempt to manage ecosystems	Management of ecosystems represents people's attempts to effect change in plant and animal systems which may be beneficial and constructive, rather than destructive to their environments.	• Kinder Scout, Studland, Ladies Spring Wood
	For successful management, it is necessary to understand fully the workings of ecosystems, the likely causes and effects of change, and the concept of sustainable yield.	
	• Local study. • Regional study. • National study. • Global study. • An overview of either grasslands as difficult to manage or forests as difficult to conserve.	• Examples of small-scale studies such as Studland Heath or Kinder Scout • National park issues, or issues in managing Korup rainforest, or savanna in Ivory Coast • Policies towards ecosystems, e.g. Cameroon policies towards rainforest conservation • Issues regarding preservation of rainforests or grasslands • Political, social, and environmental issues in managing cattle movements in Ivory Coast • Solving the problems of salinity in Australia • Comparison of issues in Korup and Kilum

Key ideas	Explanation	Examples
5 Agriculture may have important impacts on ecosystems	Agriculture is significant in creating major impacts on the functioning and structure of natural environments and systems, and on the appearance of landscapes and ecosystems.	• How agriculture changes an ecosystem,e.g. use of fertilizers in lowland and upland farms in the UK
	Use of land for agriculture may result in conflict with natural ecosystems and with other space-using human activities:	• Norfolk Broads and lowland UK farming
	• landscape change	• Landscape change in the Yorkshire Dales
	• the impact of agrochemicals	• The impact of chemicals on lowland ecosystems and farming in Norfolk
	• the impact of clearance	• The impact of clearing woodland (Chapter 8) or of clearing and altering grasslands (Chapter 9)
	• the impact of intensification.	• Changing practice and its effects in the Yorkshire Dales, lowland UK, the Murray–Darling river, and Pakistan • The impact of increasing cattle numbers in Ivory Coast
	Some forms of agriculture are more in harmony with natural ecosystems.	• Attempts to manage sustainable farming in Korup
6 Policy-making at a range of scales influences agriculture	Economic and social policies decided at national and international level have an increasingly significant impact on agricultural systems, on the people whose lives depend on such systems, and on the environments that they occupy.	• National park policies and farming in upland Britain
	Issues of quotas, subsidies, over-production, and farm diversification.	• The impact of EU policies, milk quotas, subsidies and set-aside land

Glossary

Agrochemicals Chemicals manufactured for agricultural purposes, namely fertilizers and pesticides.

Agroforestry Any system where trees are deliberately left, planted, or encouraged, on land where crops are grown or animals grazed.

Air mass A very large volume of air (measured in thousands of km^2) with uniform conditions of temperature and humidity.

Albedo The proportion of the Sun's radiation that is reflected from any surface.

Anafront Where the temperature difference between air streams is great and Polar and Equatorial airstreams mix very actively, causing the fronts to change and move rapidly.

Antecedent rainfall The amount of rainfall/water already present in the soil before a particular rainfall.

Aquifer Rock which will hold water and allow water to pass through.

Artesian water Water held in an aquifer under pressure through saturation of the rock. If a borehole is sunk at depth to tap the water an artesian well is formed from which the water will naturally flow upwards.

Baseflow Slow seepage of water through soil and rocks into a river.

Beach nourishment Sand is imported to rebuild the beach, replacing sand that has been extracted or eroded.

Biochemical oxygen demand (BOD) Quantity of oxgen used in five days for the partial oxidation of a sample effluent under standard conditions.

Biodiversity A measure of species' richness and natural genetic variation.

Biomass A measure of the amount of living material, e.g. the gross weight of plants present in an area.

Biomes A global scale ecosystem with broadly uniform climate and vegetation, covering a wide area, e.g. tundra.

Chlorofluorocarbons (CFCs) Manufactured chemicals consisting mainly of chlorine, fluorine, and carbon.

Climax community An ***ecosystem*** which is fully adapted to its climatic conditions. It is the final stage of ecological succession.

Coriolis force The force produced by the rotation of the Earth causing moving air to deviate from a straight-line path.

Corrosion or solution The chemical effect of salts or acids held in water, dissolving or corroding alkaline rocks such as limestone or chalk.

Cost–benefit analysis Projects are valued in terms of social and environmental consequences as well as economic profit and loss.

Cyclogenesis An atmospheric process where surface pressures fall rapidly as polar and subtropical air converge. Warm air spirals upwards.

Cyclonic rainfall Precipitation associated with the passage of a depression.

Decomposers The organic breakdown of vegetation, debris, animal excrement, and corpses, by bacteria and insects to release nutrients.

Discharge The volume of water in a river channel.

Diurnal range The daily difference between minimum and maximum values in 24 hours, e.g. temperature.

Dynamic equilibrium A state of balance which persists despite changes taking place.

Ecosystems A community of plants and animals together with the environment in which they live. The abiotic parts of the ecosystem interact with the biotic parts in a dynamic system.

Ecotourism Nature-based tourism.

Edaphic Produced or influenced by the soil.

El Niño Southern Oscillation (ENO) (opposite, La Niña) The warm ocean current which appears in late December off the South American coast every 3–8 years.

Enhanced greenhouse effect The increased tendency of the lower atmosphere to trap outgoing infra-red radiation caused by greenhouse gases from human activities.

Equinox Day and night are of equal length.

Evapotranspiration Water from the Earth's surface via evaporation and transpiration (water loss from plants).

Extensive farming Farming which has low inputs of investment and labour, usually with low yields.

Ferrel cell One of the three convectional cells which make up the global circulation model according to Ferrel and Rossby.

Fetch Distance travelled by a wave.

Flood abatement Management of a river's flow through land use changes in the ***river basin.***

Flood proofing Designing new buildings to withstand flood damage.

Flood return period The frequency with which flood levels are likely to be repeated.

Floodplain zoning A management strategy for reducing floodplain development.

Front The boundary between two air masses of very different temperatures and humidity.

General Circulation Models (GCMs) Computer models which process large banks of climatic data.

Geostrophic wind A high altitude wind which flows round the Earth within the troposphere.

Gleying Where air is excluded from soil through saturation of water. Ferric oxide is reduced to ferrous oxide forming a greenish-grey colour.

Global warming The increase in global mean temperature.

Greenhouse effect The processes by which heat is trapped in the lower atmosphere.

Groynes A breakwater built to stop the movement of beach material by longshore drift.

Hadley cell The central convectional cell which develops from intense heating at the Equator.

Hydrograph A graph which shows rainfall and river discharge plotted against time. Bars indicate rainfall and the line graph shows river discharge.

Infiltration Water entering soil.

Infiltration capacity or rate The rate at which water can infiltrate the soil.

Intensive farming Farming which has a high level of input – investment and labour – and high yields.

Inter-tropical convergence zone (ITCZ) The low-pressure system which moves north and south with the overhead Sun. It is the area where two airstreams meet. Along the ITCZ a **monsoon** front forms.

Jet stream A meandering flow of strong westerly winds, blowing at over 100km/h, in the upper air.

Katafront Slow-moving, less active front where upper-air movement is not strong and there is little difference in temperature between air streams.

Leaching Process by which chemicals in soil are dissolved and removed out of reach of plant roots by soil water.

Maritime Weather conditions affected by the influence of the sea.

Monsoon A seasonal wind system giving rise to a distinct rainy season in southern Asia and West Africa.

Montane Of a mountain, as in montane forest which is found in the uplands of the tropics.

Occlusion The process by which the cold front of a depression eventually catches up with a slower-moving warm front.

Permafrost Areas of soil and rock which are permanently frozen.

Permeability The extent to which water can pass through rock pores, cracks, joints.

Photochemical smog Fog formed by the chemical reaction of pollutants with oxygen, forming ozone and nitrogen dioxide.

Photosynthesis The process by which green plants manufacture organic compounds using light energy.

Pioneer species Plants which are able to colonize open or exposed sites and form the first stage in plant ***succession***.

Podzols A layer of ashy-coloured soil just below the surface where clay, humic acids, iron, and alluvial compounds have ***leached*** down.

Polar front Where Polar maritime air meets Tropical maritime air, over the North Atlantic and North Pacific.

Pollution dome Polluted air in or above an urban area, formed when polluted warm air is prevented from rising by a ***temperature inversion*** above it.

Precipitation Moisture from the atmosphere in the form of rain, dew, hail, frost, fog, sleet, or snow.

Prevailing wind The wind direction which occurs most often.

Primary producer A green plant which can synthesize carbohydrates using carbon dioxide and energy from the Sun.

Primary productivity The rate at which green plants produce organic matter (***biomass***) using the Sun's energy. Gross primary productivity (GPP) is the total amount of energy fixed by plants measured in kilograms per square metre per year. The GPP minus the energy lost through respiration is called the net primary productivity (NPP) of the ***ecosystem***.

Random sampling Sampling which selects survey points or interviewees at random.

River basin The area drained by a river and its tributaries. Also known as drainage basin.

River regime The pattern of discharge in a river over time.

Rossby waves Large global meanders of upper air flows caused by differences in pressure between Polar and Equatorial air. They are created by large mountain barriers, such as the Rockies, by differential heating of the atmosphere, and by smaller-scale atmospheric changes.

Savanna Tropical grassland.

Siltation The build-up of silt deposits.

Soil salinity Level of salts in the soil; increased when poor drainage prevents the removal of excess salts.

Stratified sampling Sampling where the area of study is divided into different segments.

Subaerial processes Processes which involve weathering and mass movement of rocks on the face of the cliff.

Subclimax community The stage in plant succession immediately before climax.

Subtropical high-pressure belt A belt of almost permanent high pressure around 30° N and 30° S.

Succession A sequence of vegetation in an area from first invasion to ***climax community***.

Sustainable A lifestyle which maintains biological productivity for succeeding generations, without depleting natural resources or damaging the systems which produce them.

Sustainable development Development which uses resources at a low steady rate so that the environment is not damaged.

Symbiosis An association between two different organisms through which both benefit.

Synoptic chart A map which summarizes meteorological conditions at a given time.

Systematic sampling Sampling which selects survey points or interviewees in a regular pattern.

Systems A group of components consisting of input , stores, and output, and linked by processes.

Temperate There are few or no extremes of temperature.

Temperature inversion Where air temperature increases as it rises rather than decreases as is normally the case.

Throughflow The flow of water down a slope through the soil.

Tidal range The difference in water-level between high and low tides.

Trade wind A wind which blows from the tropical high-pressure zones to the Equatorial low-pressure zones.

Transect A line which cuts across. In fieldwork transects are used to observe changes in vegetation, for example in a study of sand dunes.

Transpiration The process by which plants give off and lose water vapour. The moisture is then recycled as part of the hydrological cycle.

Trophic level A feeding level in a food chain. All herbivores live at one trophic level, all primary carnivores at another, and so on.

Urban heat island A large town or city which is warmer than the surrounding rural areas as a result of the heat stored by the buildings and derived from the Sun.

Vortex (vortices) Rotation currents found within violent updraughts of air within, for example, hurricanes.

Walker cell Major movement of air from east to west in the tropics.

Watershed The boundary between two river systems or drainage basins.

Water table The level below which the ground is saturated.

Weathering The disintegration or decay of rocks at the point where they are located.

Wind shear Zones of strong horizontal air movement.

Index

Page numbers in italics indicate that the terms appear in Theory or Technique boxes.